Optoelectronic Devices and Systems for Processing

Critical Reviews Series: Recent Titles

Optoelectronic Devices and Systems for Processing (CR65, Bahram Javidi, Kristina M. Johnson)

Inorganic Optical Materials (CR64, Paul Klocek)

Polymers in Optics: Physics, Chemistry, and Applications (CR63, Roger A. Lessard, Werner F. Frank)

Optoelectronic Interconnects and Packaging (CR62, Ray T. Chen, Peter S. Guilfoyle)

Standards for Electronic Imaging Technologies, Devices, and Systems (CR61, M. C. Nier)

Standards and Common Interfaces for Video Information Systems (CR60, K. R. Rao)

Technologies for Advanced Land Combat (CR59, G. R. Lane, T. A. DuBois, E. H. Vandiver)

Distributed Interactive Simulation Systems for Simulation and Training in the Aerospace Environment (CR58, T. L. Clarke)

Digital Signal Processing Technology (CR57, P. Papamichalis, R. Kerwin)

Defining the Global Information Infrastructure: Infrastructure, Systems, and Services (CR56, S. F. Lundstrom)

Adaptive Computing: Mathematics, Electronics, and Optics (CR55, S.-S. Chen, H. J. Caulfield)

Millimeter and Microwave Engineering for Communications and Radar (CR54, J. C. Wiltse)

Glass Integrated Optics and Optical Fiber Devices (CR53, S. I. Najafi)

Handbook of Critical Dimension Metrology and Process Control (CR52, K. Monahan)

64- to 256-Megabit Reticle Generation: Technology Requirements and Approaches (CR51, G. Hearn, R. Naber)

Fiber Optics Reliability and Testing (CR50, D. K. Paul)

Diffractive and Miniaturized Optics (CR49, S. H. Lee)

Photorefractive Materials, Effects, and Applications (CR48, P. Yeh, C. Gu)

Optical Technologies for Aerospace Sensing (CR47, J. E. Pearson)

Optical Inspection and Testing (CR46, J. D. Trolinger)

Integrated Optics and Optoelectronics (CR45, K.-K. Wong, M. Razeghi)

Fiber Optic Sensors (CR44, E. Udd)

Optomechanical Design (CR43, P. R. Yoder, Jr.)

The Aviation Security Problem and Related Technologies (CR42, W. H. Makky)

Lens Design (CR41, W. J. Smith)

Optical Pattern Recognition (CR40, J. L. Horner, B. Javidi)

Infrared Thin Films (CR39, R. P. Shimshock)

Infrared Optical Design and Fabrication (CR38, R. Hartmann, W. J. Smith)

Critical Reviews of
Optical Science
and Technology

Volume CR65

Optoelectronic Devices and Systems for Processing

Bahram Javidi
University of Connecticut

Kristina M. Johnson
University of Colorado/Boulder
Editors

Proceedings of a conference held
8–9 August 1996
Denver, Colorado

Sponsored by
SPIE—The International Society for Optical Engineering

SPIE OPTICAL ENGINEERING PRESS

A Publication of SPIE—The International Society for Optical Engineering
Bellingham, Washington USA

Library of Congress Cataloging-in-Publication Data

Optoelectronic devices and systems for processing: 8–9 August 1996, Denver, Colorado / Bahram Javidi, Kristina M. Johnson, chairs/editors; sponsored and published by SPIE—the International Society for Optical Engineering.
p. cm.—(Critical reviews of optical science and technology ; v. CR65)
Includes bibliographical references.
ISBN 0-8194-2257-6 (softcover)
1. Optical data processing—Congresses. 2. Optoelectronic devices—Congresses. I. Javidi, Bahram. II. Johnson, Kristina M. III. Society of Photo-optical Instrumentation Engineers. IV. Series.
TA1630.07 1996 96-23000
621.36'7—dc20 CIP

Published by
SPIE—The International Society for Optical Engineering
P.O. Box 10
Bellingham, Washington 98227-0010 USA
Telephone 360/676-3290 (Pacific Time) • Fax 360/647-1445

Printed in the United States of America.

Contents

Conference Committee

Conference Chairs

Bahram Javidi, University of Connecticut
Kristina M. Johnson, University of Colorado/Boulder

Session Chairs

1 Nonlinear Filtering for Pattern Recognition
David P. Casasent, Carnegie Mellon University

2 Optical Processing Systems
Joseph L. Horner, Rome Laboratory

3 Devices for Optical Processing I
Kristina M. Johnson, University of Colorado/Boulder

4 Optical Memory and Applications
Yunlong Sheng, COPL/Université Laval (Canada)

Nonlinear Filtering for Pattern Recognition

Optical Processors for Scene Analysis

David Casasent

Carnegie Mellon University
Department of Electrical and Computer Engineering
Center for Excellence in Optical Data Processing
Pittsburgh, PA 15213

ABSTRACT

A suite of different image processing algorithms can be optically implemented using different filter functions on one common optical architecture (a correlator). This allows one optical system to address all major operations required in general computer vision. These operations, the associated filters, examples of the use of each, and real-time optical hardware are reviewed. For specific image processing applications, only selected operations are necessary; but the same basic architecture suffices.

Keywords: Correlation, image processing, morphological processing, pattern recognition, wavelet processing.

1. INTRODUCTION

We consider image processing as a major use of optical processing. Since the input data are 2-D, since shift-invariance is required, and since multiple filter functions (operations) are necessary, optical processing is very attractive. Optical systems can inherently process 2-D data, they perform correlations (shift-invariant operations), and they allow space and frequency multiplexing (hence multiple filter functions are easily handled on the same architecture in parallel).

The basic optical architecture considered is a correlator (Fig. 1), other architectures besides the frequency plane correlator of Fig. 1 are possible. In Fig. 1, the input (P_1) and filter (P_2) are assumed to be adaptive, with the filter used (and operation performed) being dependent on the output correlation plane (P_3) results from previous operations. When the number of filter functions required at P_2 is not enormous, one can employ the architecture of Fig. 2. In this case, many different filters are recorded in different spatial regions of P_2 (spatial multiplexing) and different spatial regions of P_2 (filters) are selected by activating different laser diodes in P_0. Each spatial location in P_2 can contain several filters frequency-multiplexed; in the case shown, four filters are stored at each P_2 location and the P_3 output for one spatially localized P_2 filter set consists of four separate correlation planes (in the 4 quadrants of P_3). These 4 quadrants can be read-out in parallel by available split CCD detector arrays at P_3. The number of detector elements in P_3limit the number of frequency-multiplexed filters to 4 or 9 for most applications.

One version of the operations required in a general scene analysis problem is shown in Fig. 3. It is typical to divide the operations or functions required into different stages[1,2,3] as shown. Clutter reduction refers to removing non-object data, this operation

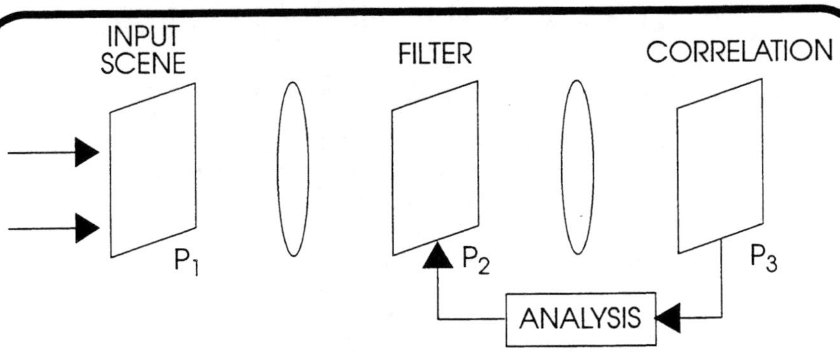

Fig. 1: Optical correlator with adaptive filter.

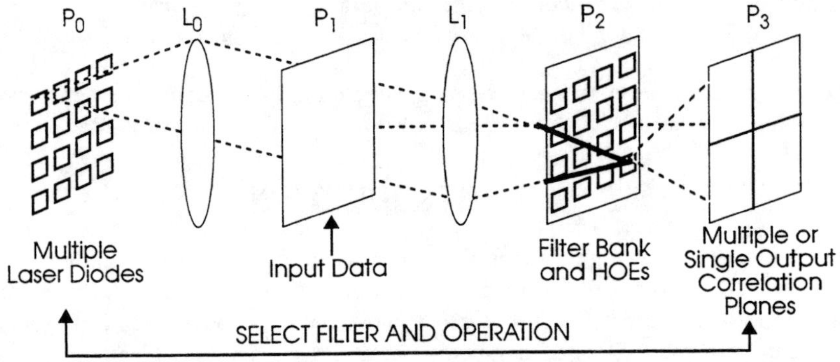

Fig. 2: Optical correlator with a fixed filter bank (space and frequency-multiplexed). The P_0 laser diode activated selects a set of four frequency-multiplexed filters at one spatial P_2 location. The correlation of the P_1 input with the set of 4 filters appears in the quadrants of the P_3 output plane.

generally alters object data and thus it is not necessarily always a separate stage (except for removal of salt and pepper etc. noise). The first major stage is detection. It must locate all objects or regions of interest (ROI) in the scene when several objects in different classes and at any 3-D aspect distortion or contrast are present. A high detection rate (P_D) is necessary, since ROIs not located in this stage will never be subsequently analyzed. This must be achieved when the scene contains significant clutter and thus false alarm rates (P_{FA}) must be low, since every ROI detected must be processed. We include a separate processing stage that reduces false alarms (P_{FA}). Shift-invariance and distortion-invariance are essential in detection, since object location is not known and since multiple objects can be present in the scene. Once ROIs have been located, each such local scene region is enhanced, features of each ROI are extracted and these are fed to a classifier that determines the class (which type of object) of the contents of each ROI (the classifier should have the ability to reject clutter and designate the ROI as a non-object region) and the pose (aspect view) of each object in each ROI. Subsequent sections address the operations required (and their optical realization) for each of these stages of a general scene analysis system. Each discussion is brief but references to more complete descriptions are provided.

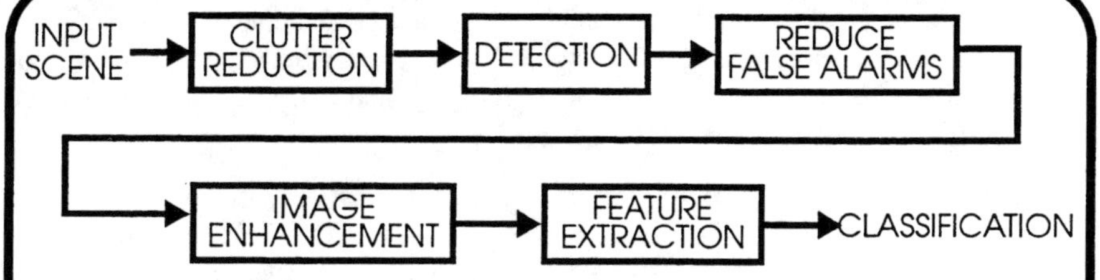

Fig. 3: Hierarchical inference levels of computer vision.

The database considered consists of a number of different scenes such as Fig. 4 containing multiple objects of different classes, aspect views and contrasts with significant clutter present. There are 8 different objects with 21 different aspect views of each object (covering 360° of aspect) in a variety of contrasts and with different backgrounds present. The objective is to locate each object (ROI) with high rates (P_D) and few clutter regions (P_{FA}) and to then classify each ROI region (object or clutter) and to denote the class and aspect view of each object.

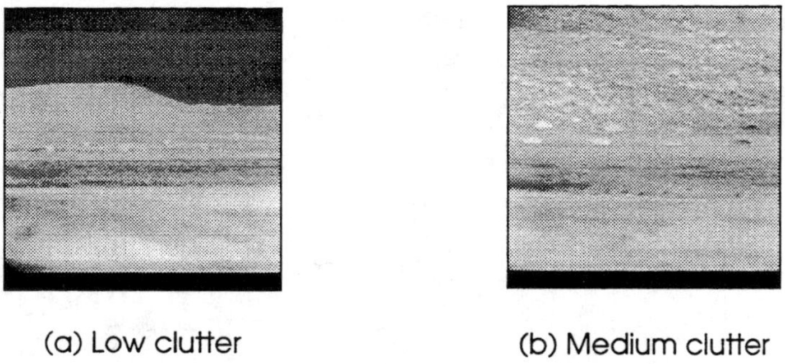

Fig. 4: Example scenes from the database.

2. GENERAL OPERATIONS ACHIEVABLE

We first note 3 general classes of operations achievable on the optical architectures of Fig. 1 and 2: morphological, wavelet, and distortion-invariant filtering.

2.1 Morphological Processing

The basic morphological operations are erosion and dilation. They are implemented on optical correlators with nonlinear input P_1 and nonlinear output P_3 correlation plane thresholding.[4] We use intermediate output thresholds and thus implement more general rank-order optical filters[5] to handle distorted objects. The two basic morphological operations are used in sequence to produce closing (dilation plus erosion) and opening (erosion plus dilation) morphological functions. Simple structuring element (SE) filters (e.g. disks) are used at P_2 of Figs. 1 or 2 and they are easily implemented optically, since their phase in the Fourier transform (FT) domain is inherently binary.[5]. In general, gray-scale morphology is necessary on gray scale imagery and a new and efficient optical correlator architecture to implement this has been advanced.[5]

Fig. 5 shows the architecture for an efficient optical gray-scale morphological processor. It consists of an optical correlator (P_1 to P_3) with a thresholding SLM at P_1 and P_3 and an added readout stage (P_3 to P_4). A gray scale input image is level-sliced at N different gray levels to produce N different binary images. Each of these is fed sequentially to P_1, correlated with the same structuring element filter at P_2, N binary morphological correlation plane outputs are produced sequentially at P_3. The SLM at P_1 with electrical or optical control can provide the N binary inputs. The optically-addressed SLM at P_3 produces the correlation plane thresholding and the desired binary morphological output for each of the N input images. Electrical or optical control of this P_3 SLM allows these binary outputs to be either erosions, dilations or rank-order morphological results. These are each formed optically in parallel. They are read out and imaged onto a CCD detector at P_4. For N sequential binary sliced inputs, N binary morphological results are produced sequentially at P_3 and they are all summed at P_4. The gray scale P_4 output is the desired gray-scale morphological result. It is the sum of N nonlinear morphological image processing operations. This architecture is very attractive as fast binary input P_1 SLMs can be used, a fast optically-addressed P_3 SLM is employed, and it avoids all electronic output interface problems since the P_4 electrical output is only required after N cycles of the P_1 or P_3 device.

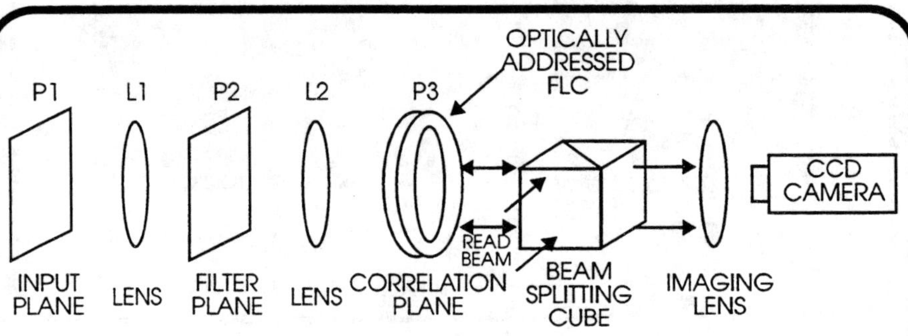

Fig. 5: Optical gray-scale morphological processor.

Morphological operations are nonlinear local image processing functions, well-founded in image algebra mathematics.

2.2 Wavelet and Gabor Transforms

Optical systems can also implement these operations with different filter functions at P_2 of Figs. 1 or 2. If the P_2 filter function is the conjugate Fourier transform (FT) of

$$g_n = e^{-\pi(x^2/a^2 + y^2/b^2)} e^{-j2\pi r\omega\cos(\theta - d)} , \qquad (1)$$

then the P_3 output is one component GT_n of the Gabor transform (GT) of the input data (f) where the GT parameters are

$$n = (a, b, \omega, \phi) . \qquad (2)$$

The $GT_n = f \otimes g_n$ is a 2-D analog output where the pixel value at each point tells the amount of the radial spatial frequency ω at an orientation ϕ present in a local region (set by a and b) around each input pixel. It thus provides information on the size and orientation of input scene data in each local region of the scene. We select $\omega = 1/2d$, where d is the width of g_n in direction ϕ. With this choice, the real part of the Gabor function (RGF) is an excellent blob or object detector and its imaginary part (IGF) is an excellent edge detector.[6]

Fig. 6 show the 1-D real and imaginary Gabor filter functions with $\omega = 1/2d$ used. As seen, the real Gabor filter (Fig. 6a) has a central positive and two negative sidelobes. The central sidelobe detects objects (blobs), the negative sidelobes detect the background around the object blob; thus, this blob (object) detection filter is similar to a hit-miss transform (HMT) optical detection filter.[5] The Gaussian shape of the function (vs. the rectangular shape in the HMT filters) makes this filter more attractive for distorted object detection. The imaginary Gabor filter with $\omega = 1/2d$ is shown in Fig. 6b. It is zero mean and has only one positive and one negative sidelobe. Thus, it is an excellent edge detector as other work has shown.[17] Section 4 shows detection results with such filters.

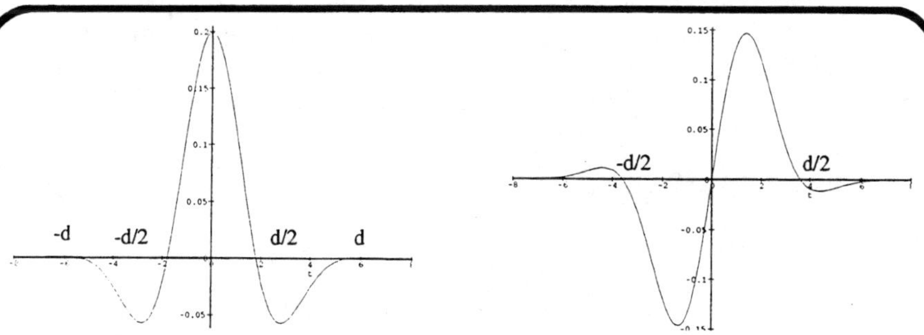

(a) Real Gabor function (b) Imaginary Gabor function

Fig. 6: Real and imaginary parts of the Gabor filter function.

We now demonstrate the ability of the real parts of such filters to recognize different distorted versions of one object, and we detail how selection of the d parameter affects this. We considered detection of all distorted versions of the TRIM-2 database of objects (they cover a range of about 10-40 pixels in width). We used a simple rectangular model of the object with different widths representing different object aspect views. We correlated these rectangular model-based objects with several different Gabor function filters whose horizontal widths were varied. Fig. 7 shows the correlation peak height obtained for Gabor filters of width 15 and 25 pixels in x when correlated with rectangular objects of widths 10 to 45 pixels. As seen, the larger width Gabor function performs best (it's width is about in the center of the range of object widths we wish to detect) with all correlation peak values within 0.3/0.45 ≈ 70% of the maximum.

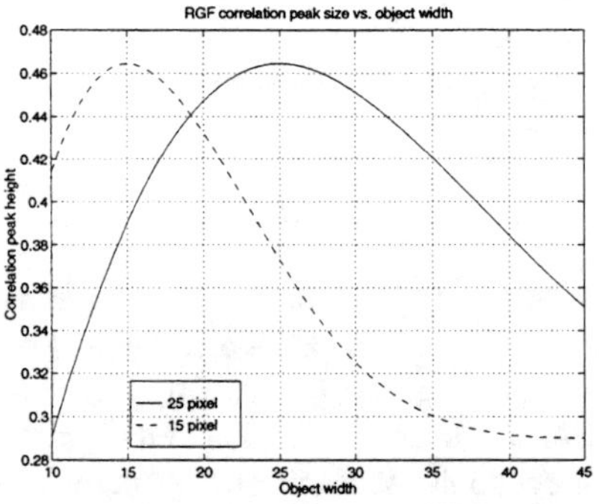

Fig. 7: 1-D correlation peak height vs. the width of a rectangular object.

By varying the parameter ϕ in the filter, we can detect object widths and heights plus horizontal or vertical object edges. We have shown that such filters can detect a large range of object widths[6,7] and can detect object edges that are smooth and sharp with the same magnitude output.[8] Gabor filters allow an easy search of different scale factors to detect objects and clutter particles in a wide range of sizes. We have also detailed how such filters can detect clutter regions of a scene (where object detection confidence is low).[8] We achieve this by use of a GF edge detection filter and wide area Gaussian integration (plus other operations). This produces a 2-D analog output (clutter map C) with larger values in clutter regions and lower values in non-clutter regions. Since clutter particles are larger, smaller and similar in size to objects, this approach was found to be very effective. These filters further broaden the repertoire of operations achievable on the same optical correlation architecture (with different filter functions).

2.3 Distortion-Invariant Filters

These filters utilize linear algebra to design a constrained optimization filter that can recognize different distorted versions of one object. There are many variations of such filters.[9,10] They represent some of the major contributions of optics to pattern recognition. Different sets of such filters have been shown[2,3] to provide detection and classification of the database considered.

The basic concept (Fig. 8) is to use representative distorted object-aspect views to produce a distortion-invariant filter h that will recognize all distorted versions of one object. The filter is formed in the Fourier transform (FT) domain as $\underline{H}$, it is a combination of the FTs of different training set images (the columns of a data matrix $\underline{\underline{X}}$), all 2-D images and $\underline{H}$ are lexicographically ordered and described as vectors. Many versions of such filters exist. Most require correlation peak constraints, i.e. $\underline{H}$ is required to give correlation peak values u = 1 for all training set images in $\underline{\underline{X}}$, i.e.

$$\underline{\underline{X}}^{+}\underline{H} = \underline{u} = [1 \cdots 1]^{T} . \tag{3}$$

The filter is also required to minimize an objective function such as

$$\begin{aligned} E &= E_S + cE_N \\ &= \underline{H}^{+}\underline{\underline{S}}\underline{H} + c\underline{H}^{+}\underline{\underline{N}}\underline{H} \end{aligned} , \tag{4}$$

where E_S (E_N) are the correlation plane energy due to signal (noise).

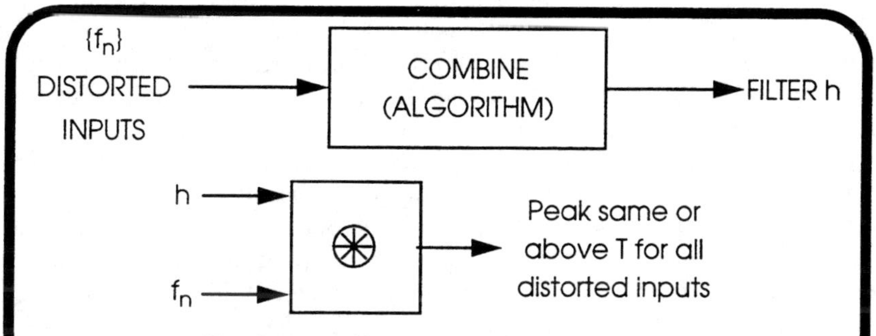

Fig. 8: Distortion-invariant filter concept.

Minimizing E_S reduces correlation plane sidelobes, this is achieved by minimizing $\underline{H}^{+}\underline{\underline{S}}\underline{H}$, where $\underline{\underline{S}}$ is a diagonal matrix whose elements are the average $|FT|^2$ of the training set. Minimizing E_S reduces false alarms due to other objects etc. by emphasizing high spatial frequencies. Minimizing E_N improves recognition P_C and detection P_D by emphasizing low spatial frequencies. The constrained optimization solution that satisfies the peak constraints in (3) and minimizes the objective function in (4) is

$$\underline{H} = \underline{\underline{T}}^{-1}\underline{\underline{X}}\left(\underline{\underline{X}}^{+}\underline{\underline{T}}^{-1}\underline{\underline{X}}\right)^{-1}\underline{u} , \tag{5}$$

where $\underline{\underline{T}}$ is a diagonal matrix whose elements are the maximum of S or cN at each spatial frequency. This solution can be shown to be a preprocessed version of a linear combination of preprocessed training images, with $\underline{\underline{T}}$ defining the type of preprocessing done (e.g. emphasizing low or high or specific spatial frequencies). One can minimize E_S or E_N as noted above; one can also minimize the correlation plane energy due to distortions in the input $E_D = \underline{H}^{+}\underline{\underline{D}}\underline{H}$, where the distortion spectrum $\underline{\underline{D}}$ is a diagonal matrix whose elements are the maximum value (for any training image) at each spatial frequency minus the average spectrum of all training images. Use of $\underline{\underline{N}}$ has been shown and found to be a simple and useful model for $\underline{\underline{D}}$. One can also minimize the correlation plane output due to clutter in the input $E_C = \underline{H}^{+}\underline{\underline{C}}\underline{H}$, where the elements of the diagonal matrix $\underline{\underline{C}}$ are the spectra of correlated noise at each spatial frequency. Such a correlated noise model for clutter has been shown to be very good for real IR and for visible data over a reasonable range of correlation lengths.

To achieve best results, a combination of several objective functions is minimized as in (4), where c is a free parameter that determines which of two objective functions we most desire to minimize. Large (small) c values emphasize minimizing noise (signal) energy and emphasize low (high) spatial frequencies. Fig. 9 shows an example of the use of c to improve recognition of a number of different versions of the title PROF (Fig. 9a). With a low value of c and only the top 2 versions of PROF included in the filter, the filter recognizes two of the 7 versions of the input (Fig. 9b shows the correlation plane output with large peaks for only the first 2 versions of PROF). When c is increased, more variations of the input are recognized until eventually all are recognized with correlation peaks ≥ 0.5 etc. (Fig. 9c). Fig. 10 shows a version of such filters that achieves all levels of computer vision (Fig. 1). The input (Fig. 10a) contains 13 distorted versions of several different classes of objects (Fig. 10b) in low SNR background. A filter using a high c value and trained only on the smallest (ZSU) objects detects all objects with only one false alarm (Fig. 10c); it thus achieves detection. If we wish to find only large objects, we synthesize a filter formed only from large objects with a smaller c value, we correlate it with the input and look only at the correlation plane regions in Fig. 10c determined from the detection filter; the result (Fig. 10d) locates only the positions of the large objects. To locate only one specific class of large objects (e.g. Scuds), we form a filter from only Scuds with a still smaller c value, correlate it versus the input, and look at the output where the macro-class filter (Fig. 10d) has outputs. The result (Fig. 10e) locates only the Scud objects and achieves classification.

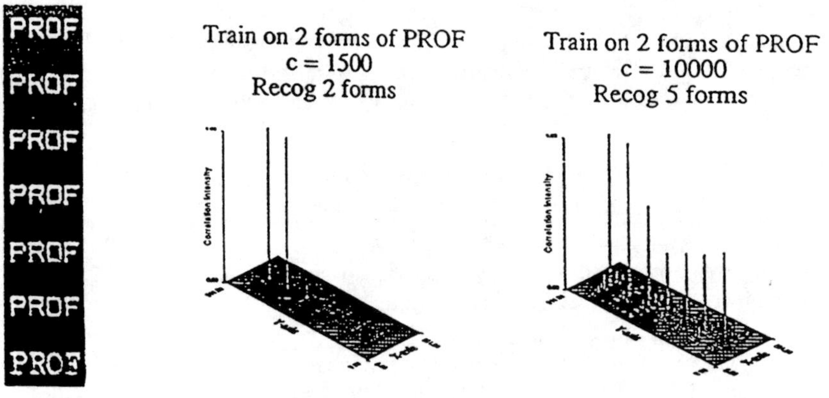

(a) Input (b) Output (high c) (c) Output (low c)

Fig. 9: Use of c control to allow recognition of more input objects.

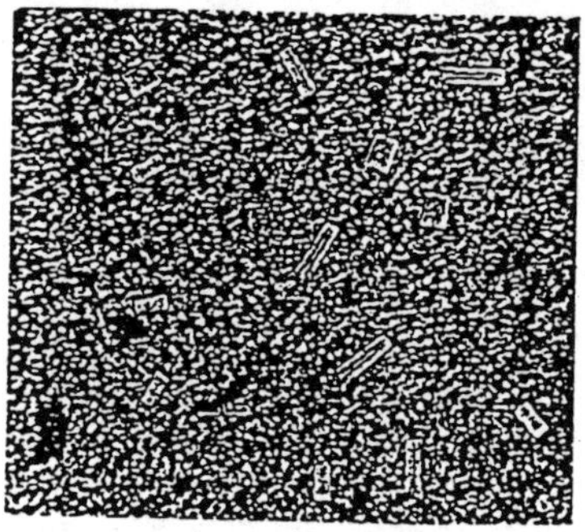

(a) Input

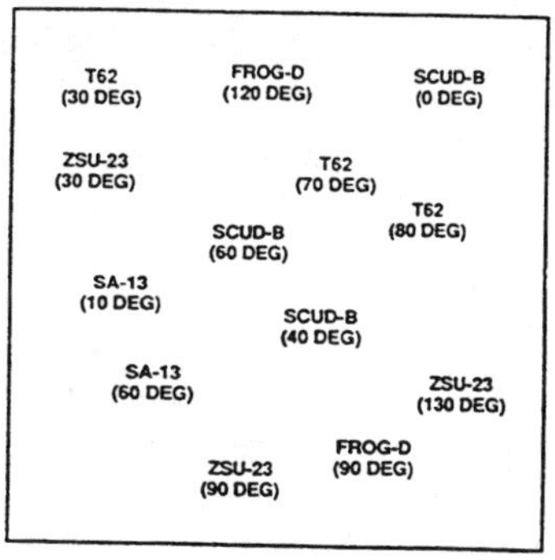

(b) Vehicle Code

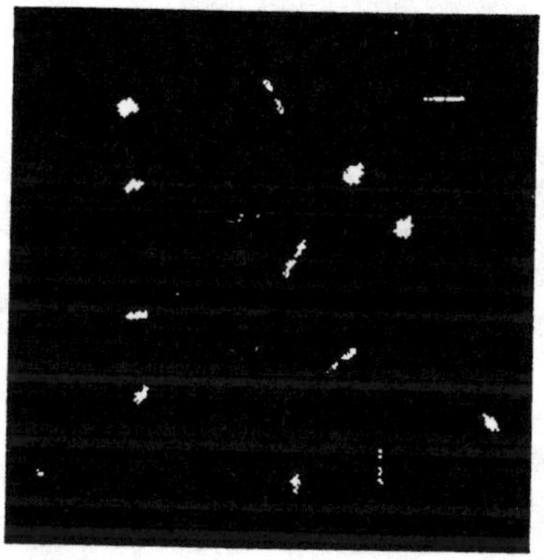

(c) Recognition example (c = 0.005)
(Scuds/Frogs vs. rest)

(d) Detection (one FA), c = 0.1
36 ZSU-23 training (smallest object)

(e) Identification example (c = 0.001)
Locate Scuds only (training only on Scuds)

Fig. 10: Use of a suite of distortion invariant filters to achieve detection, recognition, and classification.

3. ENHANCEMENT

Once local regions of interest (ROI) have been detected, one can improve these local ROI images using morphological etc. operations.[11,12] Fig. 11a shows an input object with background noise and holes on the object (at the wheel treads). The erosion of the image followed by a dilation removes the background noise (Fig. 11b). The structuring element filter used was a 5×5 disk (this size was chosen to omit noise regions smaller than 5×5 but to not remove object features (e.g. the gun) larger than 5 pixels in 1-D. The erosion removes background regions as well as some of the boundary pixels on the object and it enlarges the black holes on the object. The dilation is used to restore those boundary regions and to restore the tread holes to their original shape. To fill in holes on the object, we apply a dilation followed by an erosion to Fig. 11b. The result (Fig. 11c) has nearly all object holes filled in. The size of the structuring element filter used (a 5×5 disk) determines the size of the holes on the object that will be filled in. As before, a sequence of two morphological operations are used; the dilation fills in holes, but also enlarges the size of the object (its boundary increases); the erosion step restores the original boundary of the object.

(a) Input (b) E and D output (opening) (c) D and E output (closing)

Fig. 11: Morphological enhancement of ROI.

Some enhancement operations can also be used to preprocess an image prior to detection. Fig. 12 shows an example of gray-scale morphological processing to achieve this using CMO (close minus open) operations. The input image (Fig. 12a) contains four objects (2 are hot, white, and 2 are cold, dark) and the background level varies locally quite considerably. The top (sky) is dark and then bright (mountains) while different gray levels are present in different strips in the foreground. To *remove local background variations*, we use the CMO operation. This removes any local background variations larger than the largest object of concern (it is also useful for removing nonuniform lighting variations in robotics, inspection, etc.). We choose the structuring element to be the largest object-aspect view of interest, i.e. the largest thing in the image we do not want to remove (in this case, the size of the structuring element was chosen to equal the height and width of the broadside view of a tank). We first close the input image (this omits dark objects smaller than the size of the structuring element, i.e. it removes the 2 dark tanks). We then open the input image (this omits bright objects smaller than the size of the structuring element, i.e. it removes the 2 bright tanks). The common information present in the closed and opened outputs is thus background regions larger than the size of the structuring element filter. The subtraction (CMO, close minus open) thus removes all local

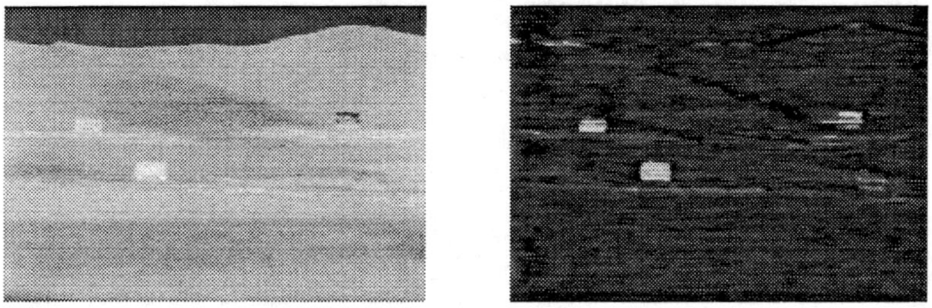

(a) Input (b) Output

Fig. 12: Morphological CMO result (background removal).

background variations (Fig. 12b). The minus of the opened image (it contains the 2 dark tanks) converts these to bright tanks (Fig. 12b). Thus, this operation also normalizes the polarity of all objects to bright. If the application is one where the objects of interest are only bright or only dark or have different structure, other variations of the basic algorithm can be used.

4. DETECTION

Three detection algorithms and the associated filters required are now discussed. The purpose of detection is to locate all objects in a scene, when the objects can be from different classes and with different distortions and contrasts present and when clutter (Fig. 4) exists.

The Gabor transform[6] detection filter uses several RGF functions combined (by neural net techniques[13]) into one macro RGF that detects all objects in all aspect views. It also uses an edge detection IGF filter to detect vertical object edges. It then fuses the results of these real and imaginary Gabor function outputs to determine if an ROI is an object (an object blob (RGF) output and edge outputs at the proper locations must both be present to determine that the input ROI is an object). Fig. 13 shows the concept, Fig. 14 shows how the RGF filter detects object blobs and the IGF filter detects object edge data and Fig. 15 shows the results for a specific scene.

An improved version of this filter (the macro Gabor filter, MGF) is under development.[7] It uses 6 different Gabor functions, two real Gabor functions (RGFs) that detect object widths and heights, and four imaginary functions (IGFs) that detect the left, right, top and bottom edges of an object. We now define a Gabor function as

$$g(\underline{x}) = \frac{w}{2\pi|\underline{A}|^{-1/2}} \exp\left[j\underline{\omega}_0^{\,T}(\underline{x}-\underline{x}_0)\right] \exp\left[-\frac{1}{2}(\underline{x}-\underline{x}_0)^T \underline{A}^{-1}(\underline{x}-\underline{x}_0)\right] . \quad (6)$$

This is a complex sinusoidally modulated Gaussian function. It is actually a weighted GF with weight w. The normalization factor $1/(2\pi|\underline{A}|^{-1/2})$ keeps the area under the Gaussian part of Eq. (6) constant as we change $\underline{A}$ and also results in a more convenient

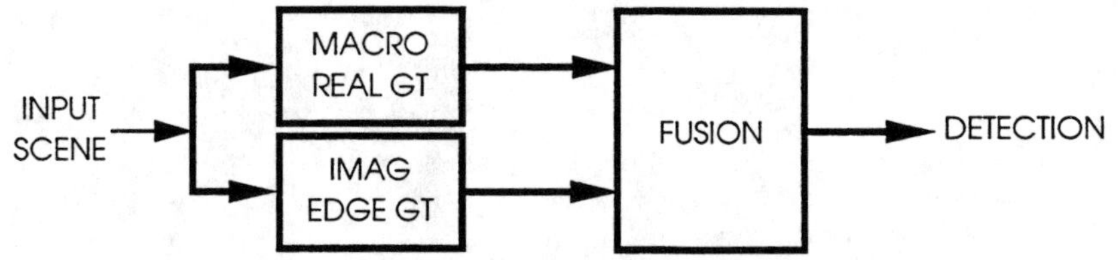

Fig. 13: Gabor transform detection concept.

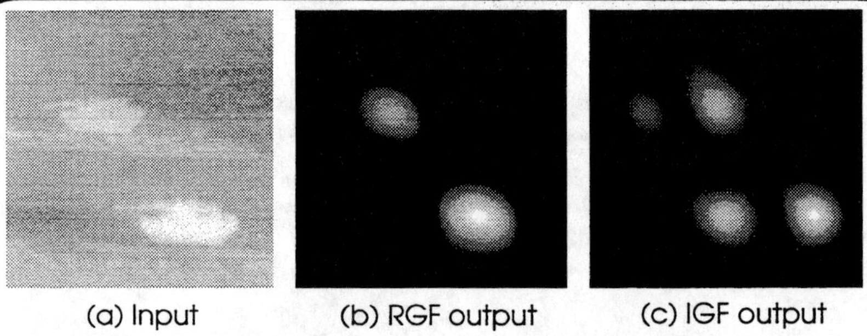

Fig. 14: Gabor transform detection example.

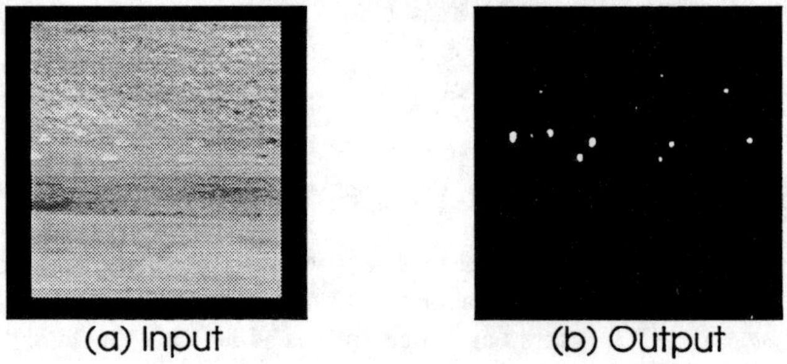

Fig. 15: Gabor transform scene results.

representation in the frequency domain. The underlined lower-case variables represent vectors. The vector $\underline{x} = [x_1,x_2]^T$ is a 2 dimensional vector representing the spatial coordinates (horizontal and vertical). The vector $\underline{x}_0$ shifts the function with respect to $\underline{x}$. $\underline{A}$ is a matrix describing the shape of the Gaussian part of the function and is defined as

$$\underline{A} = \begin{bmatrix} a^2 & 0 \\ 0 & b^2 \end{bmatrix}, \tag{7}$$

where a and b determine the variance of the Gaussian part in the horizontal (x_1) and vertical (x_2) directions. The modulation frequency is $\underline{\omega}_0$ and is set to

$$\underline{\omega}_0 = \begin{bmatrix} \pi/2a \\ 0 \end{bmatrix}. \tag{8}$$

The spatial modulation frequency is only in one direction (the x_1 or x_2 axis) since we use GFs to detect only object height or width (RGFs) or edges in one direction (IGFs). As seen, the modulation frequency scales with the shape of the Gaussian part of the function in x_1 (described by a). The particular choice used in (8) ensures that the RGFs and IGFs have the shapes in Fig. 6. To obtain GFs with different orientations, we apply the coordinate transformation

$$\tilde{\underline{x}} = \underline{J}\underline{x}, \quad \tilde{\underline{x}}_0 = \underline{J}\underline{x}_0 \tag{9}$$

with

$$\underline{J} = \begin{bmatrix} \cos(\theta) & -\sin(\theta) \\ \sin(\theta) & \cos(\theta) \end{bmatrix}. \tag{10}$$

This rotates the GF through an angle θ into $g(\tilde{\underline{x}})$ with any new desired orientation. We use $\underline{J}$ with $\theta = 0$ or $\theta = \pi/2$ to produce horizontally or vertically oriented GFs. We collectively refer to the 4 parameters controlling the shape of Eq. (6) as a vector, $\underline{\alpha} = \left[a\ b\ w\ \underline{x}_0^T\right]^T$, noting that $\underline{x}_0$ is a vector. Different choices of $\underline{\alpha}$ result in different GFs.

In these new filters, note that the position $\underline{x}_0$ of the output peak is a parameter. For each of the 6 different Gabor functions, we select $\underline{\alpha}$ (this defines each GF to optimize each GF *separately*). We require each GF to give a specified correlation function shape (Gaussian) for the training set and to give a zero output for clutter input particles. We have detailed the steps in this optimization of the parameters $\underline{\alpha}$ of each GF separately.[7] An attractive new aspect of this is that we also now require the peak value for each GF output *to occur in the same location* (in the center of the object); this is very different from the GT output in Fig. 14c, where the IGF peak output occurs on the edges of the object. This allows us to now combine all six GF functions into one MGF filter.[7] We now only need to apply one filter, implement one pass of the input image and analyze only one output GT plane for peaks for objects. This has proven successful and preferable to other Gabor detection filters in initial tests. Fig. 16 shows the six GF outputs for one input object (Fig. 16a); as seen each GF output is suitably shifted and has a peak in the central region of the object. Thus, when all GFs are combined into a single MGF, its output will have a large peak in the center of the object. The GF parameters are optimized to detect the height, width, etc. of all distorted versions of all 8 objects in our TRIM database.

The second detection algorithm we consider is the morphological wavelet transform (MWT).[8,14-16] It consists of the CMO morphological step (Fig. 12) and a Gabor wavelet filtering step that produces a clutter map (C) output. The CMO minus C difference image when thresholded yields the final output. Fig. 17 shows the concept. Fig. 12 showed one example of the CMO morphological step (object contrast is improved). Fig. 18 shows the MWT result on one scene. We now discuss several aspects of the wavelet filtering that produces the clutter map output C.

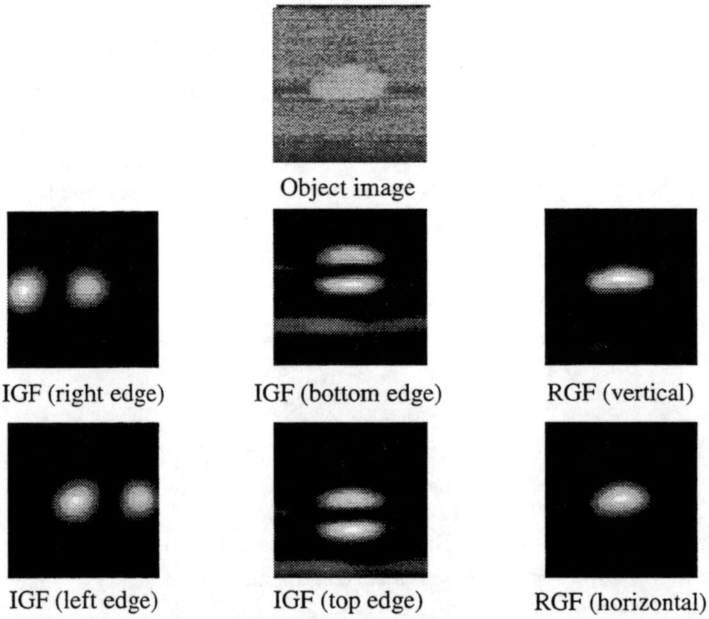

Fig. 16: Illustration of the individual GF correlation outputs for a typical object image (M2 tank at a 60° aspect angle). Note that all 6 GF outputs have peaks at the center of the object. The 4 IGFs are each shifted by $\underline{\mathbf{x}}_0$ to achieve this. The additional peaks in the IGF outputs are from the other edge of the object at the same orientation.

For a scene with significant clutter (Fig. 4b), we consider ways to locate the high clutter scene regions (regions with low P_C confidence in target detections), i.e. techniques to produce a clutter map output C (with high values in high clutter regions) and with low outputs where targets are. For the case considered, the clutter particle sizes are larger, smaller and comparable in size to that of the objects. The only difference between clutter and target regions lies in the density of particles per area. Thus, the technique we selected to locate clutter is to edge enhance the input image and low pass filter the resultant image. The edge enhancement (with magnitude detection) gives more outputs in regions of higher clutter. The Gaussian wide area smoothing integrates the number of such edges over the area of the Gaussian and produces an output C whose pixel value at each point is proportional to the number of edges (clutter particles) per area. The edge enhancement technique we chose,[8] used a 1-D horizontal imaginary Gabor edge detection filter. The imaginary Gabor filter has been shown to be an excellent edge enhancement filter.[17] We showed[8] that the area magnitude output from such a filter was invariant to whether the edge was sharp or smooth. We also addressed how selection of the order (scale) of the filter affected the range of clutter particle sizes that were detected and the effect of noise on the different types of wavelet filters at different scales. Wavelet filters are also attractive, since they allow easy scale searches. Figs. 19 and 20 show how the clutter map adapts for different types of imagery.

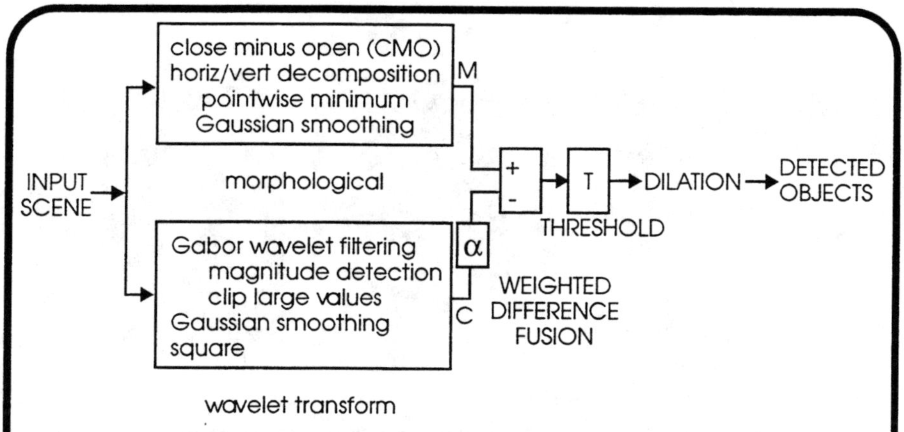

Fig. 17: MWT Concept.

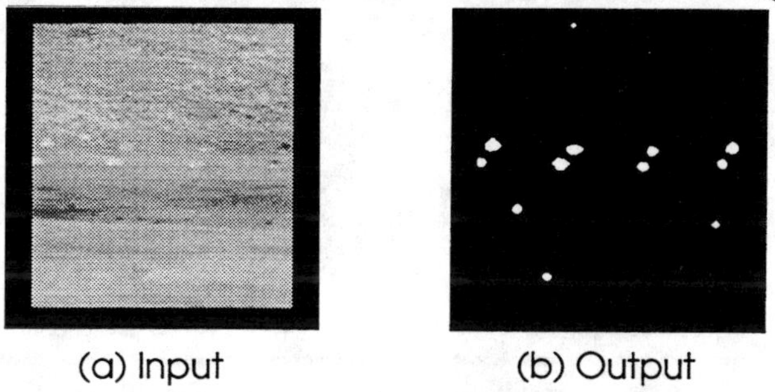

Fig. 18: MWT representative detection result.

Our final MWT algorithm[15] is very robust and preferable to the original algorithm.[14]. The size of the morphological structuring elements in the CMO algorithm are several pixels larger than the height and width of the largest object; we perform a CMO step with a horizontal and with a vertical slit structuring element and then combine the results (this directional decomposed morphology is faster, since it only requires 1-D operations, and it performs better, since more background is omitted); we Gaussian smooth the morphological output to remove structuring element structural patterns.[16] We can predict the CMO output pixel level expected on an object of a given contrast; this allows us to specify the amount of the clutter map C subtracted from the CMO output; we can thus similarly predict the pixel level expected on objects of a given contrast in the final MWT output.[16] The algorithm is very robust; it requires only the height and width of the largest object and the lowest contrast object we wish to detect; the MWT output threshold for a given object contrast is known.

The third detection algorithm we consider is a Gabor basis function (GBF) filter.[18] The filter $\underline{H}$ in the frequency domain is a linear combination (weights $\underline{w}$) of the GBFs of N_C different object aspect views and N_U typical clutter particles, i.e. $\underline{H} = \underline{M}\ \underline{w}$, where the columns of the matrix $\underline{M}$ are the Fourier transform (FT) of the Gabor functions of the $N_T = N_C + N_U$ images. We require the correlation of $\underline{h}$ and all N_C distorted object aspect view

(a) input (b) clutter map (c) MWT output

Fig. 19: MWT detection algorithm with adaptive clutter map (low clutter input).

(a) input (b) clutter map (c) MWT output

Fig. 20: MWT detection algorithm with adaptive clutter map (high clutter input)

images to be 1 (this is the peak constraint for object detection). We also require the correlation plane energy E_N due to the N_U clutter images to be reduced (this makes the filter selective and reduces P_{FA}) and we require the correlation function for the N_C Object images to have a Gaussian shape $\underline{f}$ (with Fourier transform $\underline{F}$). The constrained optimization solution, using Lagrange multipliers, is[18]

$$\underline{H} = \underline{M}\left[\underline{M}^{+}\underline{D}\underline{M}\right]^{-1} \times$$

$$\left[\underline{M}^{+}\underline{A}\underline{F} + \underline{M}^{+}\underline{X}\left\{\left[\underline{X}^{+}\underline{M}\left(\underline{M}^{+}\underline{D}\underline{M}\right)^{-1}\underline{M}^{+}\underline{X}\right]^{-1}\left[\underline{u} - \underline{X}^{+}\underline{M}\left(\underline{M}^{+}\underline{D}\underline{M}\right)^{-1}\underline{M}^{+}\underline{A}\underline{f}\right]\right\}\right] . \quad (11)$$

In (11), the columns of $\underline{M}$ are the FTs of all N_T Gabor functions, the elements of the diagonal matrix are sum of FTs of the N_T images, and the columns of data matrix $\underline{X}$ are FTs of N_C images. The is the second best algorithm (after the MWT). We add new object aspect views of training set objects to the filter, until all of the training set object aspect views give peaks ≥ 0.75. For the TRIM database (126 low contrast object-aspect views, 21 aspect views $\times$ 6 object classes), we found that only $N_C = 3$ object aspect views were necessary to recognize the entire training set. Use of only the GF (size etc.) for the N_C objects provides the good detection noted of different objects, since only partial object information is present in one Gabor function. We add N_C different clutter particle scenes

(ϕ in the GF for these is orthogonal to that for the N_C object GFs). This allows us to control the amount by which the E_N and Gaussian shape objective functions are minimized, i.e. *we control the degrees of freedom* by use of only a few N_U images. Only $N_U = 3$ clutter particle images are required for the TRIM database.

5. FEATURE EXTRACTION

Once regions of interest (ROI) where targets may occur have been selected, the contents of each ROI are enhanced. Features are then obtained from local ROI detection regions. The simplified version of Fig. 1 shown in Fig. 21 with different filter functions as computer generated hologram (CGH) filters produce different output feature spaces.[3,19] Table 1 notes some of the different optical feature spaces that can and have been produced with different filter functions. This provides an even larger set of operations achievable on the same basic optical architecture (Fig. 1 or 2). In general, all of the different functions and steps or levels of computer vision are needed. However, in certain cases (e.g. robotics, product inspection, etc.), feature extraction can be applied to the input scene and the results fed directly to a classifier. Fig. 22 shows this concept for a product inspection application.

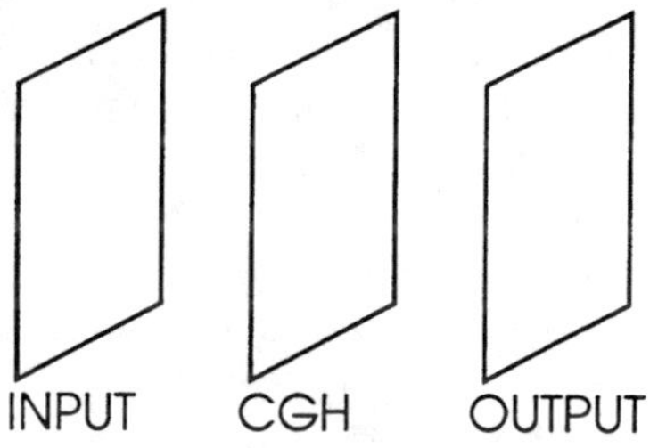

Fig. 21: Optical feature generation.

Table 1: Optically generated features.

Feature Space	P_2 filter CGH
Fourier Transform (FT)	Lens $\exp[jk(x^2+y^2)/2f_L]$
Wedge Ring FT Samples	Wedge/ring shaped FT elements
Moments	Polynomial functions
Hough Transform	Sum of cylindrical lenses
Polar Mellin Transform	Log polar coordinate transform
Wavelet Features	Wavelet/Gabor filter
Chord Distributions	Correlation of wedge ring data

6. P_{FA} REDUCTION (ALGORITHM FUSION)

No one detection algorithm is adequate (in terms of P_D and P_{FA} performance). To reduce the number of false alarms from the detection stage, we apply several different

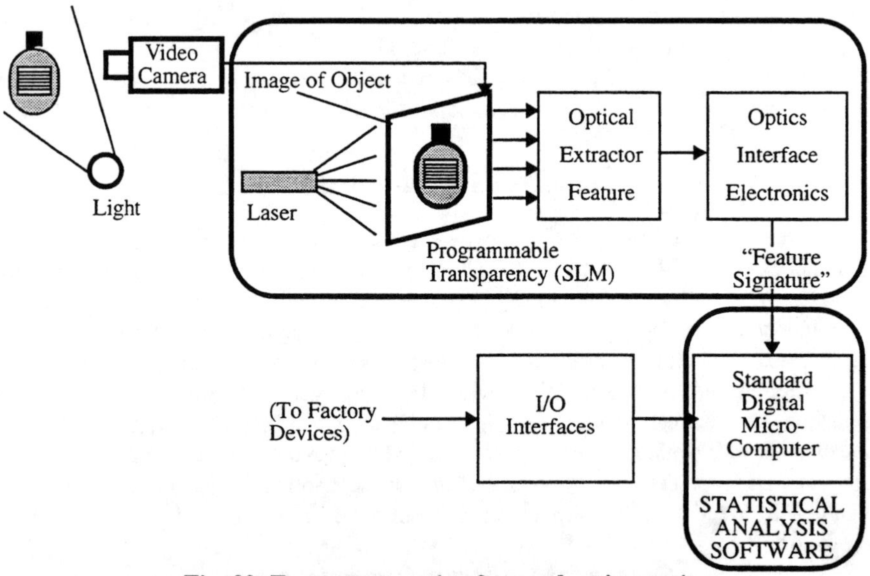

Fig. 22: Feature extraction for product inspection.

detection algorithms to the same scene and fuse the results from the different outputs. This technique has been shown[15,20] to be very attractive (P_{FA} is reduced by a factor of 10-20) since different algorithms give different false alarms and a false alarm occurs in the final fused output only if a false alarm occurs in the same pixel location in different detection algorithm outputs. Fig. 23 shows the basic concept. Fig. 24 shows a typical result.

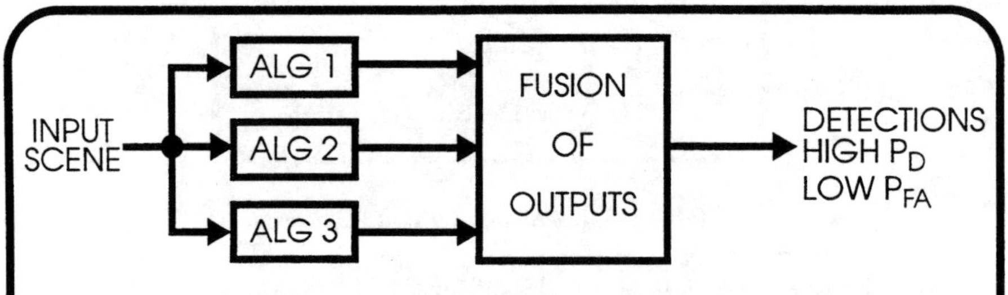

Fig. 23: Algorithm fusion concept.

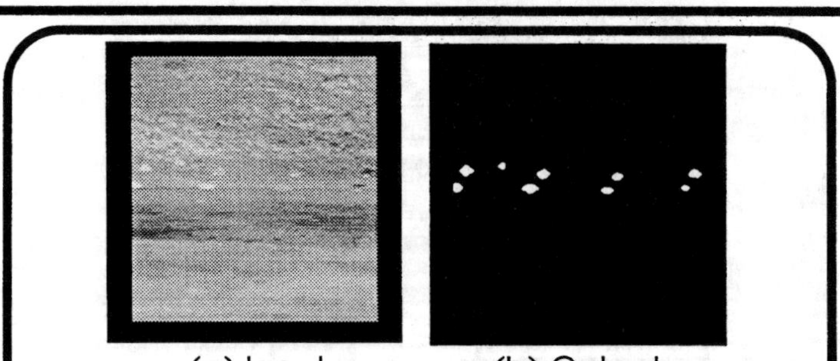

(a) Input (b) Output

Fig. 24: Typical algorithm fusion result.

Advanced fusion techniques[21] achieve improved results. An attractive technique that improves P_D and P_{FA} is to produce a gray scale (rather than a binary) fused output plane. In this case, we reduce the threshold, when a peak is encountered, we record its location and remove an elliptical region of pixels around each peak location (we use a 30×80 ellipse for the TRIM database), we then reduce the threshold until the next peak (candidate ROI) is located. This procedure is repeated until the threshold corresponding to the lowest contrast object is reached (or until many peaks, i.e. false alarms, occur). Fig. 25 shows this new detection technique, we refer to as peak sorting. It greatly improves results on individual detection algorithms and in fused data (clutter false alarms near objects are omitted, only one clutter peak is recorded in a 30×80 region, vs. many separate peaks, etc.).

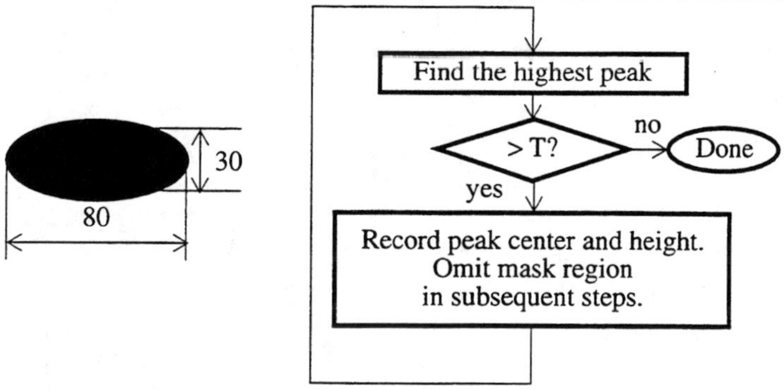

(a) region mask (b) peak sorting method

Fig. 25: Flow graph for the peak sorting detection scoring algorithm.

Another very attractive technique (Fig. 26) is to apply different mapping functions to the gray scale outputs from different detection algorithms (this uses fuzzy logic membership function methods).[21] These mapping functions produce similar target output levels from different detection algorithms.[21] We also use different mapping functions for objects with different contrasts (our peak sorting detection algorithm first detects high contrast objects, then low contrast ones, etc.). Table 2 gives typical data for 2 different detection algorithms and for their fusion. We note that the classifier we applied to these detection data ROIs (object and clutter) was able to reject all but one clutter false alarm[11] (thus, the final P_{FA} is < 0.01 per scene). The P_D values given are the percent of test objects detected (the test objects differed in aspect view by 15° from training set objects and included > 50% very low contrast objects). The P_{FA} listed is the average number of false alarms per scene.

Table 2: Improved performance with algorithm fusion.

MWT Algorithm		GBF Algorithm		Fusion	
P_D	P_{FA}	P_D	P_{FA}	P_D	P_{FA}
98.7%	17.4	98.7%	69	98.7%	10.3
97.4%	10.5	96.0%	55	97.4%	9.2
—	—	—	—	93.1%	4.0
91%	4.1	90%	43	90.6%	3.1
—	—	—	—	80.0%	0.9

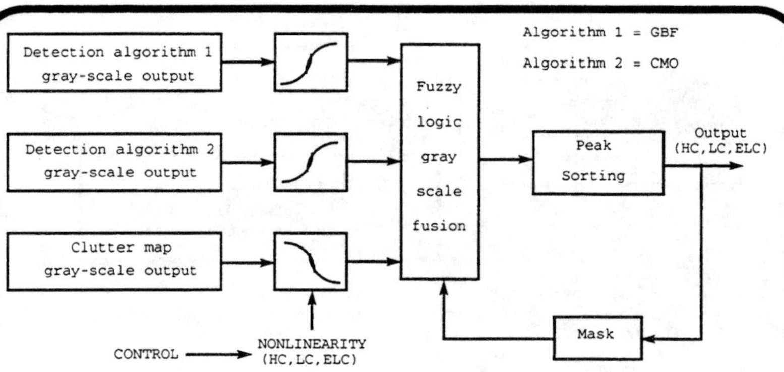

Fig. 26: Block diagram for hierarchical algorithm fusion.

7. REAL TIME HARDWARE

Several real-time 2-D optical correlators have been fabricated in small compact hardware under the TOPS program.[22] The systems have real time input (P_1) and filter (P_2) devices and operate at ≈1000 correlations per second. Present P_1 and P_2 devices are 256×256 (but are binary); other devices with gray-scale operating at 30 frames/sec are being integrated into similar small rugged hardware; devices with gray-scale and faster (1000 frame/sec) rates are in advanced research. All filter functions and operations described are realizable on the same optical correlator hardware.

Fig. 27 shows the schematic and a photograph of the 2-D correlator developed by Martin Marietta-Denver and Boulder Nonlinear Systems Inc. In Fig. 27a, the laser diode's output is collimated, reflects from the beam splitter and then from the input SLM-1. The reflected input passes back through the beam splitter, reflects from a mirror and its Fourier transform is produced by FT lens-1 at a plane where the filter SLM-2 is placed. The reflected light leaving the filter is Fourier transformed by FT lens-2 and the correlation is sensed on the 2-D detector array. The drive cards shown in Fig. 27a provide electrical input to the two SLMs and the laser diode. Fig. 27b shows a photograph of the final system delivered.

Two other well-engineered 2-D optical correlator hardware systems should be noted. The system fabricated by Litton Data Systems and Carnegie Mellon University for Eglin Air Force Base uses new magneto optic SLMs operating in reflection mode. These provide much smaller SLMs (pixels on 22 μm centers for 21 line pair/mm resolution) about 3 mm on a side (for 128×128 pixel SLM devices). The fabricated hardware is shown in Fig. 28. The laser diode source is collimated, addresses the input SLM, the reflected light strikes a mirror and then addresses the filter plane SLM. The reflected light then strikes a second mirror and the correlation appears on the CCD detector array. Curved slightly toric mirrors (with two radii of curvatures) produce the Fourier transforms. The SLMs used in the initial system are 128×128 (with extension to 512×512 pixel devices being near term efforts). The system occupies only 25 in^3, weighs less than 40 pounds and consumes 160 W of power. It is quite rugged with temperature, shock, and vibration effects allowed. It will be tested in a remotely piloted vehicle to identify three military targets at all aspect angles over a range of 20°-50° of elevation angles. The SLMs support only binary phase data. However, the major achievement of this system is again its optical engineering and the fact that the SLMs used provide operation at 1000 frame/sec rates. Thus, they are quite competitive with digital signal processing chips and processors.

The final real-time 2-D optical correlator hardware system noted was fabricated by Sandia Laboratories.[23] It is quite different from the prior ones since it implements a noncoherent correlation and uses a 1-D acousto-optic SLM and a 1-D vertical array of N light emitting diodes (LEDs) as the input data transducers. The output from each LED is expanded horizontally and imaged through the acousto-optic device at different angles and onto separate rows of a 2-D output detector array. The 2-D reference image is fed time-sequentially to the LED inputs while one row of the 2-D input image is fed to the acousto-optic device. The output on the 2-D detector array is N 1-D correlations. These output data are shifted vertically, the reference data is repeated to the LEDs with the second row of the input image fed to the acousto-optic device, and the new output is added (by charge integration) to the prior shifted result. This process continues for the different rows of the input image and the 2-D correlation of the input and reference results. The present hardware can produce about 1000 correlations per sec on 1000×1000 point data using 32×32 point reference images. A major attractiveness of this processor is that it allows use of gray-scale input and reference functions. It occupies less than 1 ft^3 of volume.

8. CLASSIFIERS

This last stage of a general scene analysis system operates on ROIs or features extracted from ROIs to determine the class etc. of the contents of each ROI. Neural net classifiers are the most attractive since they are data driven, adaptive, robust and parallel. Optical interconnections are useful here, but the need for optical processing is of less concern. The feature space trajectory neural net (FST NN)[24] is very attractive as it overcomes small training set size problems, ad hoc parameter selection, problems when the number of hidden layer neurons is chosen too large or too small, etc. It also allows non target data to be rejected and it provides pose estimates of each object.

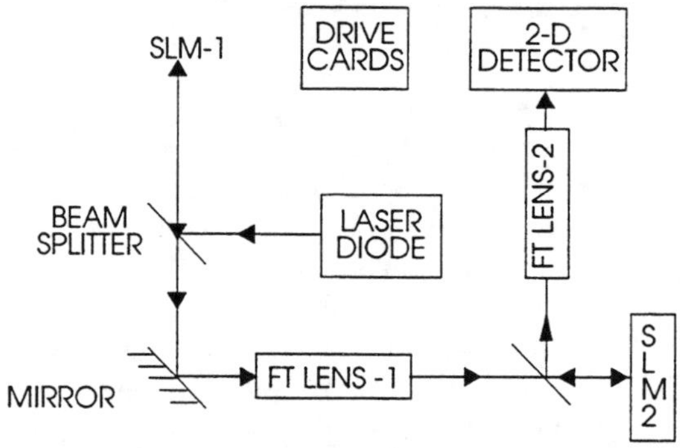

(a) Schematic

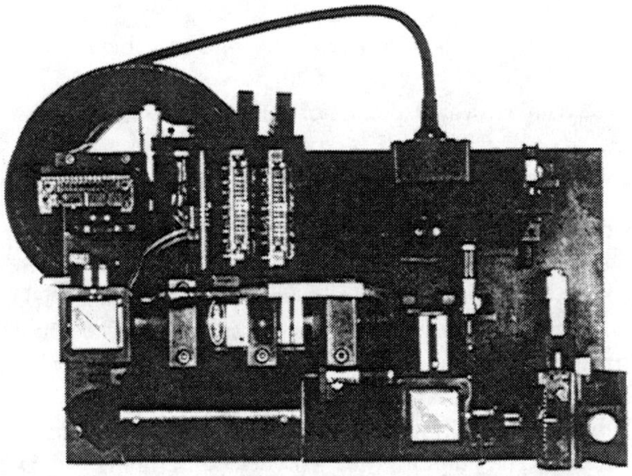

(b) Photograph

Fig. 27: Martin Marietta/Boulder nonlinear systems 2-D optical correlator.

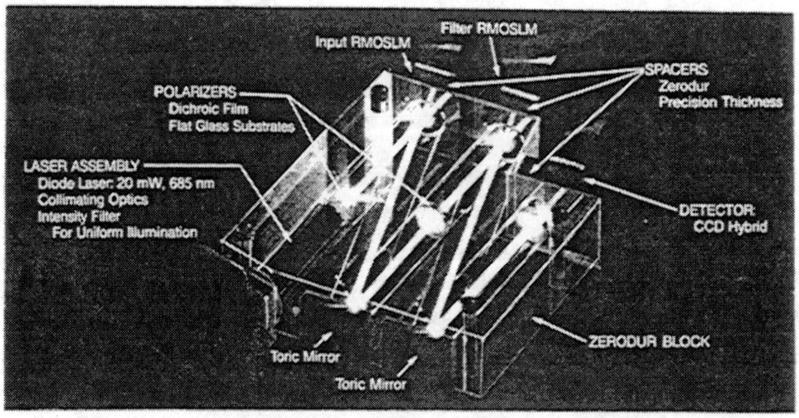

Fig. 28: Eglin Air Force Base/Litton Data Systems compact 2-D optical correlator.

9. SUMMARY

New algorithms have been detailed that greatly increase the number of operations achievable on a common optical correlator architecture by use of different filter functions. We have discussed many of these, provided examples of many, and have described their use in a general scene and system. Optical real-time hardware to achieve these operations is also noted.

ACKNOWLEDGMENT

The support of the optical realization of the algorithms noted and much of the Gabor filter work by an ASSERT Grant from the Office of Naval Research and the support of the detection algorithms by a contract from the Advanced Research Project Agency monitored by the Naval Surface Warfare Center (China Lake) are gratefully acknowledged.

REFERENCES

1) B. Bhanu, "Automatic target recognition: a state of the art survey", *IEEE Trans. Aeros. Elec. Sys.*, vol. AES-22, pp. 364-379, 1986.
2) D. Casasent, "An optical correlator, feature extractor, neural net system", *Opt. Engr.*, vol. 31, no. 5, pp. 972-978, May 1992.
3) D. Casasent, "General-purpose optical pattern recognition image processors", *Proc. IEEE*, vol. 82, no. 11, pp. 1724-1734, November 1994.
4) D. Casasent and E. Botha, "Optical symbolic substitution for morphological transformation", *Applied Optics*, vol. 27, pp. 3806-3810, September 1988.
5) R. Schaefer and D. Casasent, "Nonlinear optical hit-miss transform for detection", *Applied Optics*, vol. 34, no. 20, pp. 3869-3882, 10 July 1995.
6) D. Casasent and J.S. Smokelin, "Real, imaginary and clutter Gabor filter fusion for detection with reduced false alarms", *Opt. Engr.*, vol. 33, no. 7, pp. 2255-2263, July 1994.
7) D. Casasent, D. Weber and M. Sipe, "Nonlinear fusion of Gabor wavelet filters for locating objects, edges and clutter", *Proc. SPIE*, vol. 2762, April 1996.
8) D. Casasent, "Gabor wavelet filters and fusion for distortion-invariant multi-class object detection", *Proc. SPIE*, vol. 2491, pp. 430-440, April 1995.
9) *Optical Pattern Recognition VI*, D. Casasent and T.H. Chao, eds., *Proc. SPIE*, vol. 2490, April 1995.
10) *Optical Pattern Recognition VII*, D. Casasent and T.H. Chao, eds., *Proc. SPIE*, vol. 2752, April 1996.
11) "L. Neiberg, D. Casasent and A. Talukder, "Feature space trajectory neural net classifier: confidences and thresholds for clutter and low contrast objects", *Proc. SPIE*, vol. 2760, April 1996.
12) D. Casasent, "An inference approach to optical scene analysis" (invited paper), *Euro-American Workshop on Optical Pattern Recognition*, La Rochelle, France, June 1994.
13) D. Casasent and J-S. Smokelin, "Neural net design of macro Gabor wavelet filters for distortion-invariant object detection in clutter", *Opt. Engr.*, vol. 33, no. 7, pp. 2264-2271, July 1994.
14) A. Ye and D. Casasent, "Morphological and wavelet transforms for object detection and image processing", *Appl. Opt.*, vol. 33, no. 35, pp. 8226-8239, 10 December 1994.

15) D. Casasent and A. Ye, "Detection filters and algorithm fusion for ATR", *IEEE Trans. on Image Processing*, 1996.
16) A. Ye and D. Casasent, "Detection filters using wavelets, Gabor, morphology, and fusion", *Proc. SPIE*, vol. 2490, pp. 23-32, April 1995.
17) R. Mehrotra et al, "Gabor filter-based edge detection", *Pattern Recognition*, vol. 25, pp. 1479-1494, 1992.
18) A. Iyer and D. Casasent, "New basis function distortion-invariant detection filters", *Proc. SPIE*, vol. 2297, pp. 4-9, July 1994.
19) D. Casasent, "Computer generated holograms in pattern recognition: a review", *Opt. Engr.,* vol. 24, pp. 724-730, September/October 1985.
20) D. Casasent, A. Ye, J.S. Smokelin and R. Schaefer, "Optical correlation filter fusion for object detection", *Opt. Engr.*, vol. 33, no. 6, pp. 1757-1767, June 1994.
21) D. Casasent, A. Ye, and A. Talukder, "Algorithm fusion for detection with reduced P_{FA}", *Proc. SPIE*, vol. 2752, April 1996.
22) *Transition of Optical Processors into Systems 1993*, D. Casasent, ed., *Proc. SPIE*, vol. 1958, April 1993; *Transition of Optical Processors into Systems 1994*, D. Casasent, ed., *Proc. SPIE*, vol. 2236, April 1994; *Transition of Optical Processors into Systems 1995*, D. Casasent, ed., *Proc. SPIE*, vol. 2489, April 1995.
23) P.A. Molley and K.T. Stalker, "Acousto-optic signal processing for real-time image recognition", *Optical Engineering*, vol. 29, no. 9, pp. 1073-1080, September 1990.
24) D. Casasent and L. Neiberg, "Classifier and shift-invariant ATR neural net networks", *Neural Networks*, vol. 8, no. 7/8, special issue on Automatic Target Recognition, pp. 117-1129.

Distortion invariant pattern recognition using fourier plane nonlinear filters

Bahram Javidi and Dean Painchaud

University of Connecticut
Department of Electrical Engineering
Storrs, Connecticut 06269-3157

ABSTRACT

The use of nonlinear techniques in the Fourier plane of pattern recognition correlators can improve the correlators' performance in terms of discrimination against objects similar to the target object, correlation peak sharpness, and correlation noise robustness. Additionally, filter designs have been proposed which provide the linear correlator with invariance properties with respect to input signal distortions and rotations. In this paper, we propose simple modifications to presently known distortion invariant correlator filters which enable these filter designs to be used in a nonlinear correlator architecture. These Fourier plane nonlinear filters can be implemented electronically, or may be implemented optically using a nonlinear joint transform correlator. Extensive simulation results are presented which illustrate the performance enhancements that are gained by the unification of nonlinear techniques with these filter designs.

1. INTRODUCTION

Optical techniques have a great potential for solving pattern recognition problems. The proposal of the frequency plane optical correlator[1] and the joint transform optical correlator[2] in the 1960's gave us the optical techniques necessary to attack the problem of pattern recognition. In the correlation approach to solving the pattern recognition problem,[3,4] the correlation between a known reference signal and an unknown "scene" signal is computed in hopes that a high degree of correlation will be obtained at points in the "scene" signal where the reference signal is present. The correlation approach requires the design of a reference signal, or "correlation filter."[3,4] The design of filters for these correlators has led to a wide selection of filters. Each new filter design that has been proposed has attempted to improve correlator performance with respect to specific performance criteria (e.g. noise immunity, distortion invariance, discriminability, etc). Optimal filter designs with respect to these criteria have been proposed and tested[5-7] which illustrate the tradeoffs encountered between the various optical correlator performance criteria. Additionally, a constraint often placed on the filter design is laboratory feasibility, i.e. the ability to implement the filter using presently available electronic or optical technology.

It has been shown that the use of nonlinearities in the Fourier plane of pattern recognition correlators can improve correlator performance.[8-10] One way of implementing these techniques is through the use of the joint transform correlator (JTC), although electronic realizations are also possible. When these nonlinearities have been applied to the pattern recognition problem, these techniques have proven themselves useful in improving the output correlation's peak sharpness and noise immunity. Additionally, it has been demonstrated that presently available optical spatial light modulators are able to implement these Fourier plane nonlinearities.[11]

Much work has been done to improve the performance of the frequency plane correlator when subjected to input signal distortions (e. g. signal rotations and signal views from different perspectives with respect to the reference signal)[3,4]. As a result of this work, many "composite" filter designs have been proposed which attempt to provide distortion invariance to the frequency plane correlator. When used with nonlinear correlators such as the nonlinear JTC, however, these filter designs may not perform as well as in the frequency plane correlator, since they were conceived and designed to be used in a linear optical correlator. In response to this incompatibility, we propose simple modifications to these filter designs which enable these filter designs to be used in the nonlinear JTC; we shall refer to these specially designed filters as Fourier plane nonlinear filters. In this paper, we will extensively examine the performance of the nonlinear correlator in terms of its immunity to both in-plane distortions (e.g. signal rotations) and out-of-plane distortions (e.g. views of an object from different angles) when these specially designed distortion invariant filters are employed.

We begin with a review of the relevant results from the theory of the Synthetic Discriminant Function (SDF) filters[5,12]. Subsequently, a brief discussion of the nonlinear joint transform correlator[8] is undertaken. The seemingly disjoint topics of nonlinear correlation and SDF filter design are finally brought together as the previously presented SDF filters are extended for use in the nonlinear JTC. We note that we are not proposing optimal filter designs here, but rather are proposing filter modifications which enable SDF filters to be used in the nonlinear JTC. Computer simulation results are presented which demonstrate that these modifications provide filters which demonstrate performance exceeding that of their linear counterparts when used in the nonlinear JTC. The results of simulations using the linear counterparts to the above "nonlinearly transformed" filters, in a linear correlator, are also presented to provide a baseline for comparison. A description of the test images used and the correlator performance metrics employed to evaluate the filters is included. We conclude the paper with a discussion of the simulation results and a comparison of the results with the relevant theory.

2. BACKGROUND

2.1. Distortion invariant pattern recognition

For any pattern or object recognition system to be useful, it must recognize patterns and objects viewed from various angles and perspectives. To the correlator, these perspectives can be categorized as signal distortions. Several composite filter designs have been proposed to perform distortion invariant pattern recognition, but here, we concentrate on the classic equal correlation peak (ECP) Synthetic Discriminant Function (SDF) filter[12] and one of its derivatives, the minimum average correlation energy (MACE) SDF filter[5]. We note that major improvements in terms of SDF filter performance have been achieved since the development of the ECP and MACE SDF filters[4]. We work here, however, with the ECP and MACE SDF filters because they are computationally simpler than newer SDF filters such as the Optimal Tradeoff (OT) SDF filter[7]. At the same time, the ECP and MACE designs can be used to clearly illustrate the benefits of the nonlinear techniques we are proposing.

In the ECP SDF filter, proposed by Hester and Casasent[12], the objective is to use a linear combination of reference (or training) images to form a composite image, to then cross-correlate unknown inputs with this image, $h(x, y)$. Weights for this linear combination could be selected such that the cross-correlation output origin value would be the same for all images belonging to one class. A key assumption to this filter design is that each class of images that must be discriminated is characterized by a linear subspace spanned by the N training images available for that class.

By sampling the training images $s_1(x, y)$, $s_2(x, y)$, ..., $s_N(x, y)$ we can obtain matrices with $m \times n$ pixels in each. d-dimensional column vectors $\mathbf{s}_1$, $\mathbf{s}_2$, ... , $\mathbf{s}_N$ are then generated from these matrices by taking the rows of a matrix and linking their transposes end to create a $d = m \times n$ vector. This image scanning is performed from left to right and from top to bottom. We will refer to this technique as "lexicographical scanning" throughout the rest of this paper, in keeping with the previous literature. Additionally, we set $h(x, y) = \mathbf{h}$, a d-dimensional column vector representing the digital composite SDF filter. Using such vectors, we can define a matrix $\mathbf{S}$ with vector $\mathbf{s}_i$ as its ith column. Using such notation, the ECP SDF filter has been given as

$$\mathbf{h}_{\text{ECP}} = \mathbf{S}(\mathbf{S}^{+}\mathbf{S})^{-1}\mathbf{c}^{*}, \tag{1}$$

where $\mathbf{S}^{+}$ is the complex conjugate transpose of $\mathbf{S}$, the vector $\mathbf{c}$ contains the desired cross-correlation output origin values for each training image vector $\mathbf{s}_i$, and the notation $\mathbf{c}^{*}$ denotes the complex conjugate of $\mathbf{c}$. This resulting $\mathbf{h}_{\text{ECP}}$ vector can be reverse lexicographically scanned for use in the JTC.

The MACE SDF filter[5,13] was designed to produce a sharp peak in the correlation plane, while reducing the extraneous peaks not occurring at the origin. This is critical to discriminating the various image classes, but is also critical to locating the correlation peak in noisy target location applications, where the output correlation plane contains many large peaks. A major tradeoff made in the design of MACE filters, however, is that the filters exhibit poor noise tolerance. The MACE SDF filter is more aptly described in the frequency domain, and seeks to minimize the average correlation plane energy

$$E_{\text{ave}} = \frac{1}{N}\sum_{i=1}^{N}\sum_{n=1}^{d}\left|\hat{\mathbf{h}}(n)\right|^{2}\left|\hat{\mathbf{s}}_i(n)\right|^{2}, \tag{2}$$

where Parseval's theorem has been used in deriving the frequency-domain equation for E_{ave}, and $\hat{\mathbf{h}}$ and $\hat{\mathbf{s}}_i$ are d-dimensional lexicographically scanned vectors representing $H(f_x, f_y)$ and $S_i(f_x, f_y)$, the filter and training image 2-dimensional Fourier transforms respectively.

Minimizing E_{ave} is an attempt to minimize the sidelobes as low as possible. These sidelobes can be thought of as peaks in the correlation plane other than the true cross-correlation output origin value. Letting $\hat{\mathbf{S}}$ denote a $d \times$ N matrix with $\hat{\mathbf{s}}_i$ as its ith column, the resulting MACE filter[5] expression is given as

$$\hat{\mathbf{h}}_{\text{MACE}} = \hat{\mathbf{D}}^{-1}\hat{\mathbf{S}}(\hat{\mathbf{S}}^{+}\hat{\mathbf{D}}^{-1}\hat{\mathbf{S}})^{-1}\mathbf{c}^{*}, \tag{3}$$

where $\hat{\mathbf{D}}$ is a $d \times d$ diagonal matrix, with diagonal elements given by

$$\hat{\mathbf{D}}(r,r) = \frac{1}{N}\sum_{i=1}^{N} \hat{\mathbf{s}}_i(r)\hat{\mathbf{s}}_i^*(r), \quad r = 1, \ldots, d \tag{4}$$

Note that the inversion of the $\hat{\mathbf{D}}$ matrix is a simple task, since it is diagonal. As above, this resulting $\hat{\mathbf{h}}_{\mathrm{MACE}}$ vector can be reverse lexicographically scanned, and inverse Fourier transformed for use in the JTC.

2.2. Nonlinear joint transform correlation

The advantages of using nonlinearities in spatial filtering have been known for many years[9]. These techniques have recently been applied to the joint transform correlator[2] (JTC) where it has been shown that the performance of the JTC can be significantly improved through the use of nonlinearities in its Fourier plane[8]. This performance improvement includes a sharpening of the correlator output cross-correlation peak when the target signal is present in the scene (i.e., when a "match" is found between the signal and the reference). Also improved by these techniques is the correlator's discriminability, or its ability to distinguish between the presence of a target signal, and the presence a signal which is similar to the target signal, but which should not be detected as the target signal (i.e. low output cross-correlation peak is desirable). Additionally, correlator light efficiency and correlator noise immunity[14] can be improved through the use of these nonlinear techniques. When the nonlinear JTC has been compared to the linear frequency plane correlator, the performance of the nonlinear JTC has been shown to be superior for the test cases presented[15]. The JTC affords us an additional degree of freedom in that the nonlinearities can be applied to the input signal as well as to the reference signal, although electronic implementations would also provide this extra degree of freedom.

The nonlinear JTC applies a nonlinearity $g(E)$ to each point of the JTC Fourier plane cross-power spectrum before this spectrum is inverse Fourier transformed to generate the output correlation. For the specific case of the odd k^{th} law nonlinearity, described mathematically by

$$g(E) = \mathrm{sgn}(E)|E|^k, \qquad k \leq 1 \tag{5}$$

the resulting nonlinearly transformed joint power spectrum (JPS) can be expressed as an infinite sum of harmonic terms[8], the first of which is proportional to

$$g_1(E) \propto \left[|H(f_x,f_y)||S(f_x,f_y)|\right]^k \exp\{j[\phi_H(f_x,f_y) - \phi_S(f_x,f_y)]\}, \tag{6}$$

where $H(f_x, f_y)$ and $S(f_x, f_y)$ are the filter and input signal Fourier plane magnitudes, and $\phi_H(f_x, f_y)$ and $\phi_S(f_x,f_y)$ are the filter and input signal Fourier plane phase distributions, respectively. We note here that Eq. (6) describes a portion of the JPS minus the DC terms normally found in it. Our interest in this specific type of nonlinearity arises due to its analytical tractability. In addition, some real world optical devices approximately exhibit this nonlinear transfer function from input to output[11]. Various other types of nonlinearities can also be used as well. It is clear that the term given by Eq. (6) has the same phase as the linear JTC, but has distorted the amplitude of the linear JTC by raising it to the kth power. For $k = 1$, we obtain the same correlation output as the linear JTC, while for $k = 0$, i.e. a binarizing nonlinearity[16,17], we see that the nonlinearity has removed all amplitude information from the Fourier plane intensity and the resultant first harmonic distribution contains only phase information. This ease of gradually changing from a linear to a binary correlator is another attractive feature of the nonlinearity given by Eq. (6). We will focus on this first order harmonic term throughout the remainder of this paper, and will assume that the higher order harmonics[8] are spatially separated (diffracted off-axis) from the first order term in the correlation output plane.

3. EXTENSIONS OF SDF FILTERS FOR USE IN THE NONLINEAR JTC

As we can see from above, composite filter designs for use in correlators may be designed to optimize specific correlator output performance critera. However, the use of these filters in a nonlinear JTC requires the modification of the composite filter design to take into account the effects of the specific nonlinearity employed. Here, we attempt to unify, to some extent, the Synthetic Discriminant Function approach to distortion invariant pattern recognition and the use of nonlinearities in the Fourier plane of the joint transform correlator. As the use of nonlinearities has been shown to improve the discriminability of the JTC, this performance improvement does not come without a cost. The cost here is that the improvement in discriminability is offset by a degradation in the distortion invariance of the JTC. Preliminary studies of the nonlinear JTC's sensitivity to target signal in-plane rotations with respect to the reference signal have confirmed this degradation[18]. Consequently, we see the motivation for wanting to apply SDF based filter designs to the nonlinear JTC. In this section, we propose modifications for the previously described ECP and MACE SDF's. These modifications are made in an attempt to improve the distortion invariance of the kth law JTC, while retaining the optimization of the specific performance criteria associated with each of the SDF filter functions.

3.1. The kth law ECP SDF filter

We present the extension of the ECP SDF first; the theory is simple and straightforward. It can be easily shown that the frequency domain expression for $\mathbf{h}_{ECP}$ is given by

$$\hat{\mathbf{h}}_{ECP} = \hat{\mathbf{S}}(\hat{\mathbf{S}}^{+}\hat{\mathbf{S}})^{-1}\mathbf{c}^{*}, \tag{7}$$

where, as before, the notation $\hat{\mathbf{f}}$ denotes the lexicographically scanned frequency domain version (actually the Discrete Fourier Transform, or DFT) of $\mathbf{f}$, a spatial domain entity.

We can also define the following vector operation for any complex vector $\mathbf{v}$,

$$\mathbf{v}^{k} \triangleq \begin{bmatrix} |\mathbf{v}[1]|^{k} \exp[j\phi_{\mathbf{v}}[1]] \\ |\mathbf{v}[2]|^{k} \exp[j\phi_{\mathbf{v}}[2]] \\ \vdots \\ |\mathbf{v}[d]|^{k} \exp[j\phi_{\mathbf{v}}[d]] \end{bmatrix}. \tag{8}$$

Now, we can sample and lexicographically scan $S_n(f_x, f_y)$ for each of the N training images to form $\hat{\mathbf{S}}^k = [\hat{\mathbf{s}}_1{}^k \quad \hat{\mathbf{s}}_2{}^k \quad \cdots \quad \hat{\mathbf{s}}_N{}^k]$ using the above defined vector operator. Comparing the expressions for $\hat{\mathbf{h}}_{\mathrm{ECP}}$ and the kth law JTC Fourier plane intensity expression, given by Eq. (6), we see that we must replace $\hat{\mathbf{S}}$ in the $\hat{\mathbf{h}}_{\mathrm{ECP}}$ expression with $\hat{\mathbf{S}}^k$ defined above, since the desired filter function $\hat{\mathbf{h}}^k$ operates on the distorted signal Fourier transforms, $\hat{\mathbf{s}}_n{}^k$, instead of the undistorted transforms $\hat{\mathbf{s}}_n$. We reason, therefore, that the kth law ECP SDF is given by

$$\hat{\mathbf{h}}_{\mathrm{ECP},k}{}^k = \hat{\mathbf{S}}^k((\hat{\mathbf{S}}^k)^+\hat{\mathbf{S}}^k)^{-1}\mathbf{c}^* . \tag{9}$$

Finally, we define the following vector operation for any complex vector $\mathbf{v}$,

$$\mathbf{v}^{\frac{1}{k}} \triangleq \begin{bmatrix} |\mathbf{v}[1]|^{\frac{1}{k}} \exp[j\phi_{\mathbf{v}}[1]] \\ |\mathbf{v}[2]|^{\frac{1}{k}} \exp[j\phi_{\mathbf{v}}[2]] \\ \vdots \\ |\mathbf{v}[d]|^{\frac{1}{k}} \exp[j\phi_{\mathbf{v}}[d]] \end{bmatrix} \tag{10}$$

and we arrive at the frequency domain k^{th} law ECP SDF as

$$\hat{\mathbf{h}}_{\mathrm{ECP},k} = (\hat{\mathbf{S}}^k((\hat{\mathbf{S}}^k)^+\hat{\mathbf{S}}^k)^{-1}\mathbf{c}^*)^{\frac{1}{k}} , \tag{11}$$

where $(\ )^{-1}$ denotes the matrix inverse. The inverse discrete Fourier transform of $\hat{\mathbf{h}}_{\mathrm{ECP},k}$ above will provide the spatial domain k^{th} law ECP SDF,

$$\mathbf{h}_{\mathrm{ECP},k} = \mathrm{F}^{-1}\left\{(\hat{\mathbf{S}}^k((\hat{\mathbf{S}}^k)^+\hat{\mathbf{S}}^k)^{-1}\mathbf{c}^*)^{\frac{1}{k}}\right\} . \tag{12}$$

Reviewing Eq. (12), one can see that the training image Fourier transforms must be "pre-warped" using the above defined vector operator $\mathbf{v}^k$ before being used in the computation of $\mathrm{h}_{\mathrm{ECP},k}$. $\mathrm{h}_{\mathrm{ECP},k}$ is then "un-warped" using the above defined vector operator $\mathbf{v}^{1/k}$, reverse lexicographically scanned, and inverse discrete Fourier transformed before being placed in the input plane of the nonlinear JTC. We must work in the frequency domain due to the nonlinear spectral distortion of the kth law JTC. By the fact that the training images used to construct the filter are real valued images, and by the symmetry properties of Fourier transforms, we can see that the spatial domain k^{th} law ECP SDF filter will be a real valued filter.

3.2. The kth law MACE SDF filter

By inspection, with the above discussion in mind, we can write the spatial domain k^{th} law MACE SDF filter as

$$\mathbf{h}_{\text{MACE},k} = \mathrm{F}^{-1}\left\{\left[(\hat{\mathbf{D}}^k)^{-1}\hat{\mathbf{S}}^k((\hat{\mathbf{S}}^k)^{+}(\hat{\mathbf{D}}^k)^{-1}\hat{\mathbf{S}}^k)^{-1}\mathbf{c}^{*}\right]^{\frac{1}{k}}\right\}, \tag{13}$$

where $\hat{\mathbf{D}}^k$ is a $d \times d$ diagonal matrix given by

$$\hat{\mathbf{D}}^k(j,k) = \begin{cases} \sum_{i=1}^{N} \left|\hat{\mathbf{s}}_i^{\,k}(k)\right|^2 & , \quad j = k = 1, \ldots, d \\ 0 & , \quad j \neq k \end{cases}. \tag{14}$$

In other words, the diagonal of $\hat{\mathbf{D}}^k$ contains the sum of the kth law transformed training image spectral magnitudes. As above, by the fact that the training images used to construct the filter are real valued images, and by the symmetry properties of Fourier transforms, we can see that the spatial domain k^{th} law MACE SDF filter will also be a real valued filter.

Based on the above two k^{th} law SDF filter designs presented, it is clear that these nonlinear design techniques can be used to implement any classical composite filter. Specifically, these techniques can also be extended to the MV SDF and OT SDF filters previously mentioned.

4. COMPUTER SIMULATIONS

4.1. Correlator performance metrics

To properly evaluate the performance of a filter in a correlator, several criteria must be compared. As we have seen, correlation filters may be designed to optimize particular correlator output performance criteria, e.g. peak location error, correlation energy, peak sharpness, or peak variance in the presence of input noise. Kumar and Hassebrook presented a review of the most popular metrics[14], while Horner has suggested an improved metric which was shown to be independent of the correlator output plane mean value[19]. Although we do not present any algorithms in this paper for deciding whether a target is present or absent in an input scene based on the correlator output plane, we shall select a set of metrics which measure correlator output characteristics important to such a decision. All metrics are computed based upon an output correlation plane with its axes centered upon the theoretical correlation output function peak value (i.e. value set to a constant in the SDF filter formulation above).

When evaluating the joint transform correlator as a useful pattern recognition system, we shall use the following key criteria and metrics to measure and quantify the performance of the proposed SDF filter designs. We work with correlation plane intensity only in these metrics, since most real-world detectors can only detect optical signal intensity, but not optical signal phase. The notation $|\ \ |$ is used to denote magnitude, and E{ } denotes the expected value, computed where multiple simulations with noise are performed.

Correlation Output Peak Intensity:

To denote the maximum output correlation value, the following shall be used,

$$R_o^2 = \max_{x,y}\left(|C(x,y)|^2\right), \tag{15}$$

where R_o^2 is the output correlation peak value, and $C(x, y)$ is the output correlation value at position (x, y). Ideally, this value should be located at $(x, y) = (0, 0)$, the value forced to a constant value in the SDF filter formulation.

Correlation Output Maximum Sidelobe Intensity:

We compute the maximum side lobe intensity in the correlation output plane based upon R_o^2 above by computing the maximum correlation output plane value outside an annular region centered at the position of R_o^2. The radius of this annular region is given by the average radius at which the output correlation function drops to $R_o^2/2$,

$$SL = \max_{\substack{x,y>\text{circular region}\\ \text{centered at position}\\ \text{of } R_o(x,y) \text{ and with}\\ \text{diameter given by}\\ R_o^2(x,y)/2}} \left(|C(x,y)|^2\right). \tag{16}$$

Correlation Output Peak-to-Sidelobe Ratio:

We desire that the ratio,

$$R_o SL = \frac{R_o^2}{SL}, \tag{17}$$

be as large as possible.

Correlation Output Plane Peak-to-Correlation Energy Ratio:

One measure of correlator efficiency, and peak sharpness, is the peak-to-correlation-energy metric proposed by Horner[19]. As detailed by Horner, this metric was shown to be independent of the correlator output plane mean value. This peak-to-correlation-energy is given by

$$PCE = \frac{\left(R_o^2 - \overline{|C(x,y)|^2}\right)}{\left[\sum\limits_{\substack{\text{all } x,y\\ x,y\neq 0}} \frac{\left(|C(x,y)|^2 - \overline{|C(x,y)|^2}\right)^2}{(N_x N_y - 1)}\right]^{\frac{1}{2}}}, \tag{18}$$

where

$$\overline{|C(x,y)|^2} = \sum_{\text{all } x,y} \frac{|C(x,y)|^2}{N_x N_y} \tag{19}$$

is the correlation output plane mean value, and N_x and N_y are sizes of the output correlation plane in the x and y directions respectively.

Target Discriminability:

Additionally, a means of determining the discriminability of the correlation filter under evaluation is required. Here, "discriminability" refers to the filter's ability to correctly distinguish between the target to which it has been trained and a similar, but not identical target. For this purpose, since there could be many "similar, but not identical targets", but only one true target, we require a large output correlation peak in response to the true target, and small output correlation peaks in response to any other targets, including "similar, but not identical targets". This criteria is embodied in the discriminability ratio

$$DR = \frac{\mathrm{E}\left\{R_{o,\text{target}}\right\}}{\mathrm{E}\left\{R_{o,\text{non-target}}\right\}}, \tag{20}$$

where $R_{o,\text{target}}$ and $R_{o,\text{nontarget}}$ are the output correlation peaks of the target and non-target images respectively. For a true measure of the capabilities of the correlator, the target and non-target images should be very similar to each other. Good performance would be indicated by a large DR.

Correlation Output Peak Signal-to-Noise Ratio[14]:

Finally, we require a metric to measure the variability of the output correlation peak in response to noise in the input scene of the the correlator. To compute this correlation peak signal-to-noise ratio, a large number of noise samples must be taken and the output correlation peaks due to a single input signal plus each of these samples must be recorded. The signal-to-noise ratio is thus given by

$$SNR = \frac{\mathrm{E}\left\{R_o^{\,2}\right\}}{\sqrt{\mathrm{E}\left\{\left(R_o^{\,2} - \mathrm{E}\left\{R_o^{\,2}\right\}\right)^2\right\}}}, \tag{21}$$

where

$$\mathrm{E}\left\{R_o^{\,2}\right\} = \sum_s \frac{R_{o,s}^{\,2}}{N_s} \tag{22}$$

is the mean value of the output correlation peak, $R_{o,s}^{\,2}$ is the output correlation peak value from simulation s, and

$$E\left\{\left(R_o^{\,2} - E\left\{R_o^{\,2}\right\}\right)^2\right\} = \sum_s \frac{\left(R_{o,s}^{\,2} - E\left\{R_o^{\,2}\right\}\right)^2}{N_s} \tag{23}$$

is the variance of the output correlation peak due to the input noise, and N_s is the number of noise ensembles used to compute the expected values.

4.2. The simulation environment, training images

Computer simulations were performed to evaluate the performance of the linear and nonlinear versions of both filter designs presented in the previous sections. We present here a description of the simulation test bed which was used. The correlation metrics described above were applied to the simulated output correlation plane, and the results are presented below.

Details of the Simulation Environment:

Because of the high degree of computational power required to perform the simulations, filters were limited to 128 × 128 pixels in size. In order to elimate the effects of circular convolution/correlation which arise when multiplying two DFT's, it was necessary to zero pad both the filter and signal images to increase their respective sizes to 256 × 256 pixels. This amount of zero padding is the minimum necessary to eliminate circular convolution /correlation effects. In performing the simulations, the filter and signal images were centered in these zero padded 256 × 256 pixel input images, and a 2-d Fast Fourier Transform (2-d FFT) algorithm was employed to compute the Discrete Fourier Transform of both of the signal and reference images which would be present in the JTC input. At this point, the kth law nonlinearity was applied to the normalized amplitudes of these transforms. These two Transforms were then multiplied together and the 2-*d* FFT algorithm applied to this result to compute the output plane. Without loss of generality, this technique simulates the intensity of only one of the identical nonlinear JTC output lobes in the JTC output correlation plane (the first harmonic only), and relieves the computational burden of the simulations substantially. In order to compute the performance metrics for the case in which the input signal is embedded in additive white noise, thirty simulations were run for each input signal. For each simulation, a new sample of pseudo-random noise was used, and the results of the thirty simulations were used to compute the statistical averages required by the above performance metrics.

A brief discussion concerning the training image and signal image pixel sizes is in order here. With the requirement that filter images (and consequently the signal images) be limited to 128 × 128 pixels, we found that training images needed to be scaled down in size to an average maximum size of 32 × 32 pixels before the SDF filters could be computed using these training images. Recall that the SDF filter designs are computed based on the samples of the training image Fourier Transforms, i.e. the training image DFT values. Since Fourier Transforms of these images may be rapidly varying in amplitude and phase over a relatively small number of samples, it was necessary to oversample the training image Fourier Transforms in order to adequately represent these variations. Experimentally, we found that zero padding the training images by a factor of 4 seemed to work well in our simulations. Zero padding by a factor of 4, coupled with the requirement that filters be limited to 128 × 128 pixels in size, resulted in the previously stated training image size of 32 × 32 pixels. We emphasize that the size chosen for the input images is not a consequence of nonlinear correlation, but is only due to a need to limit simulation computation time. Since we will not be designing SDF filters to be invariant to scale changes in a signal, (although we could just as easily as for the in-plane rotation case), the training image size of 32 × 32 pixels required that our correlator input signals be the same size as their corresponding training images.

A Note on Normalization and Quantization

As an initial step in computing all filters, all training and signal images were quantized to a value from 0 to 255. These images were then scaled such that each training and signal image had a maximum value of 1 and minimum value of 0. These normalized training image sets were used in computing the SDF filters using the expressions given above. As a final step in computing all filters, after the inverse DFT operation, the spatial domain filters were again normalized such that all filters pixel values were between +1 and −1. We are allowing for filter values to be both positive and negative without loss of generality. All nonlinear filter designs provide real valued spatial domain representations, by virtue of the symmetry of the training image Fourier transforms, and therefore the issue of representing complex valued filter distributions is not present here. No quantization was performed on the resulting spatial domain filters. The reason for this was to eliminate a possible source of error. The effects of such quantization are left for study in future work.

Training Image Sets and Test Images

The training images which we shall work with thoughout are summarized in Table 1, and are shown in Figs. 1 and 2. These images were chosen to simulate both in-plane distortions, and out-of-plane distortions. The aircraft images in Fig. 1 depict in-plane distortions, and were created from single baseline digitized images of both Mig and F14 images. A bilinear interpolation function was used[20] to create scaled and rotated versions of these images, and these derived images were used to build both training and signal images sets. The Mig and F14 aircraft images were chosen because of their similarity to each other. These "similar but not identical" images provide a good discriminability test of the correlation filter designs for the in-plane distortion case. A set of training images was built using the Mig images in 5° increments of rotation over an entire quadrant (0° to 90°, for a total of 19 images in the set). Two members of this set (0° and 45°) are shown in Fig. 1. This image set was used as training images to create the linear and nonlinearly transformed filters. An additional set of images was built using this image in 0.5° increments of rotation between 30° and 50°, for a total of 41 images in this set. This image set was used as the "signal" image in the simulations, in order to compute the various correlator performance metrics. Figure 1 also contains an example of Mig imagery embedded in normally distributed, zero mean, additive white noise with $\sigma = 0.2$. These noisy inputs were used to test filter performance. Additionally, a set of F14 images was built using images in 0.5° increments of rotation between 30° and 50°, for a total of 41 images as above for the Mig image. This image set was used to test discriminability ratio (*DR*) of the various filters, and was used only as the "signal" image in the simulations. The member of this set at 37.5° is shown in Fig. 1.

Table 1. Summary of training/test image sets.

Training Image Set	Figure	Used As	Tests Used In
Mig, rotated in 5° increments from 0° to 90°	1a, 1b	Filter References	in-plane dist. invariance, correlator performance, discriminability
Mig, rotated in 0.5° increments from 30° to 50°, embedded in σ = 0.2 noise	1d	Input Signals	in-plane dist. invariance, correlator performance
F14, rotated in 0.5° increments from 30° to 50°, embedded in σ = 0.2 noise	1c (shown w/o noise)	Input Signals	in-plane discriminability
Lamborghini, rotated in 5° increments from 0° to 90°	2a, 2b	Filter References	out-of-plane dist. invariance, correlator performance, discriminability
Lamborghini, rotated in 1° increments from 30° to 40°, embedded in $\sigma = 0.2$ noise	2d	Input Signals	out-of-plane dist. invariance, correlator performance
Ferrari, rotated in 1° increments from 30° to 40°, embedded in σ = 0.2 noise	2c (shown w/o noise)	Input Signals	out-of-plane discriminability

The automobile images in Fig. 2 depict out-of-plane distortions, and were created by digitizing images of both a Lamborghini and Ferrari taken at varying angles of view. The resulting images were scaled to the required size for simulation using the same bilinear interpolation function as was used for the aircraft images described above. The Lamborghini and Ferrari automobile images were also chosen because of their similarity to each other. These "similar but not identical" images provide a good discriminability test of the correlation filter designs for the out-of-plane distortion case. A set of training images was built using the Lamborghini images in 5° increments of rotation over an entire quadrant (0° to 90°, for a total of 19 images in the set). Two members of this set (0° and 45°) are shown in Fig. 2. This image set was used as training images to create both the linear and nonlinearly transformed filtes for testing the out-of-plane rotation case. An additional set of images was built using these images taken at 1° increments of rotation between 30° and 40°, for a total of 11 images in this set. Due to the resolution and accuracy of the equipment used in rotating the automobiles before their respective images were digitized, we were only able to obtain images in 1° increments for the out-of-plane cases. This image set was used as the "signal" image in the simulations, in order to compute the various correlator performance metrics. Figure 2 also contains an example of Lamborghini imagery embedded in additive white noise with $\sigma = 0.2$. These noisy inputs were used to test filter performance. As above, an additional set of Ferrari images was built using these images in 1° increments of rotation between 30° and 40°, for a total of 11 images as above for the Lamborghini image. This image set was used to test the dis-

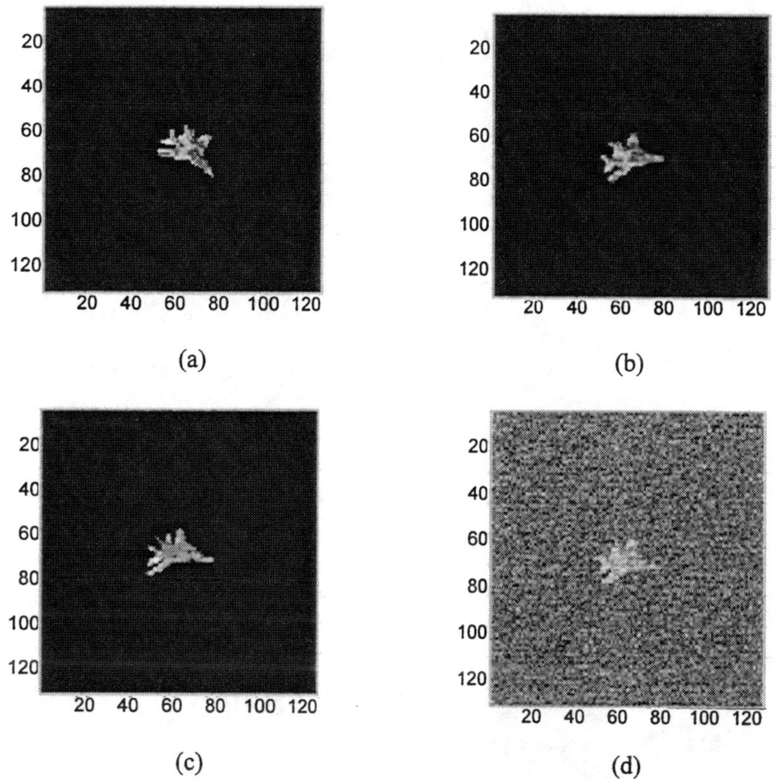

Figure 1. (a) Mig reference set member at 0° rotation. (b) Mig reference set member at 45° rotation. (c) F14 input signal set member at 37.5° rotation. (d) Mig input signal set member at 37.5° rotation with noise ($\sigma = 0.2$).

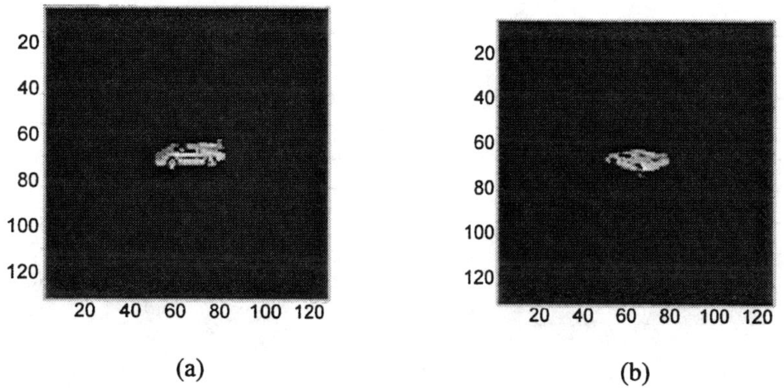

Figure 2.

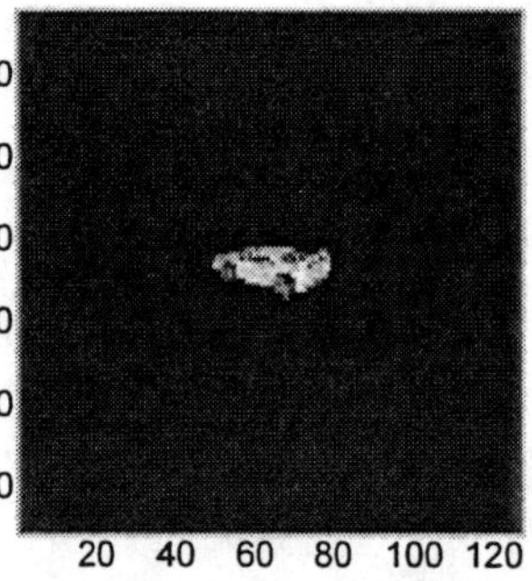

(c)

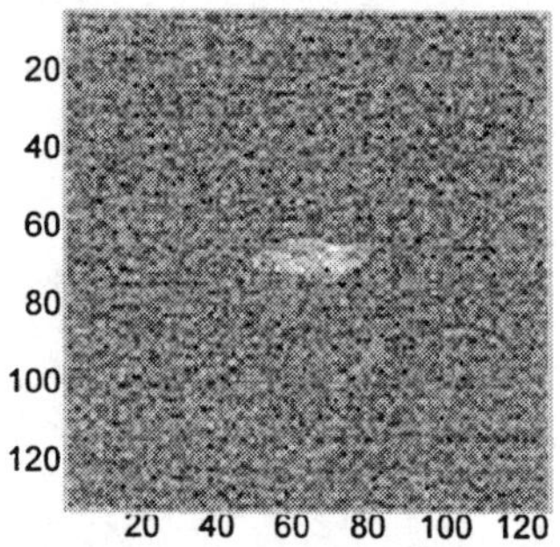

(d)

Figure 2. Lamborghini reference set member at 0° rotation. (b) Lamborghini reference set member at 45° rotation. (c) Ferrari input signal set member at 37° rotation. (d) Lamborghini input signal set member at 33° rotation with noise ($\sigma = 0.2$).

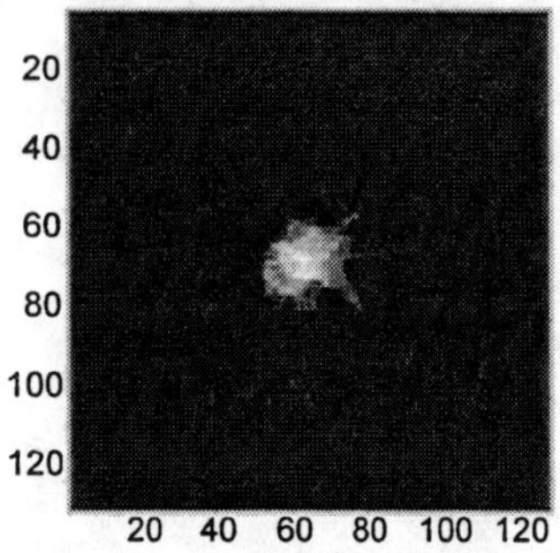

Figure 3. Mig ECP SDF filter created for $k = 1$.

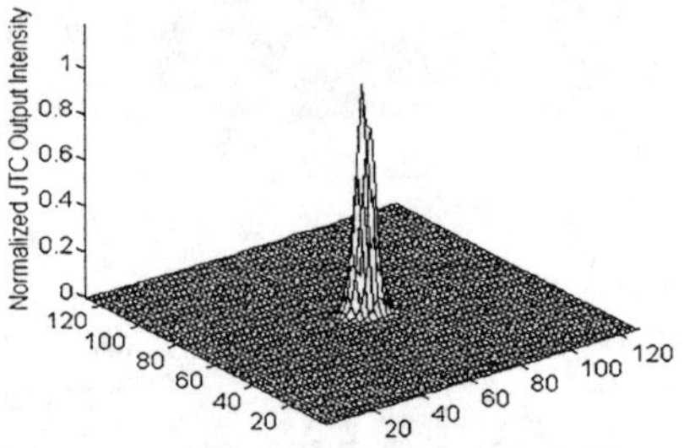

Figure 4. JTC output intensity ($k = 1$) with Mig ECP filter and Lamborghini (37.5°).

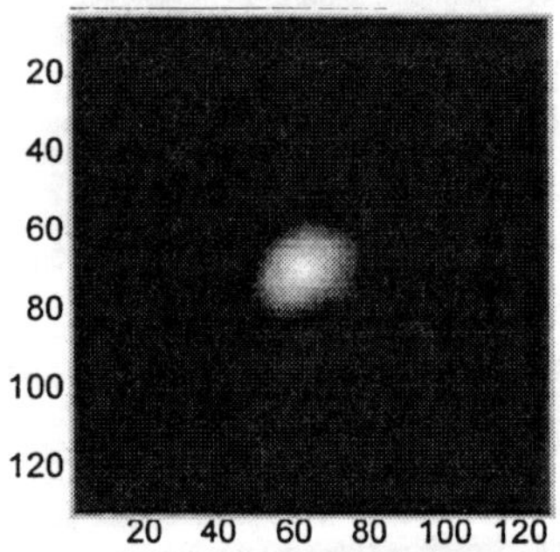

Figure 5. Mig ECP SDF filter created for $k = 0.3$.

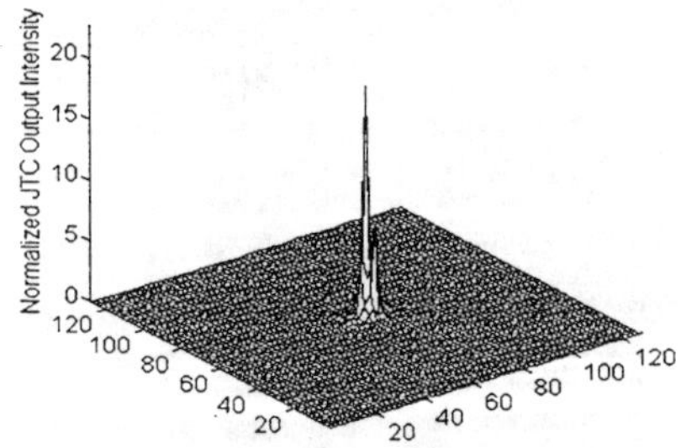

Figure 6. Same as Fig. 4 except with $k = 0.3$.

criminability ratio (*DR*) of the various filters, and was used only as the "signal" image in the simulations. The member of this set at 37° is shown in Fig. 2.

4.3. Simulation results

We present here the computer simulation results for the ECP SDF and MACE SDF filters when used in the linear and nonlinear JTC. Results are presented for both the in-plane and out-of-plane distortion cases. For good distortion invariance, we desire a correlation peak which remains relatively constant over an entire signal set (i.e. images rotated at increasing angles), even when the signal image is not one of the training images, but is an image rotated somewhere between two of the training images (i.e., signal at 37 degrees rotation with nearest training images at 35 and 40 degrees rotation). We also desire a correlation peak which remains relatively constant in the presence of input scene noise, and which is consistently higher than any other sidelobes in the output correlation plane. Also, the correlation peak location should provide an error-free indication of the location of the input signal.

Results for the kth Law ECP SDF Filter

The ECP SDF filter implemented here for the simulations is given by Eqs. (1) and (12). A one class filter was implemented, i.e. no training images from the F14 image set were included in the filter design. We have reasoned here that a true test of a correlation filter is to measure its ability to reject any arbitrary signal which is not its target image. By not including the F14 images in the training set, we can, to a certain extent, measure the discriminability of the filter against such a criterion. Figures 3 and 5 show the resulting spatial domain ECP SDF filters created using 5° spaced Mig training images, while Figs. 8 and 10 show the corresponding spatial domain ECP SDF filter which was created using 5° spaced Lamborghini training images.

Figures 4 and 6 are mesh plots given as examples of the output correlation intensity resulting from this filter when used in the linear and nonlinear JTC with noise-free Mig input images. Figures 9 and 11 are mesh plots given as examples of the output correlation intensity resulting from this filter when used in the linear and nonlinear JTC with noise-free Lamborghini input images. We have compiled the correlation output metrics over the input signal rotation range from 30° to 50° for the Mig, in-plane rotated images, and over the input signal rotation range from 30° to 40° for the Lamborghini, out-of-plane rotated images.

For the input signal with additive white noise ($\sigma = 0.2$) case, simulation results are presented in Figs. 7 and 12. We can see from the plots that the nonlinear JTC has significantly increased the peak output value, R_o, the output correlation peak-to-sidelobe ratio, the output correlation *PCE* ratio, and the output discriminability ratio. The plots show that the kth law ECP SDF has reduced the sensitivity of the nonlinear JTC to distortion as the value of k decreases[18]. This sensitivity reduction is not as evident with the out-of-plane imagery, and we attribute this to the fact that each training image contains information not present in the other members of the training image set for the out-of-plane case. A significant improvement in JTC performance is still seen in the out-of-plane case, however. For $k = 0.6$, we see significant improvements in these metrics over the linear case ($k = 1$) without the variability seen at $k = 0.3$. Even with these variations, however, the perform-

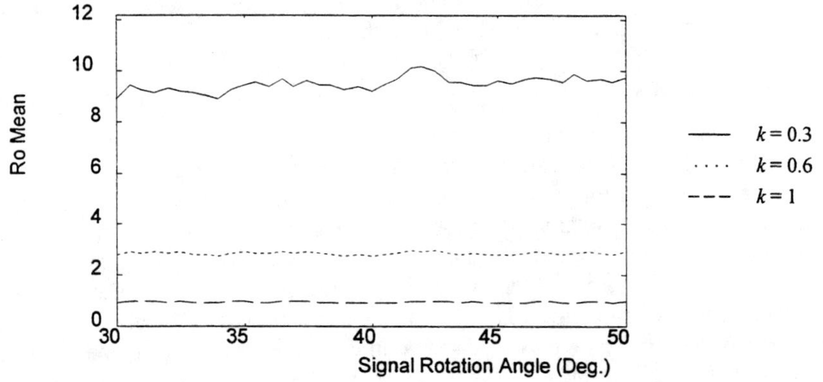
Ro Mean
12
10
8
6
4
2
0
30
35
40
45
50
Signal Rotation Angle (Deg.)
k = 0.3
k = 0.6
k = 1

(a)

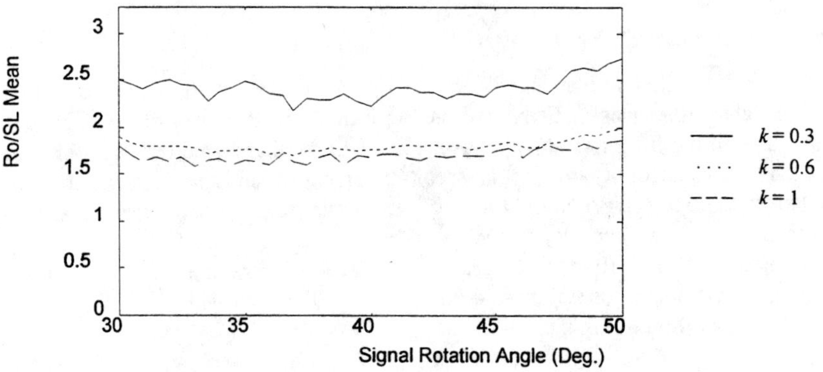
Ro/SL Mean
3
2.5
2
1.5
1
0.5
0
30
35
40
45
50
Signal Rotation Angle (Deg.)
k = 0.3
k = 0.6
k = 1

(b)

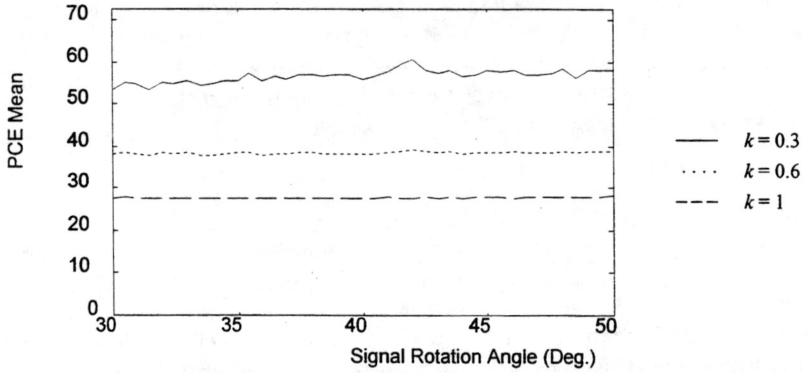
PCE Mean
70
60
50
40
30
20
10
0
30
35
40
45
50
Signal Rotation Angle (Deg.)
k = 0.3
k = 0.6
k = 1

(c)

Figure 7.

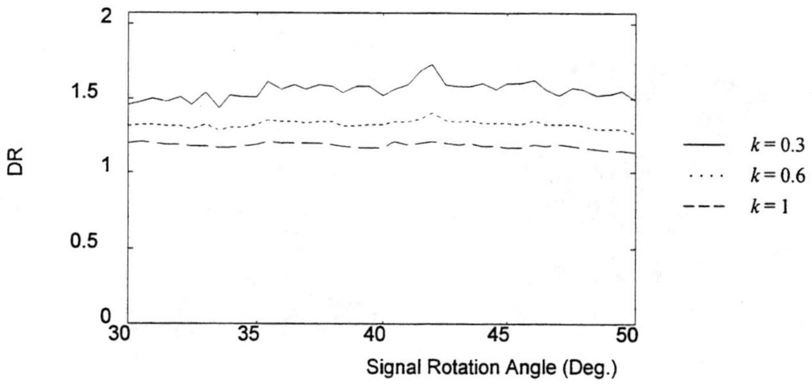

(d)

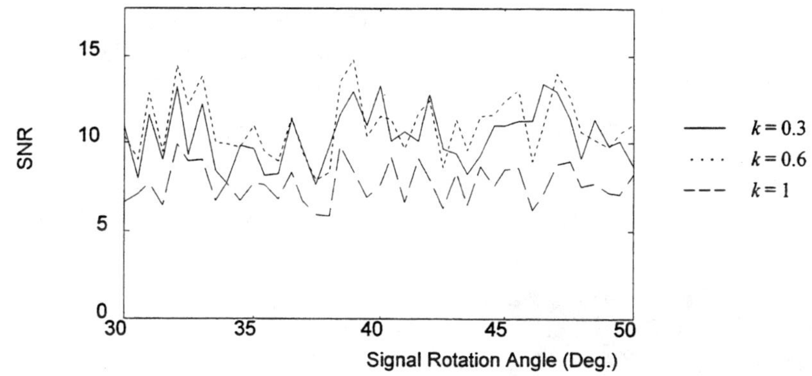

(e)

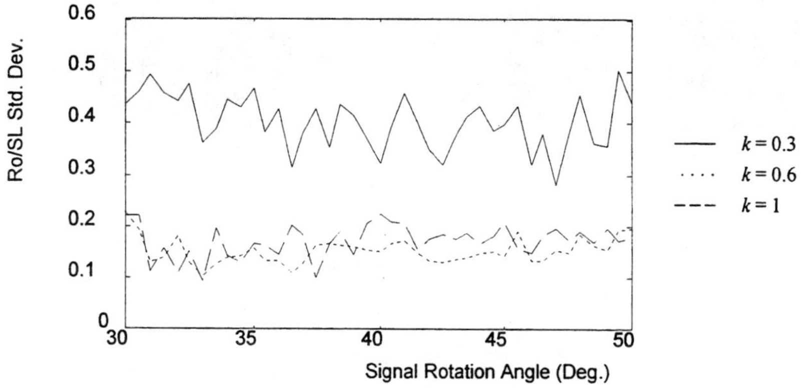

(f)

Figure 7. Mig ECP SDF filter performance with noisy input scene ($\sigma = 0.2$).

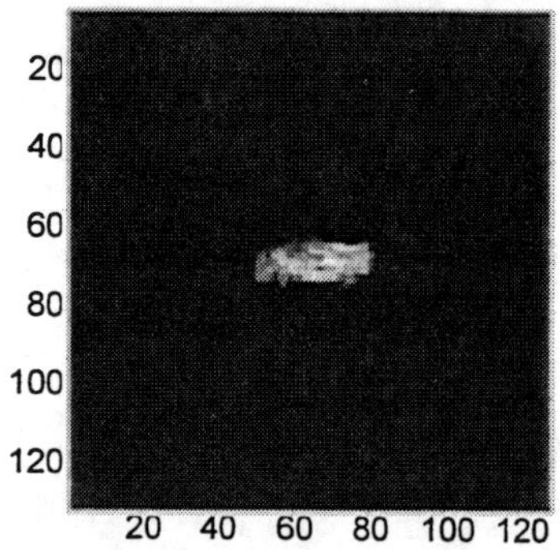

Figure 8. Lamborghini ECP SDF filter created for $k = 1$.

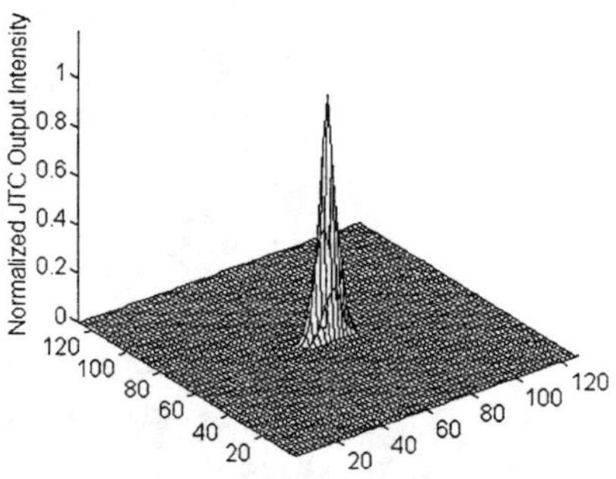

Figure 9. JTC output intensity ($k = 1$) with Lamborghini ECP filter and Lamborghini (33°).

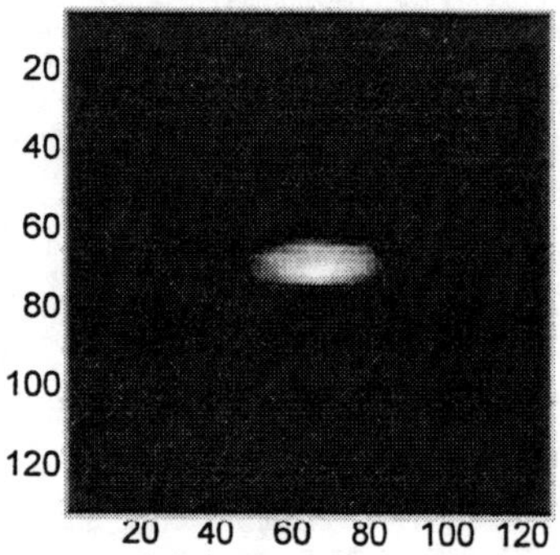

Figure 10. Lamborghini ECP SDF filter created for $k = 0.3$.

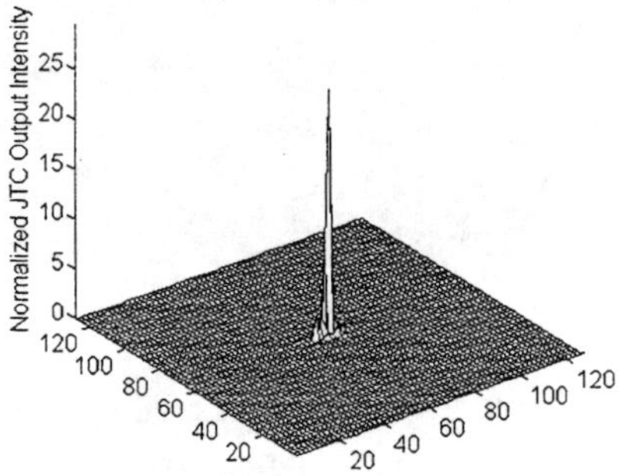

Figure 11. Same as Fig. 9 except with $k = 0.3$.

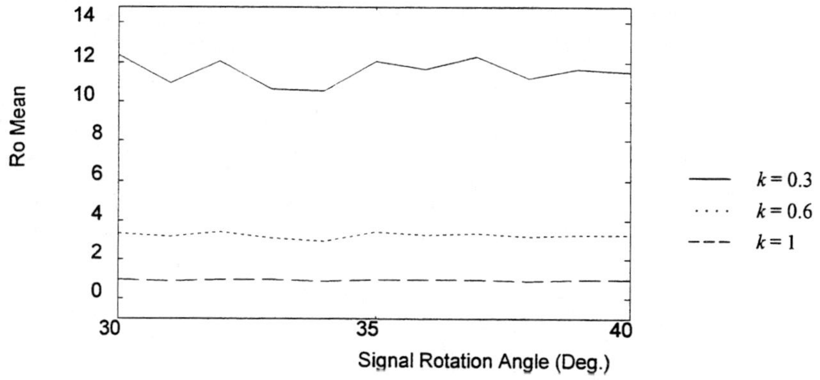

(a)

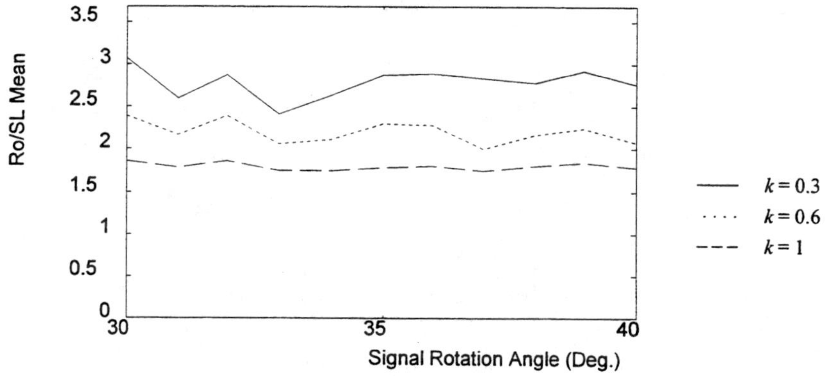

(b)

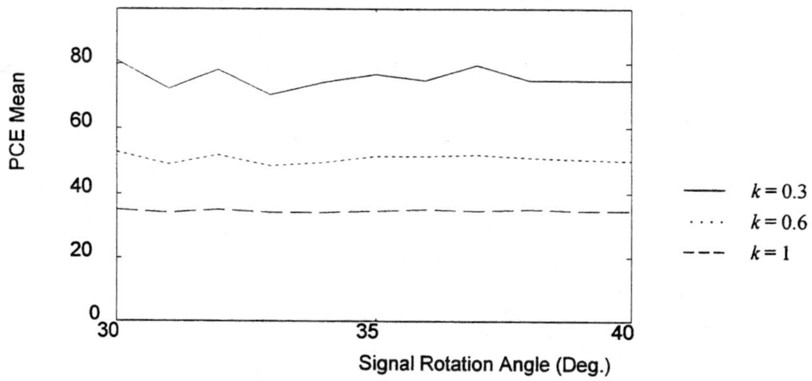

(c)

Figure 12.

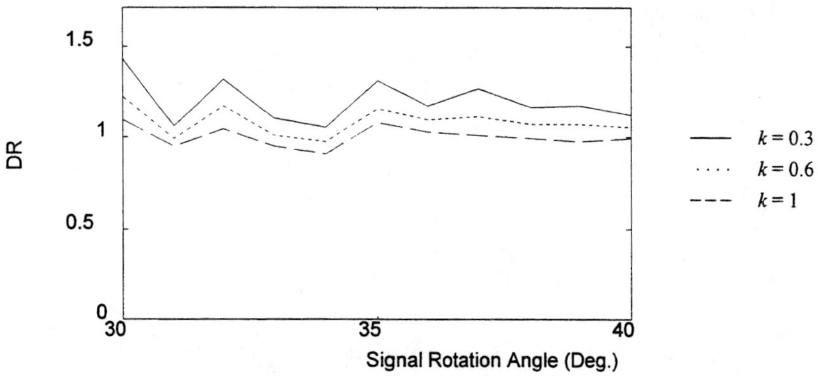

(d)

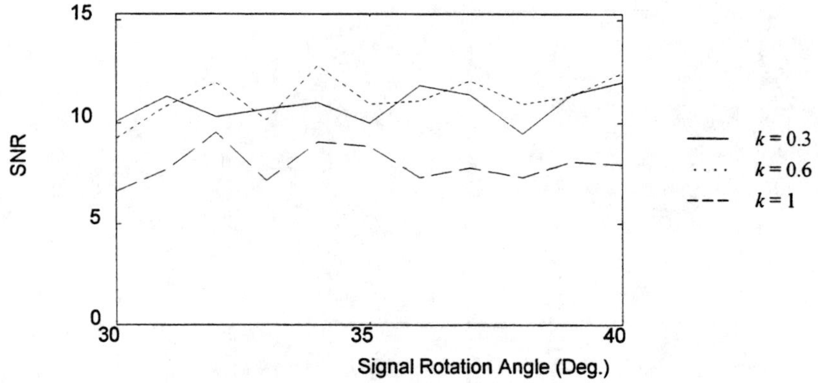

(e)

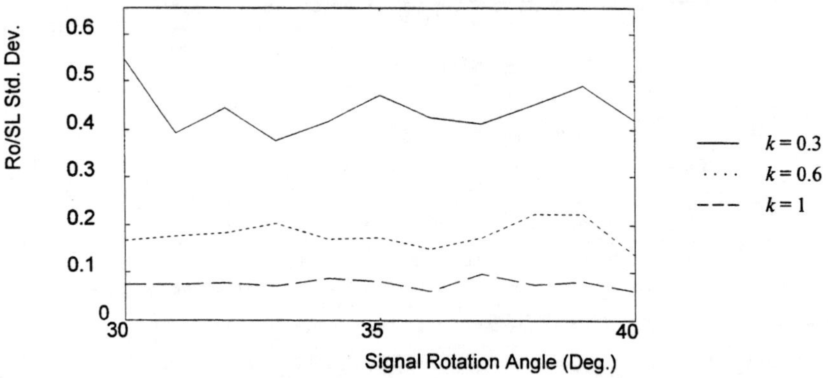

(f)

Figure 12. Lamborghini ECP SDF filter performance with noisy input scene ($\sigma = 0.2$).

ance of the correlator at $k = 0.3$ is still far better than that at $k = 1$ (linear case) for all of these metrics. For all simulations, the position of R_o was not in error from its theoretical position by more than one pixel value, and therefore positional data for R_o is not presented here. Although the R_o/SL standard deviation increases as k decreases, this increase is nullified by the significant improvement in the mean value of this metric. Additionally, we see a significant improvent in the *SNR* for the $k = 0.6$ and $k = 0.3$ cases. Similar improvements are seen for both the in-plane and out-of-plane distortion cases. For the noise free case similar results were seen for the ECP SDF filters, and therefore no results are presented here. Reviewing the performance results as a whole, the $k = 0.6$ case appears to provide the best overall improvement for both the in-plane and out-of-plane distortion cases.

Results for the kth Law MACE SDF Filter

The MACE SDF filter implemented here for the simulations is given by Eqs. (3) and (13). As above, a one class filter was implemented. Figures 13 and 15 show the resulting spatial domain MACE SDF filters created using 5° spaced Mig training images, while Figures 18 and 20 show the corresponding spatial domain MACE SDF filter which was created using 5° spaced Lamborghini training images. Note that in these plots, the gray backgrounds represents a spatial filter value of zero. The spatial domain MACE SDF filters contained both positive and negative valued pixels, and were biased for these plots, so that the full detail of the filters can be seen (the image plotting utility used required positive pixel values only).

Figures 14 and 16 are mesh plots given as examples of the output correlation intensity resulting from this filter when used in the linear and nonlinear JTC with noise-free Mig input images. Figures 19 and 21 are mesh plots given as examples of the output correlation intensity resulting from this filter when used in the linear and nonlinear JTC with noisy Lamborghini input images. Again, as for the ECP filters, we have compiled the correlation output metrics over the input signal rotation range from 30° to 50° for the Mig, in-plane rotated images, and over the input signal rotation range from 30° to 40° for the Lamborghini, out-of-plane rotated images.

For k^{th} law MACE SDF filter, for the case of a noise free JTC input signal, only the peak output value, R_o, was increased by the nonlinearities. All other metrics were neither decreased or increased by the lower values of k versus the results at $k = 1$. The performance of the filter is already excellent at $k = 1$ (optimal in fact for *PCE*), and therefore, it makes sense that no additional performance improvement would be gained for the MACE SDF filter in the noise free input case.

For the input signal with additive white noise ($\sigma = 0.2$) case, however, we see significant improvements in the performance of the MACE SDF filter through the use of nonlinearities. We can see from the plots in Figs. 17 and 22 that the nonlinear filters have a high degree of distortion invariance. As above, this distortion invariance is better for the in-plane distortion case, but is still good for the out-of-plane distortion case as well. The nonlinearities have significantly increased the mean of the peak output value, R_o, the mean of the output correlation peak-to-sidelobe ratio, and the mean of the output correlation *PCE* ratio. The output discriminability ratio is improved for the in-plane distortion case, but is degraded slightly for the out-of-plane case. For all simulations, the position of

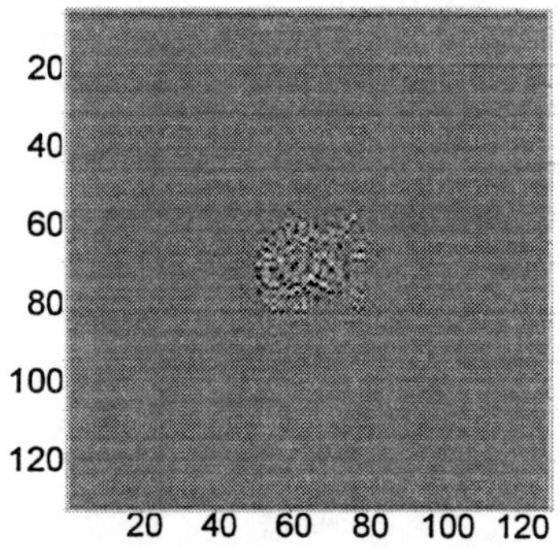

Figure 13. Mig MACE SDF filter created for $k = 1$.

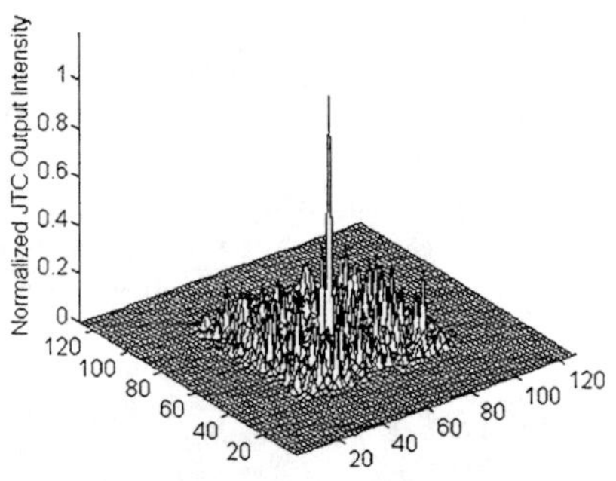

Figure 14. JTC output intensity ($k = 1$) with Mig MACE filter and Mig (37.5°, $\sigma = 0.2$).

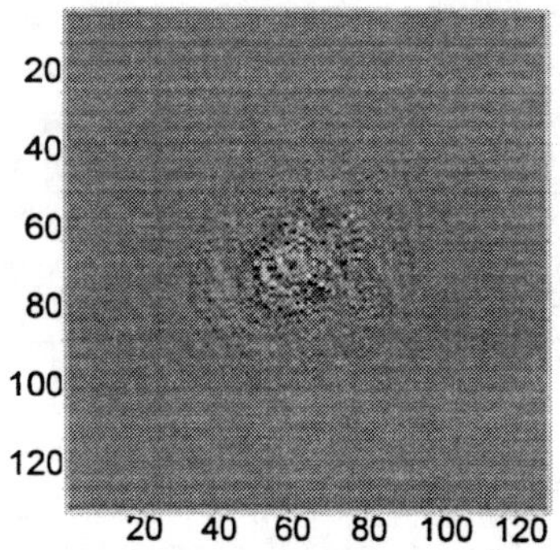

Figure 15. Mig MACE SDF filter created for $k = 0.3$.

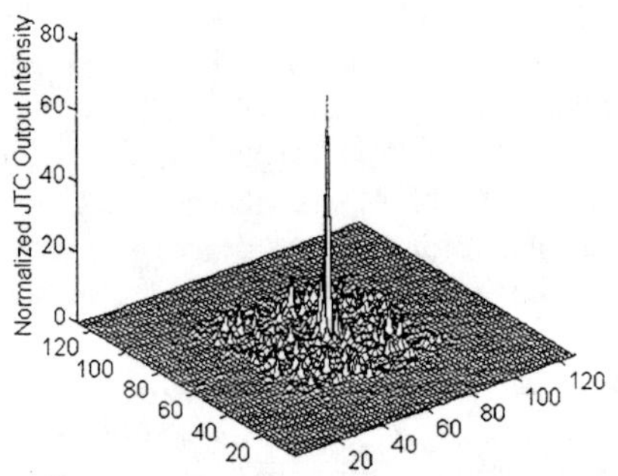

Figure 16. Same as Fig. 14 except with $k = 0.3$.

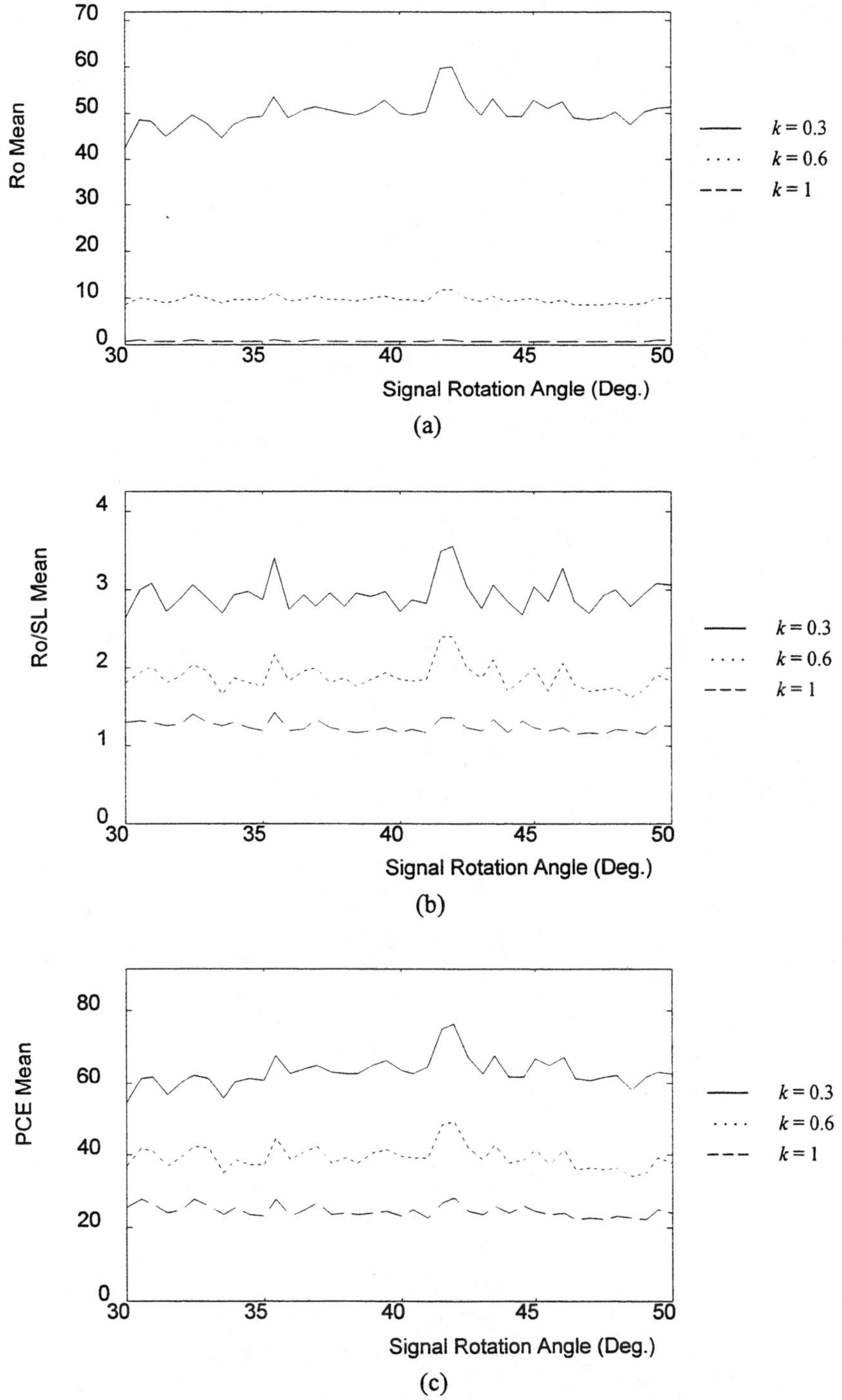
Ro Mean
70
60
50
40
30
20
10
0
30
35
40
45
50
Signal Rotation Angle (Deg.)
k = 0.3
k = 0.6
k = 1
(a)
Ro/SL Mean
4
3
2
1
0
Signal Rotation Angle (Deg.)
k = 0.3
k = 0.6
k = 1
(b)
PCE Mean
80
60
40
20
0
Signal Rotation Angle (Deg.)
k = 0.3
k = 0.6
k = 1
(c)

Figure 17.

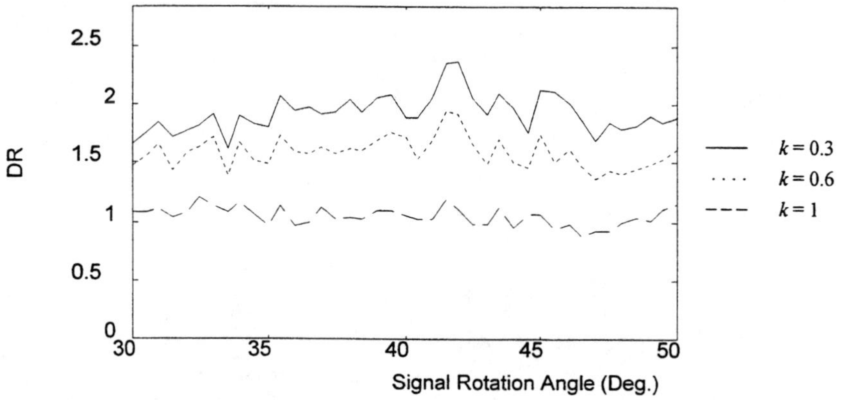

(d)

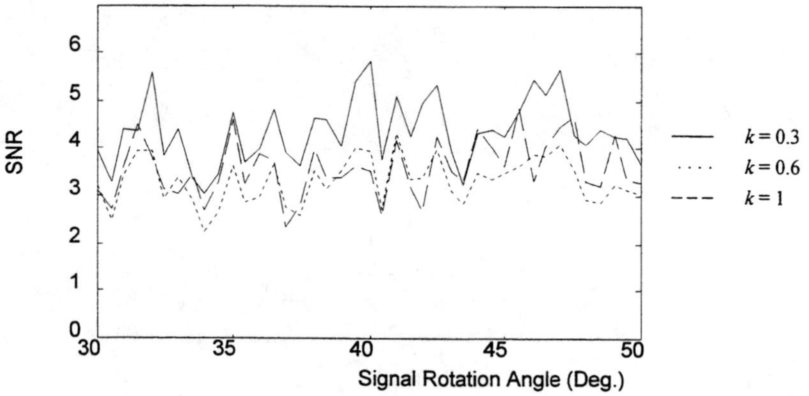

(e)

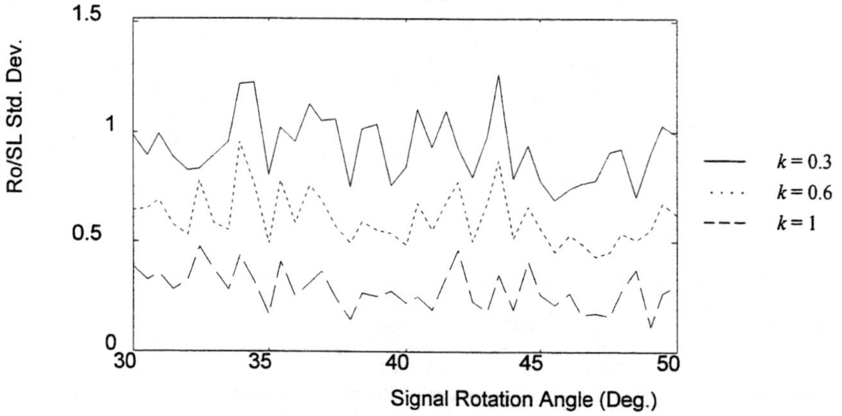

(f)

Figure 17. Mig MACE SDF filter performance with noisy input scene (σ = 0.2).

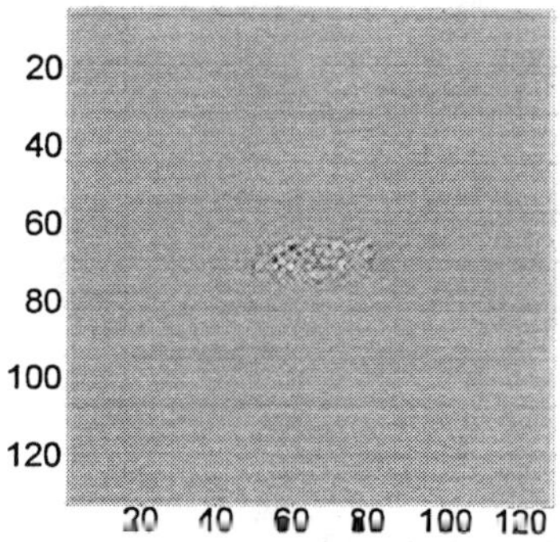

Figure 18. Lamboghini MACE SDF filter created for $k = 1$.

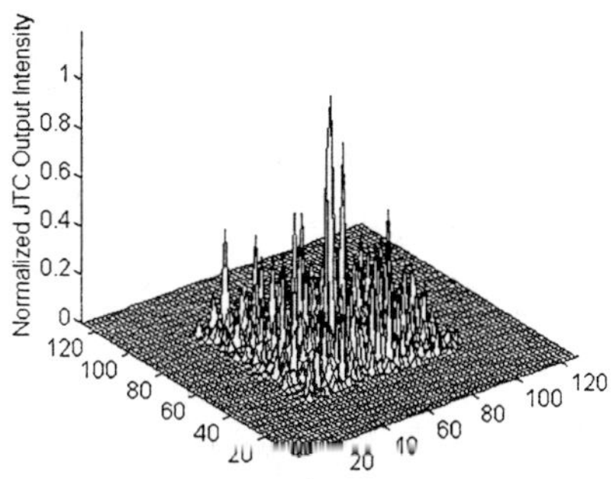

Figure 19. JTC output intensity ($k = 1$) with Mig Lamboghini filter and Lamboghini (33°,$\sigma = 0.2$)

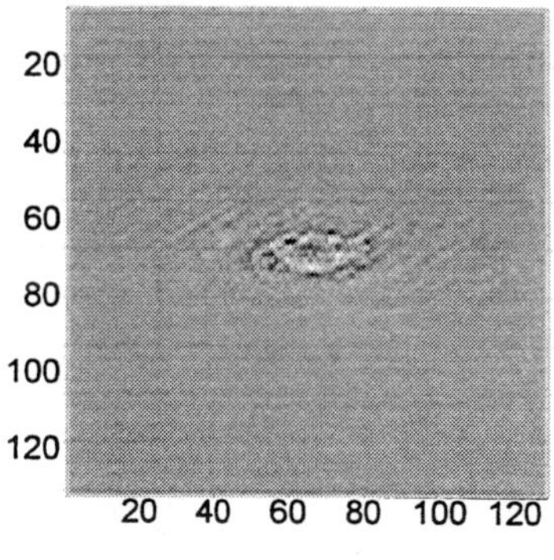

Figure 20. Mig MACE SDF filter created for $k = 0.3$.

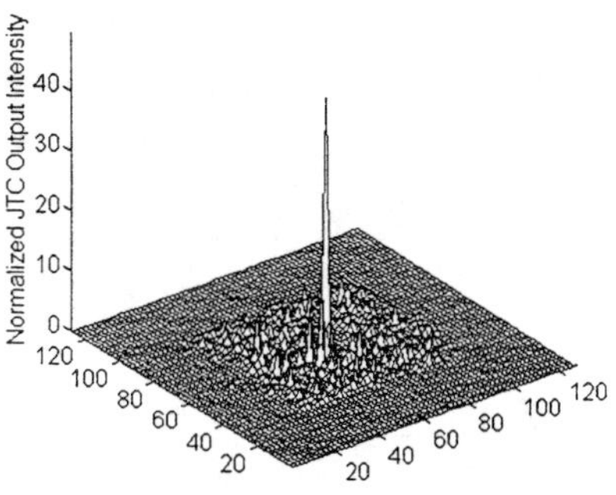

Figure 21. Same as Fig. 19 except with $k = 0.3$.

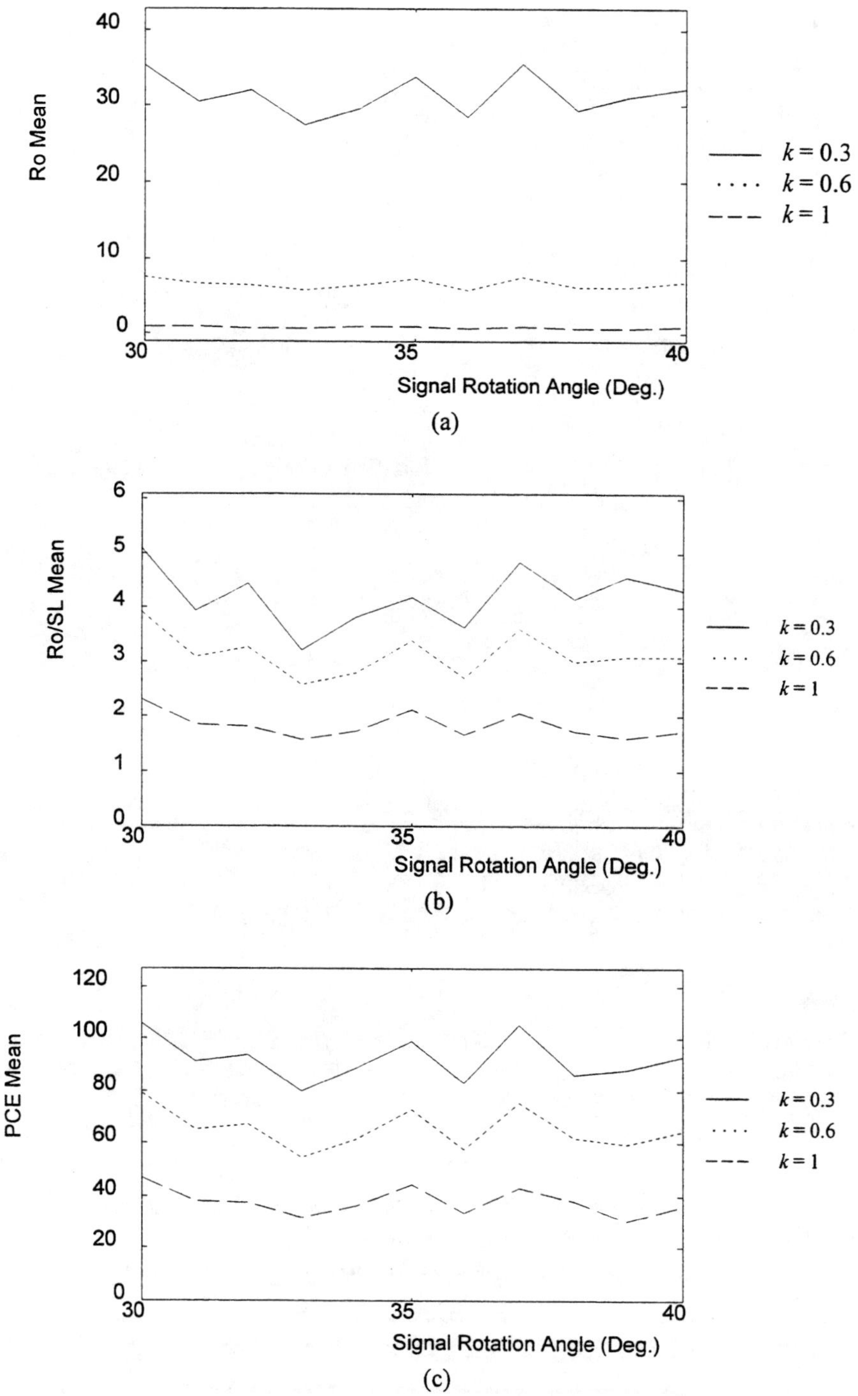

(a)

(b)

(c)

Figure 22.

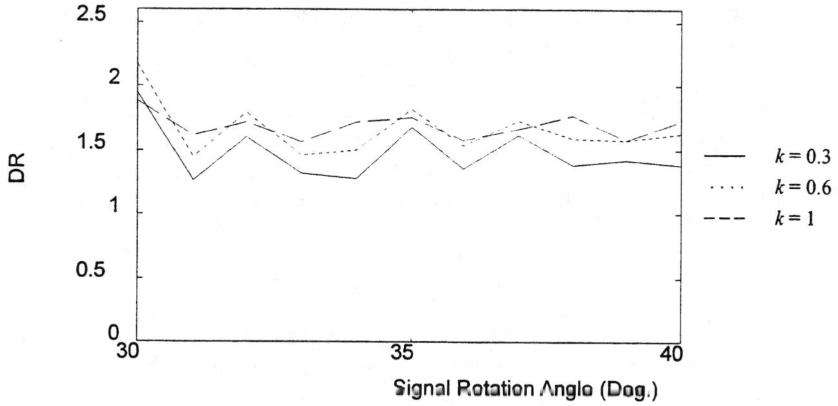

(d)

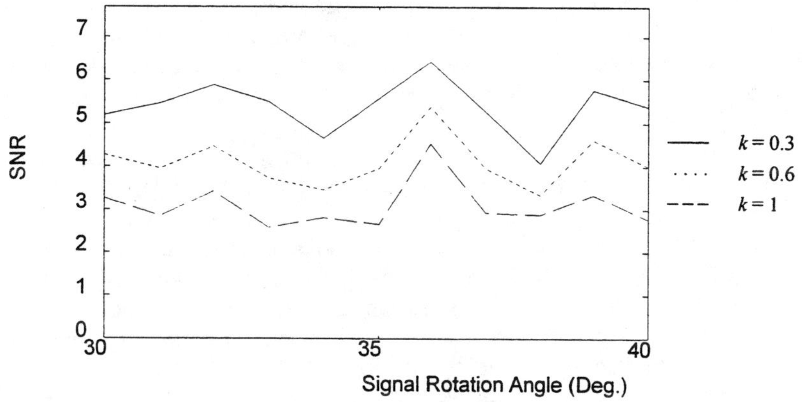

(e)

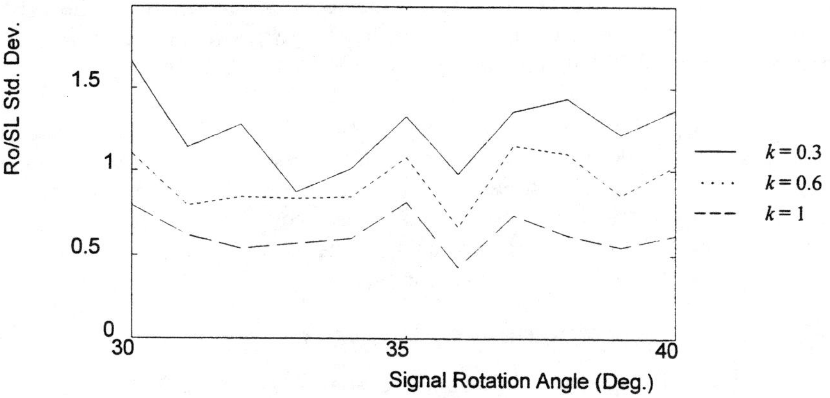

(f)

Figure 22. Lamborghini MACE SDF filter performance with noisy input scene ($\sigma = 0.2$).

R_o was not in error from its theoretical position by more than one pixel value, and therefore positional data for R_o is not presented here.

Although the R_o/SL standard deviation increases as k decreases, this increase is again nullified by the significant improvement in the mean value of this metric. We once again see a significant improvent in the *SNR* for the $k = 0.6$ and $k = 0.3$ cases, as above for the ECP SDF case. The performance results for both the $k = 0.6$ and $k = 0.3$ cases show significant improvements for the kth law MACE SDF filter.

5. CONCLUSIONS

In the above analysis and simulations, we have demonstrated the analytical tractability of designing filters for use in a Fourier plane nonlinear correlator architecture such as the nonlinear JTC. We have, we hope, further demonstrated the intriguing possibilities which result from nonlinearly manipulating the pattern recognition correlator Fourier plane. We are able to see substantial improvements in the correlator origin *SNR*, and in the mean values of various correlator performance criteria. The kth law ECP SDF filter and the kth law MACE SDF filter have been shown to outperform their linear precursors, even though these filter designs are not designed to be optimal with respect to any particular metric or performance criteria. We have seen that the main performance improvements gained from the incorporation of the kth law nonlinearity in our filter designs are a large increase in the peak output intensity, a significant improvement in the peak-to-sidelobe ratio, and a substantial improvement in the filters discriminability. Additionally, the nonlinearities improve the correlator light efficiency (*PCE* ratio), and the *SNR* of the peak value. At the same time, the SDF filter formulation has largely improved the distortion invariance of the nonlinear JTC. We have seen that for sharp nonlinearities, distortion invariance begins to degrade as a tradeoff for other criteria. We know from the work on phase-only filters[3,4] that their distortion invariance is poor, and since sharp nonlinearities shape the Fourier spectrum to be more “phase-only”, these results make sense. We have seen that there is a middle ground between “phase-only” and “linear” where nonlinear transformations can improve performance in all respects. Also, when compared to the fractional power filters[14,15] recently introduced for use in the matched filter correlator, we see that the nonlinear JTC can provide the ability to apply the nonlinearities to the signal, as well as the reference. This additional degree of freedom is key to the performance improvements gained by the kth law nonlinear filters.

Much work still remains, however. The subject of disjoint signal and scene noise has been gaining attention[21-23], and this topic should be addressed in light of this nonlinear noise estimation. New variations on the original MACE filter design have also recently been introduced[24,25] and the application of nonlinearities to these filter designs must be investigated. Additionally, the performance of these nonlinear filters in the presence of colored noise must also be investigated and understood[26].

REFERENCES AND NOTES

1. A. VanderLugt, "Signal detection by complex spatial filtering," IEEE Transactions on Information Theory **IT-10**, 139-145 (1964).
2. C. S. Weaver and J. W. Goodman, "A technique for optically convolving two functions," Applied Optics **5**(7), 1248-1249 (1966).

3. D. L. Flannery and J. L. Horner, "Fourier Optical Signal Processors," Proceedings of the IEEE **77**(10), 1511-1527 (1989).
4. B. V. K. V. Kumar, "A tutorial survey of composite filter designs for optical correlators," Applied Optics **31**(23), 4773-4801 (1992).
5. A. Mahalanobis, B.V.K.V. Kumar, and D. Casasent, "Minimum average correlation energy filters," Applied Optics **26**(17), 3633-3640 (1987).
6. B.V.K.V. Kumar, "Minimum variance synthetic discriminant functions," Journal of the Optical Society of America A **3**(10), 1579-1584 (1986).
7. Ph. Refregier and J. Figue, "Optimal trade-off filters for pattern recognition and their comparison with the Wiener approach," Optical Computing and Processing **1**(3), 245-266 (1991).
8. B. Javidi, "Nonlinear joint power spectrum based optical correlation," Applied Optics **28**(12), 2358-2367 (1989).
9. A. VanderLugt and F. B. Rotz, "The use of film nonlinearities in optical spatial filtering," Applied Optics **9**(1), 215-222 (1970).
10. Ph. Refregier, V. Laude, and B. Javidi, "Nonlinear joint-transform correlation: an optimal solution for adaptive image discrimination and input noise robustness," Optics Letters **19**(6), 405-407 (1994).
11. B. Javidi, Q. Tang, D. A. Gregory, and T. D. Hudson, "Experiments on nonlinear joint transform correlator using an optically addressed spatial light modulator in the Fourier plane," Applied Optics **30**(14), 1772-1776 (1991).
12. C. F. Hester and D. Casasent, "Multivariant technique for multiclass pattern recognition," Applied Optics **19**(11), 1758-1761 (1980).
13. A. Mahalanobis and D. P. Casasent, "Performance evaluation of minimum average correlation energy filters," Applied Optics **30**(5), 561-572 (1991).
14. B.V.K.V. Kumar and L. Hassebrook, "Performance measures for correlation filters," Applied Optics **29**(20), 2997-3006 (1990).
15. B. Javidi, "Generalization of the linear matched filter concept to nonlinear matched filters," Applied Optics **29**(8), 1215-1224 (1990).
16. K. H. Fielding and J. L. Horner, "1-f binary joint transform correlator", Optical Engineering **29**(9), 1081-1087 (1990).
17. W. B. Hahn and D. L. Flannery, "Design elements of binary joint transform correlation and selected optimization techniques", Optical Engineering **31**, 896-905 (1992).
18. Q. Tang and B. Javidi, "Sensitivity of the nonlinear joint transform correlator: experimental investigations," Applied Optics **31**(20), 4016-4024 (1992).
19. J. L. Horner, "Metrics for assessing pattern-recognition performance," Applied Optics **31**(2), 165-166 (1992).
20. W. K. Pratt, *Digital Image Processing*, 2nd Ed., (John Wiley and Sons, New York, 1991).
21. B. Javidi and J. Wang, "Limitation of the classic definition of the correlation signal-to-noise ratio in optical pattern recognition with disjoint signal and scene noise," Applied Optics **31**(32), 6826-6829 (1992).
22. Ph. Refregier, B. Javidi, and G. Zhang, "Minimum mean-square-error filter for pattern recognition with spatially disjoint signal and scene noise," Optics Letters **18**(17), 1453-1455 (1993).
23. B. Javidi, P. Refregier, and P. Willett, "Optimum receiver design for pattern recognition with non-overlapping target and scene noise," Optics Letters **18**(19), 1660-1662 (1993).

24. D. Casasent and G. Ravichandran, "Advanced distortion invariant MACE filters," Applied Optics **31**(8), 1109-1116 (1992).
25. G. Ravichandran and D. Casasent, "Minimum noise and correlation energy (MINACE) optical correlation filter," Applied Optics **31**(11), 1823-1833 (1992).
26. B. Javidi, J. Wang, and A. H. Fazlollahi, "Performance of the nonlinear joint transform correlator for images with low pass characteristics," Applied Optics **33**(5), 834-848 (1994).
27. D. Painchaud, *Distortion invariant nonlinear optical signal processing*, master's thesis, University of Connecticut, Storrs, CT, 1994.
28. H. J. Caulfield and W. T. Maloney, "Improved discrimination in optical character recognition," Applied Optics **8**, 2354-2356 (1969).

Fingerprint recognition using binary nonlinear joint transform correlators

Thomas J Grycewicz

United States Air Force Rome Laboratory
31 Grenier Street, Bedford, Massachusetts 01731

ABSTRACT

The joint transform binarization routines of frame subtraction and Fourier plane windowing are reviewed. A binary joint transform correlator (BJTC) fingerprint verification system design based on constant false alarm rate (CFAR) processing is proposed. These techniques are used to show that a BJTC based recognition system can be designed to have very good system performance. The tradeoffs between using all or part of the joint power spectrum and between displaying one or several reference prints in the input plane are addressed through simulation. It is shown that both Fourier plane windowing and multiple reference inputs can be used to significantly improve throughput at a slight cost in system performance. A recognition system with a single reference is shown to operate with a CFAR of 1 percent and a probability of false pass of less than 1 percent through a rotation range of ±3.5 degrees. A three reference system with different rotations of the same print as references obtains the same CFAR and similar false pass performance through a range of ±6 degrees.

Keywords: binary joint transform correlator, fingerprint verification, optimum threshold function, subtraction algorithm, Fourier plane windowing, constant false alarm rate

1.0. OVERVIEW

This paper reviews recently developed Fourier plane techniques for improving the performance of nonlinear[1] binary[2] joint transform correlators[3] (JTCs). The mechanism for this is to outline systems issues and design tradeoffs considered when designing a fingerprint verification system using the binary joint transform correlator (BJTC). Simulations are used to evaluate various types of Fourier plane processing and assess the performance gain in terms of the probability of incorrect fingerprint identification in a system designed to operate with a constant false alarm rate[4] (CFAR).

The binary joint transform correlator (BJTC) has been shown to be a good signal processing architecture for the identification of fingerprints, and has been suggested as a useful part of a security identification or recognition system[7]. In 1991, shortly after binary electrically addressed spatial light modulators (SLMs) became available for use in JTCs, it was realized that they provided an ideal medium for optical fingerprint identification systems[8]. JTC based systems have been shown to be robust in the presence of small rotations, partial occlusions, and the minor distortions seen between multiple printings of a finger[9,10]. Recent work has shown that the BJTC can compare an input fingerprint with four to six reference prints, and that the main practical limitation for multiple print comparisons is the space-bandwidth product of the input SLM[5,6].

The system investigated here assumes that a single SLM JTC architecture[11] with a binary SLM is used to compare fingerprints in a secure entry system. This assumption introduces both some constraints and some opportunities. The principal constraint is that both the input images and the joint power spectrum are displayed as binary images. The primary advantage is that the data is captured by a computer, simple preprocessing can be done along with thresholding between the detection of the joint power spectrum and its display in the second stage. The Fourier plane processing methods investigated are frame subtraction[12-15] and Fourier plane windowing[16-19]. The effects of using multiple reference fingerprints[5,6,12] of using only a portion of the joint power spectrum[5] are investigated. A constant false alarm rate (CFAR) verification strategy is introduced. The effect of the Fourier plane techniques on the overall system performance is analyzed using image pairs of the same fingerprint taken at different times and allowing varying amounts of rotation between the pair.

2.0. BACKGROUND: JOINT TRANSFORM CORRELATION

The BJTC is shown in Figure 1 below. The BJTC implements a two stage correlation process. The reference and scene images are displayed side by side in the input plane and a lens is used to take the Fourier transform. The intensity of this Fourier Transform is the joint power spectrum, which is captured by a CCD camera and binarized. This signal is displayed on a SLM at the second stage input. The joint power spectrum is spatially modulated by the cross-correlation between the scenes. The cross-correlation of the input images is computed by the second Fourier transform and forms a pair of correlation peaks in the output plane.

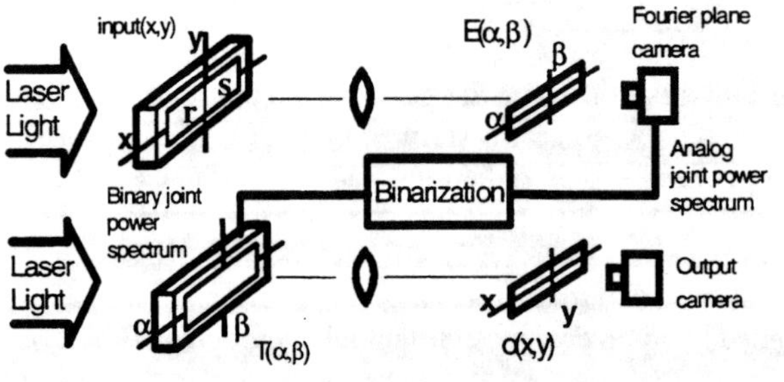

Figure 1. The binary joint transform correlator.

In a JTC, the inputs are displayed in the x-y plane of the input SLM. The scene, $s_0(x-x_0,y-y_0)$, contains a fingerprint which is to be identified. The scene fingerprint is compared to one or more reference fingerprints, $s_1(x-x_1,y-y_1)$ through $s_N(x-x_N,y_N)$. The reference images provide the templates to which the scene fingerprint is correlated. The input is given by

$$in(x,y) = s(x,y) = \sum_{n=0}^{N} s_n(x-x_n,y-y_n) \ . \tag{1}$$

A lens is used to form the two dimensional Fourier transform of in(x,y). The Fourier transforms of the input prints are defined by:

$$\mathcal{F}\{s_k(x,y)\} = S_k\left(\frac{\alpha}{\lambda f},\frac{\beta}{\lambda f}\right) \exp\left[j\phi_k\left(\frac{\alpha}{\lambda f},\frac{\beta}{\lambda f}\right)\right] = S_k(\bullet) \exp\left[\, j\phi_k(\bullet)\,\right] \ , \ and$$
$$S(\bullet) \exp\left[\, j[\phi(\bullet)\,\right] = \sum_{k=0}^{N} S_k(\bullet) \exp\left\{ j[\ \phi_k(\bullet)-\alpha x_k-\beta y_k\ ]\right\} . \tag{2}$$

Here $S_k(\bullet)$ are real quantities, and the complex phase is expressed by the exponential term. The joint power spectrum is the intensity of the Fourier transform of the input plane as captured using a square law detector (a camera). The joint power spectrum is given by:

$$E(\bullet) = |\ S(\bullet) \exp\{\ j\phi(\bullet)\ \}\ |^2 = \sum_{p=0}^{N}\sum_{q=0}^{N} S_p(\bullet)\ S_q(\bullet) \exp\left[\theta_{pq}(\bullet)\right]$$
$$= \sum_{n=0}^{N} S_n^2(\bullet) + 2\sum_{p=1}^{N}\sum_{q=0}^{p-1} S_p(\bullet)\ S_q(\bullet) \cos\left[\theta_{pq}(\bullet)\right] \ , \tag{3}$$

where:

$$\theta_{pq}(\bullet) = [\ \phi_p(\bullet)-\alpha x_p-\beta y_p-\phi_q(\bullet)+\alpha x_q+\beta y_q\ ] \ . \tag{4}$$

When two fingerprints correlate, their individual power spectra interfere in the transform plane. This results in amplitude modulation of the joint power spectrum. This modulation is expressed by the term $\cos[\theta_{pq}(\alpha,\beta)]$. The second stage output is the Fourier transform of the joint power spectrum. It contains cross-correlations between each pair of prints in the input, as well as the autocorrelations of these prints:

$$o(x,y) \propto \sum_{n=0}^{N} s_p(x,y)\star s_p(x,y) + 2\sum_{p=1}^{N}\sum_{q=0}^{p-1} s_p(x-x_p,y-y_p)\star s_q(x-x_q,y-y_q) \ . \tag{5}$$

Here $\star$ denotes correlation. The resulting output contains the autocorrelations centered on the optical axis, and cross-correlations to either side of the optical axis, displaced from the optical axis by a distance equal to the separation between the correlating prints. The output function is a proportionality rather than an equality because the energy in the output plane is determined by the illumination of the second stage SLM.

2.1. The binary JTC (BJTC)

In the BJTC, a binary SLM at the second stage input displays the joint power spectrum intensity. Binarization increases the optical throughput of the second stage and enhances the high spatial frequency components in the joint power spectrum. The binarization of the joint power spectrum results in bright, narrow output peaks.

In order to mathematically model the BJTC, the binarized joint power spectrum must first be expressed in terms of linear functions. This can be accomplished using the transform method of analysis[1,20]. The binarization function, g(E), is the sign function which has a value of 1 when the intensity of the joint power spectrum is larger than the threshold value and -1 when the joint power spectrum is less than the threshold value. The Fourier transform of this function is computed in the energy domain:

$$g(E) = sgn\ [\ E(\bullet) - V_T(\bullet)\ ]\ ;$$

$$G(\omega) = \int_{-\infty}^{\infty} g(E)\ e^{-j\omega E}\ dE = \frac{2}{j\omega}\ . \tag{6}$$

The binarized signal displayed in the Fourier plane input can be expressed in terms of the inverse Fourier transform of G(ω) :

$$T(\bullet) = \frac{1}{2\pi} \int_{-\infty}^{\infty} G(\omega)\ \exp\left\{ j\omega\left[\ E(\bullet) - V(\bullet)\ \right] \right\}\ d\omega$$

$$= \frac{1}{j\pi} \int_{-\infty}^{\infty} \frac{1}{\omega}\ \exp\left\{ j\omega\left[\ E(\bullet) - V(\bullet)\ \right] \right\}\ d\omega\ . \tag{7}$$

T(•) is the binary phase transmission function of the Fourier plane SLM. The evaluation of T(•) is simplified further when the optimum threshold function[1] is used. The optimum threshold function is defined by:

$$V_T(\bullet) = S_0^2(\bullet) + \left| \sum_{n=1}^{N} S_n(\bullet) \exp\{ j [\phi_n(\bullet) - \alpha x_n - \beta y_n] \} \right|^2 , \; or$$
$$E(\bullet) - V_T(\bullet) = 2 \sum_{p=1}^{N} S_p(\bullet)\, S_0(\bullet) \cos[\theta_{p0}(\bullet)] \; . \tag{8}$$

In order to model the BJTC in a manner which can be extended to describe Fourier plane windowing, two new variables, a constant A, and a Fourier plane phase function $\Omega(\alpha,\beta)$ are introduced[19]:

$$A > \left| E(\bullet) - V_T(\bullet) \right| \; over \; all \; (\alpha,\beta), \; and$$
$$\sin[\Omega(\bullet)] = \frac{1}{A} [E(\bullet) - V_T(\bullet)] \; . \tag{9}$$

Since $\sin[\Omega(\bullet)]$ is proportional to the difference between the joint power spectrum and the threshold:

$$sgn\{\sin[\Omega(\bullet)]\} = sgn[\Omega(\bullet)] = sgn[E(\bullet) - V_T(\bullet)] = sgn\left\{ \sum_{p=1}^{N} S_p(\bullet)\, S_0(\bullet) \cos[\theta_{p0}(\bullet)] \right\} , \tag{10}$$

and therefore all of these functions are equivalent as inputs to the thresholding operation.

The binary phase transmission function is analyzed by expanding it as a power series and evaluating the first order term of the result, which contains the majority of the output energy and has the proper phase relationship to form correlation peaks in the same locations as the JTC[1,12,21]. The higher order terms in the expansion contribute a relatively small amount of noise energy to the valid correlation regions of the output plane:

$$T(\bullet) \propto \frac{1}{j\pi} \int_{-\infty}^{\infty} \frac{1}{\omega} \exp\left\{ j\omega \sin\left[\Omega(\bullet) \right] \right\} d\omega$$
$$= \sum_{\nu=0}^{\infty} \frac{\epsilon_\nu}{j\pi} j^\nu \int_{-\infty}^{\infty} \frac{1}{\omega} J_\nu(\omega) \sin\left[\nu\Omega(\bullet) \right] d\omega \tag{11}$$
$$\approx \frac{4}{\pi} \sin\left[\Omega(\bullet) \right] .$$

This model is very flexible in terms of its ability to accommodate various threshold functions. If the optimum threshold function[12,21] is used, the output peaks are determined by the Fourier transform of $\Omega(\bullet)$:

$$o(x,y) \propto \frac{2}{A\pi} \sum_{p=1}^{N} S_p(\bullet)\, S_0(\bullet) \left[\delta(x+x_p-x_0, y+y_p-y_0) + \delta(x-x_p+x_0, y-y_p+y_0) \right]. \qquad (12)$$

The correlation peaks are in the same locations as they are for the JTC, and are weighted by their correlation intensity. The output plane energy is shared between the peaks according to the ratios of their weightings. To find the peak heights, the output intensities can be normalized using the total energy received in the correlation plane.

2.2. The frame subtraction technique for binarization

Frame subtraction applies the optimum threshold function in the Fourier plane, and has been the object of much recent study[5,6,13-15]. This process works extremely well with the single SLM JTC architecture[11], shown in Figure 2. This JTC architecture is quite popular since by using only one SLM, the system cost is greatly reduced. Given the speed of modern small computers, the pixel by pixel subtraction process is much faster than the capture of the joint power spectrum using a CCD camera.

The most efficient implementation of the subtraction process is when the modulation of the reference and scene images is in the form of a phase shift of π radians[13]. The sign of either the reference or scene image is reversed between then first input frame and the second. These two inputs result in the two joint power spectra:

$$E_{00}(\bullet) = \left| \sum_{n=1}^{N} S_n(\bullet) \exp\{ j\phi_n(\bullet) \} - S_0(\bullet) \exp\{ j\phi_0(\bullet) \} \right|^2$$

$$= \sum_{n=0}^{N} S_n^2(\bullet) + 2\sum_{p=2}^{N}\sum_{q=1}^{p-1} S_p(\bullet)\, S_q(\bullet) \cos[\theta_{pq}(\bullet)] - 2\sum_{m=1}^{N} S_m(\bullet) S_0(\bullet) \cos[\theta_{m0}(\bullet)] \; ;$$

$$E_{01}(\bullet) = \left| \sum_{n=1}^{N} S_n(\bullet) \exp\{ j\phi_n(\bullet) \} + S_0(\bullet) \exp\{ j\phi_0(\bullet) \} \right|^2 \qquad (13)$$

$$= \sum_{n=0}^{N} S_n^2(\bullet) + 2\sum_{p=2}^{N}\sum_{q=1}^{p-1} S_p(\bullet)\, S_q(\bullet) \cos[\theta_{pq}(\bullet)] + 2\sum_{m=1}^{N} S_m(\bullet)\, S_0(\bullet) \cos[\theta_{m0}(\bullet)] .$$

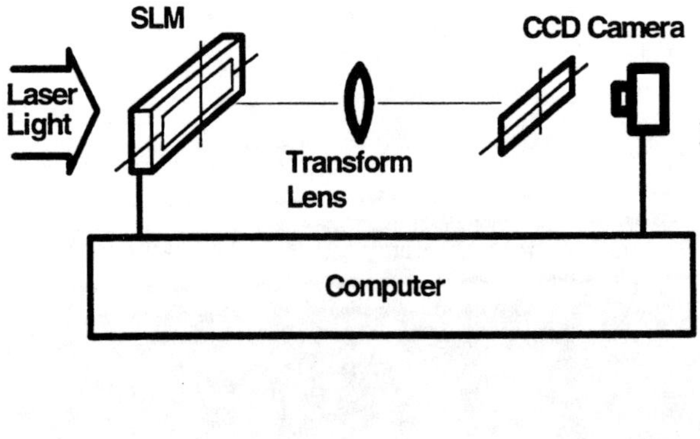

Figure 2. The single SLM joint transform correlator.

When the phase of either the reference or scene signal is reversed, the effect is to reverse the phase of the cross term. The output is found by subtracting the two power spectra:

$$E_{sub}(\bullet) = E_{00}(\bullet) - E_{10}(\bullet) = 4\sum_{m=1}^{N} S_m(\bullet)\ S_0(\bullet)\ \cos\left[\theta_{m0}(\bullet)\right] \ . \tag{14}$$

This result is equivalent to thresholding with the optimum threshold function and yields the best performance achievable using a BJTC[1,12].

2.3. The Fourier plane windowing technique for binarization

Windowing is defined as convolving the joint power spectrum with a small window mask in the Fourier plane[19]. This amplifies the high spatial frequency components in the joint power spectrum which contain cross-correlation information and rejects the low spatial frequency components which contain primarily autocorrelation information. Fairly small masks (5-25 pixels) often work well for this application[4,5,16,17]. Since filtering with a convolution mask is a local process in the Fourier plane, it does not require a large amount of computation. (The number of operations required is proportional to the number of pixels in the image multiplied by the number of pixels in the mask. For simple masks, these operations need only be adds and shifts.) When a CCD camera is used in the Fourier plane, windowing along the horizontal direction can be achieved by electrically high-pass or band-pass filtering the camera output instead of by digital filtering. The effect of windowing on the JTC output will be considered first, and then the model will be extended to the BJTC. The window mask is convolved with the joint power spectrum, and the windowed spectrum is given by

$$T_w(\bullet) = T(\bullet) \otimes w(\alpha,\beta) \ . \tag{15}$$

Here $\otimes$ represents convolution and $w(\alpha,\beta)$ is the convolution mask. The "(•)" notation will not be used with the convolution mask. This is to emphasize that the spatial values of this mask are applied in the Fourier plane, and it is the Fourier transform of the mask which applies in the output plane. The output of the windowed JTC is the product of the conventional JTC output and the Fourier transform of the windowing function, $W(x,y)$:

$$o_w(x,y) = o(x,y)\ W(x,y) \ . \tag{16}$$

To extend this analysis to the BJTC, the binarized joint power spectrum must first be expressed in terms of linear functions. This is accomplished using the transform method of analysis, as in Equations 6-10. The windowed and binarized transform plane is given by

$$T_{WB}(\bullet) \propto \frac{1}{2\pi} \int_{-\infty}^{\infty} G(\omega) \exp\left\{ j\omega[\ T_W(\bullet)\ ] \right\} d\omega$$

$$= \frac{1}{j\pi} \int_{-\infty}^{\infty} \frac{1}{\omega} \exp\left\{ j\omega \left[\sum_{n=0}^{N} S_n^2(\bullet) \otimes w(\alpha,\beta) \right] \right\} \tag{17}$$

$$\times \exp\left\{ j2\omega \left[\sum_{p=1}^{N} \sum_{q=0}^{p-1} S_p(\bullet)\ S_q(\bullet) \cos\ [\theta_{pq}(\bullet)] \otimes w(\alpha,\beta) \right) \right] \right\} \right\} d\omega\ .$$

If the weighting coefficients in the Fourier plane convolution mask have a mean of zero convolution will perform a high pass spatial filtering. The Fourier transform of this mask will then have a zero as its central value [W(0,0) = 0] and the value of W(x,y) will be small for values of x and y near the optical axis. The Fourier transform of a small convolution mask varies slowly in the output plane, and can be considered as constant across the width of an individual correlation peak. The output peaks are approximated by delta functions whose amplitude is the product of the correlation amplitude and the amplitude of the Fourier transform of the window evaluated at the correlation location. Since W(0,0) = 0 the autocorrelation terms [whose peaks are at (0,0)] are eliminated:

$$\exp\left\{ j\omega \left[\sum_{n=0}^{N} S_n^2(\bullet) \otimes w(\alpha,\beta) \right] \right\} = \exp\left(j\omega \left\{ \mathcal{F}^{-1} \left[\sum_{n=0}^{N} s_n(x,y) \star s_n(x,y)\ W(x,y) \right] \right\} \right) \tag{18}$$

$$\approx \exp\left\{ j\omega \sum_{n=0}^{N} S_n^2(\bullet)\ W(0,0) \right\} = \exp(0) = 1\ .$$

A similar approach is used with the cross-correlation terms. The correlation peaks are at coordinates $(x_0\text{-}x_1,y_0\text{-}y_1)$ and $(x_1\text{-}x_0,y_1\text{-}y_0)$. This results in the simplification:

$$\exp\left(j2\omega \left\{ \sum_{p=1}^{N} \sum_{q=0}^{p-1} S_p(\bullet)\ S_q(\bullet) \cos\ [\theta_{pq}(\bullet)] \otimes w(\alpha,\beta) \right\} \right)$$

$$= \exp\left\{ j\omega\ \mathcal{F}^{-1} \left[\sum_{p=0}^{N} \sum_{q=0}^{N} s_p(x\text{-}x_p,y\text{-}y_p) \star s_q(x\text{-}x_q,y\text{-}y_q)\ W(x,y) \right] \right\} \tag{19}$$

$$\approx \exp\left\{ j2\omega \left[R(\bullet)\ S(\bullet) \cos \left(\theta_{pq}(\bullet)\right) \right] W(x_p\text{-}x_q,y_p\text{-}y_q) \right\}\ .$$

Note that here $W(x_p\text{-}x_q,y_p\text{-}y_q)$ is a constant whose value depends on the peak location. Equations 16-18 are combined to give the Fourier plane SLM transmission function after windowing:

$$T_{WB}(\bullet) \propto \frac{1}{j\pi} \int_{-\infty}^{\infty} \frac{1}{\omega} \exp\left(j2\omega \left\{ \sum_{p=1}^{N} \sum_{q=0}^{p-1} S_p(\bullet) S_q(\bullet) \cos[\theta_{pq}(\bullet)]\ W(x_p\text{-}x_q,y_p\text{-}y_q) \right\} \right) d\omega\ . \tag{20}$$

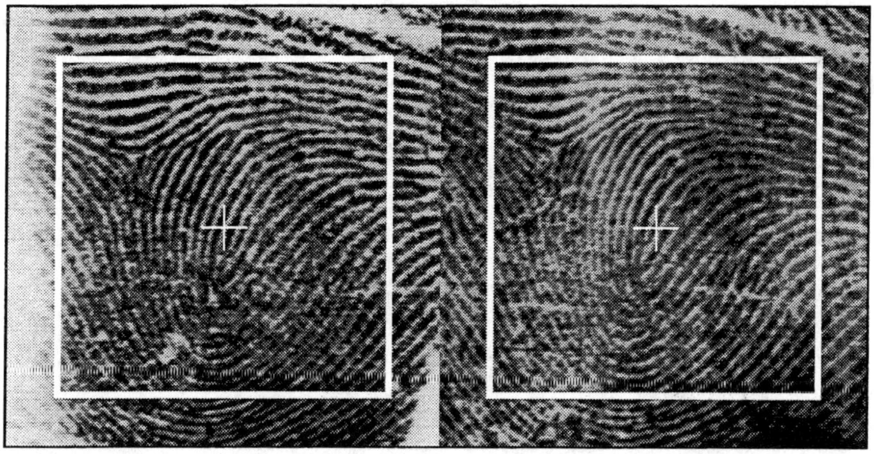

Figure 3. A pair of prints from the NIST Special Database 4 used in this study. The white boxes show the portion of the print actually used.

As with the BJTC, the Jacobian-Anger formula is used to expand the exponential term to a power series. The windowed joint power spectrum is approximated using the first order term of the series[19]. This results in the output:

$$o_{WB}(x,y) \propto \sum_{p=1}^{N} \sum_{q=0}^{p-1} \left[s_p(x,y) \star s_q(x,y) \right] \otimes \left[\delta(x+x_p-x_q, y+y_p-y_q) + \delta(x-x_p+x_q, y-y_p+y_q) \right] \times W(x_p-x_q, y_p-y_q) \; . \qquad (21)$$

The correlation peaks are in the same locations as they are for the BJTC. They are weighted by the product of the correlation intensity and the Fourier transform of the window mask at the location of the correlation peak. The output plane energy is shared between the peaks in proportion to their weightings and their correlation intensities.

3. INPUT PRINTS AND PREPROCESSING

The prints for this study were selected from *NIST Special Database 4, Fingerprint Database*[22]. The images are 512x512 pixel, 8 bit gray scale images taken from standard fingerprint cards collected by the FBI. Each entry in the database includes two sets of fingerprints collected at different times. The 25 right forefinger print pairs chosen for this study were free from major distortions. Minor distortions were present as a result of cuts and abrasions on the fingertips as well as printing differences. These included overinking, underinking, and nonuniform inking as well as slight stretching, twisting, or sliding of the finger pad as it was rolled on the card. The goal of fingerprint analysis is to compare the ridge structure on the pad of the finger with a reference. In doing this, the print can be treated as a binary image. The conversion of the gray scale fingerprint image to a binary image for display on the SLM is key.

A typical pair of fingerprints used in this study is shown in Figure 3. A 400x400 pixel area is shown, with a "+" marking the point selected as the center of the print. The 300x300 pixel area inside the boxes are the areas of the prints used to form the JTC inputs. To obtain different rotations of the print, the print is digitally rotated about the "+". Figure 3 illustrates typical differences in for the print pairs used in this study. The ridge structure is degraded by minor cuts and abrasions, which are in different locations on each print, the inking is not uniform, and the print on the right has been rolled, while the print on the left was simply pressed on the card.

The first steps in the fingerprint preprocessing are to center the print in the input window, reduce the image size to 128x128 pixels, and produce a set of gray scale images rotated in 0.5 degree increments. In the original images, the ridge to ridge spacing on the prints ranges from seven pixels to over twenty pixels. In order to make maximum use of the space-bandwidth product in the input plane of the BJTC, the minimum ridge to ridge spacing on the SLM display should not be larger than four pixels. The pixel values were converted from bytes (values of 0 to 255) to a floating point representation. A 300x300 pixel area of each print was selected, and the size of this area was reduced to 160x160 pixels using a cubic spline process. The images were then rotated about their center point, and the central 128x128 area of the rotated print binarized to form the input image. The resizing reduced the typical inter- ridge spacing to four to twelve pixels.

Binarization of the fingerprints reduces the continuous (floating point) gray scale images to binary images which capture the ridge and valley structure of the finger pad. The gray scale image is filtered using a spatial band-pass filter, and the filtered signal is binarized by thresholding at zero. The convolution mask filtering process used here is the same as described in Section 2.3 for windowing the joint power spectrum. The convolution mask filter should have a large response to the fingerprint ridge pattern. The convolution mask used in the input binarization process was

$$w_{IN}(x,y) = \begin{bmatrix} -1 & -1 & -1 & -1 & -1 \\ -1 & 1 & 1 & 1 & -1 \\ -1 & 1 & 8 & 1 & -1 \\ -1 & 1 & 1 & 1 & -1 \\ -1 & -1 & -1 & -1 & -1 \end{bmatrix} . \tag{22}$$

Figure 4. Binarized input pair resulting from the fingerprints shown in Figure 3.

Filtering by convolution with this mask amplifies the spatial frequency components in the fingerprint which make up the ridge structure. The result of applying this convolution mask to the inputs in Figure 3 is shown in Figure 4. This is a simple mask to implement digitally, requiring only add and shift operations. The same input processing was used in all of the simulations.

4.0. SIMULATION OF BJTC BASED FINGERPRINT RECOGNITION

This study compares the performance of BJTC based fingerprint identification systems using different processing algorithms to binarize the joint power spectrum. The two fingerprint correlations were done using a 256x128 pixel input; three and four fingerprint correlations used a 256x256 pixel input. (A set of the two fingerprint data was run in both 256x128 and 256x256 pixel formats to verify that the results were consistent.) For each of the BJTC configurations, the following study was performed:

a. For each of the 25 pairs of prints, the two fingerprints were correlated at through a rotation range of 15 degrees at 0.5 degree increments. (For the simulations with an extended rotation range, the increment was 1 degree and the range was 24 degrees.) The peak heights were recorded for each trial and used to build a statistical model of output peak height *vs.* rotation. For each rotation range the mean was determined, as well as boundaries for the 1, 10, and 90 percentile regions.

b. The statistics gathered in step (a) was used to determine a correlation intensity threshold which would result in a 1 percent false alarm rate with the input allowed to rotate through continuous ranges of $\pm$ 0.5 degrees to $\pm$ 6 degrees. A false alarm is defined as rejection of a correct print.

c. Three hundred simulation runs were done using unmatched print pairs. These correlation intensities were used to statistically model the unmatched print response.

d. The unmatched print data was used to determine the probability of a false pass for the system, where a false pass is defined as accepting an incorrect print. Low probability of false pass was the figure of merit used to compare the algorithms.

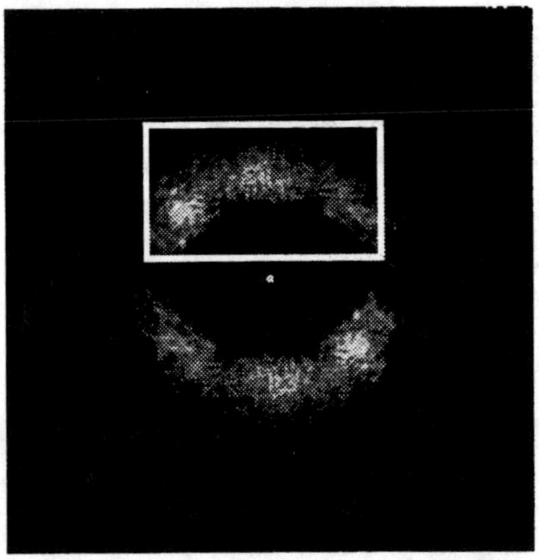

Figure 5. Histogram of the power spectra for the 25 binarized fingerprints used in the study. The box shows the area which was binarized in the joint power spectra.

The baseline for the study was a correlator with two fingerprints in the input plane and using frame subtraction as the Fourier plane binarization algorithm. Rather than binarizing the entire joint power spectrum, only the portion of the joint power spectrum inside the box in Figure 5 was used. The following variations were investigated:

a. The full joint power spectrum was binarized, rather than just the 42 percent of the transform plane which is enclosed in the box in Figure 5.

b. Fourier plane windowing was used as the binarization procedure.

c. A comparison was made using one, two, and three reference prints in the input plane. This was done both using different prints for the references, and using different rotations of the same print as references.

4.1. The baseline case: frame subtraction, two fingerprints

The frame subtraction technique was used with an input plane containing two fingerprints as the basis for comparison in this study. Since frame subtraction implements the optimum binarization threshold, other techniques are expected to have poorer performance. The portion of the joint power spectrum which was binarized and used in the second stage is the area inside the box in Figure 5. The box includes the area where the joint power spectrum contains significant energy. Half of the joint power spectrum contains all of the correlation information because of the symmetry of the transform. The central DC response of the joint power spectrum, which is at the center in Figure 5, is outside the box. This is important since the central DC spike often contains enough energy to cause serious blooming problems with most CCD cameras. This area was enlarged to form the input of the second Fourier transform using floating point arithmetic and a cubic interpolation routine. This resizing is consistent with designing an optimum single lens BJTC system, since the overall scaling of the joint power spectrum is determined by the focal length of the transform lens, and the x to y aspect ratio is determined by the camera and the setup of the frame grabber card in the computer.

Two binary 128x128 pixel fingerprints as shown in Figure 4 were displayed side by side in a 512x256 pixel input plane. The joint power spectrum was computed twice, with the pixel values for the reference print inverted in the second run to form the two inputs to the subtraction routine. Then the central 256x128 areas of the joint power spectra were subtracted, and the difference binarized by thresholding at zero. The binarized spectrum is the input to the second Fourier transform, yielding a 512x256 pixel output plane. The size of the output plane is reduced to 64x32 pixels for output display and analysis. Typical outputs for a pair of matched prints and a pair of unmatched prints are shown in Figure 6. The central DC peak normally seen in a BJTC output is eliminated because of the frame subtraction process and because the second stage SLM pixels are modulated at +1 as at -1. The largest output value in the region within two pixels of the expected output location was taken as the peak height. This requires that the original input images were aligned to within about thirty pixels. The actual alignment was within 5 pixels, assuring correct measurement of the peak value.

The fingerprints were correlated over of rotations covering a range of 15 degrees in 0.5 degree increments. in order to build the database of peak height *vs.* rotation angle, the measurements were lined up based on the assumption that the correctly aligned orientation was the one giving the largest correlation response. With this assumption there is an uncertainty in the rotation angle of 0.5 degrees. The first graph in Figure 7 plots the mean of the correlation response for each angle as well as the tenth and ninetieth percentile lines. The statistics plotted for each angle were computed based on 75 measurements including those ± 0.5 degrees away from the central measurement. The gray lines at the bottom of the graph are the mean and the tenth and ninetieth percentile lines for the 300 unmatched print samples.

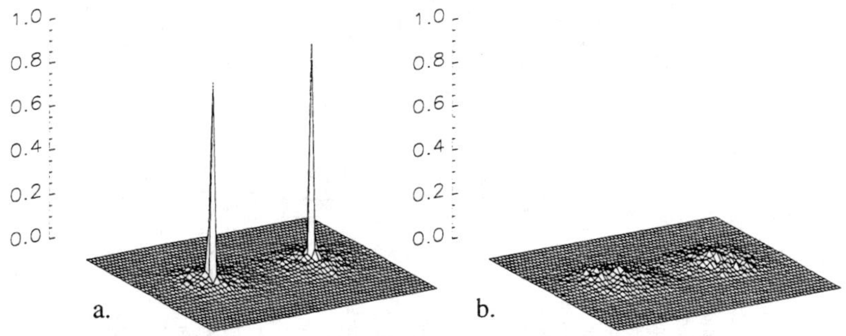

Figure 6. BJTC outputs for matched (a) and un-matched (b) prints. The central DC spike is eliminated by the frame subtraction process.

4.2. Fingerprint verification with a constant false alarm rate (CFAR) constraint

The first graph in Figure 7 is a statistical summary of the correlation data. This allows comparison of the expected correlation peak height of a matched print with an unmatched print. When the prints are rotated more than a few degrees with respect to one another, there is significant overlap between the expected peak height distributions for the matched and unmatched prints. This indicates that this processor cannot be used to build a verification system which will be reliable at these rotation angles.

To determine the performance under a CFAR constraint, first the data was regrouped into samples which covered an increasing range of angle distribution. The ±1° sample included the 75 results from the three central rotation bins, the ±2° sample included 175 results from the central seven bins, etc. This created a sample with a reasonably uniform probability of rotation throughout the range. These results were sorted in a histogram, and a threshold set at the one percentile location. Use of this threshold establishes the constant false alarm rate for that rotation range. The threshold is compared to the 300 samples of unmatched correlations, and the portion of these samples larger than the threshold determines the probability of a false pass. The second graph shows that for a rotation range of up to ±4°, this fingerprint verification system will work well, and that to include rotations of more than 4° results in performance which degrades quickly. The reason for this that as more rotation is allowed, the threshold level needed to pass 99% of the valid prints drops to a level which passes a significant number of unmatched prints. In order to obtain good verification performance while allowing more rotation of the input fingerprint, the input can be compared to several rotated versions of the reference as is described in section 4.6, or by using a composite reference input.

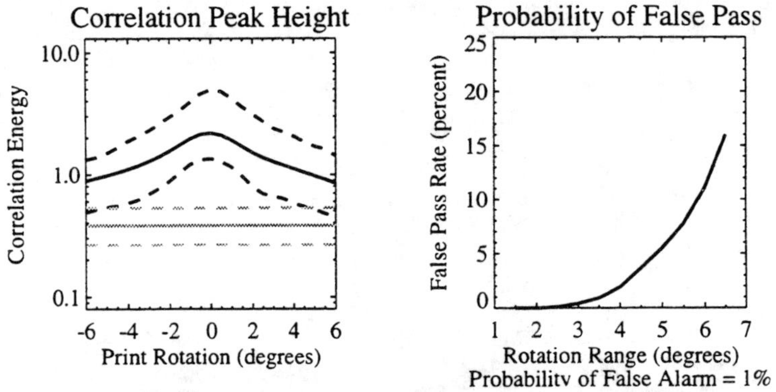

Figure 7. Correlation peak height for matched and un-matched prints as a function of the relative rotation, and the resulting probability of false pass for a verification system with a 1 % probability of false alarm. The solid black line shows the mean of 25 matched correlations and the solid grey line the mean of 300 unmatched correlations. The dashed lines show the 10 and 90 percentile boundaries.

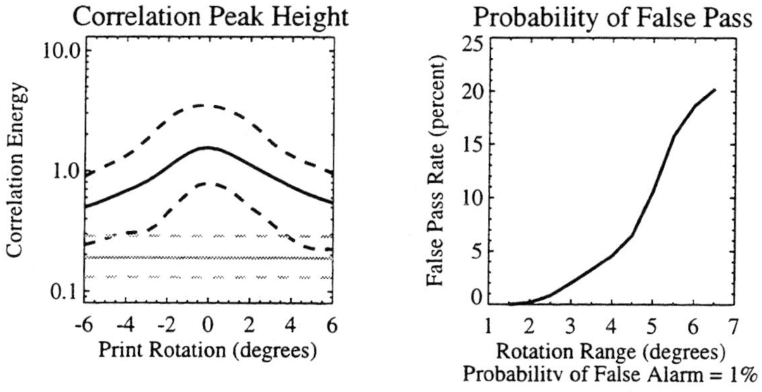

Figure 8. Correlation peak height and the resulting probability of false pass when the entire joint power spectrum is used to form the second stage input.

4.3. The effect of using a portion of the joint power spectrum

The area of the power spectrum chosen for use in the second stage of the BJTC was selected to include the area where the power spectrum is large, while avoiding regions where the expected energy is small. The reasoning for this is that in the regions where the signal is large, there will be relatively less effect from clutter noise, camera noise, quantization noise, etc., since these noise sources have energy which is reasonably evenly distributed across the joint power spectrum. The central DC spike was excluded, since in a real system, there is a great deal of noise energy near the optical axis. When this is not eliminated through use of a DC block it generally contains enough energy to cause difficulties.

In order to verify that the performance of the correlator is better when only a portion of the power spectrum is used, a simulation was done using the entire central region of the spectrum illustrated in Figure 5. For this simulation, the joint power spectrum was not resized before binarization. The portion of the power spectrum used includes the entire annular region in which the power spectrum has significant energy. The processing was the same as described in Section 4.2. Since camera noise was not addressed in this simulation, only the effects of less than perfect correlation are seen here. The results are shown in Figure 8.

The output peak intensity for both matched and unmatched peaks is lower than with use of the partial joint power spectrum. While system performance is still good when the rotation range is small, the performance falls off more quickly with increased rotation angle than when only a portion of the joint power spectrum is used.

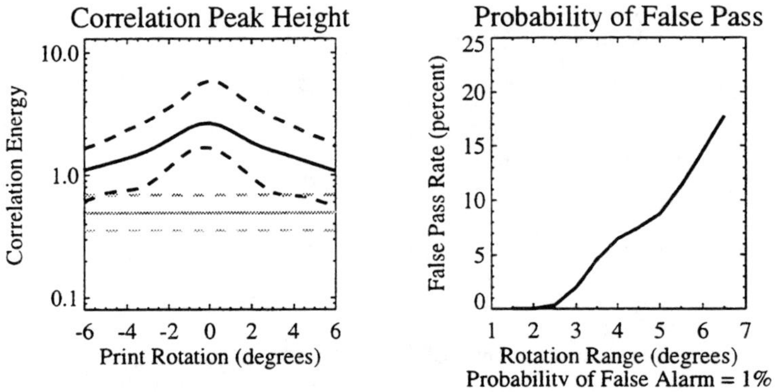

Figure 9. The correlation peak height and the resulting probability of false pass when Fourier plane windowing is used to binarize the joint power spectrum.

4.4. Fourier plane windowing

Fourier plane windowing has been shown experimentally to have correlation performance which approaches the performance obtained using Frame subtraction[6]. This study shows that its system performance in terms of the probability of false pass under a CFAR constraint approaches the performance achieved with frame subtraction. This will make this form of processing the obvious choice in many situations, since only one frame of input data is needed rather than two. The processing done by the optical system is done in two passes as opposed to three (two for the input stage and one for the output stage).

The portion of the joint power spectrum indicated by the box in Figure 5 is spatially high-pass filtered with the convolution mask defined by

$$w(\alpha,\beta) = \begin{pmatrix} -1 & 0 & 2 & 0 & -1 \end{pmatrix} \tag{23}$$

and the result is binarized by thresholding at zero. No multiplications were used when implementing this mask. Only additions and subtractions were needed. The fringes on the joint power spectrum generated by a correlating pair of prints are expected to have a spatial period of 4.4 pixels, which is very close to the peak response of this convolution filter. Figure 9 summarizes the system response for a verification system based on this BJTC. As with the system using frame subtraction, the system performance is very good when the rotation range is small. The system performance is more degraded than when frame subtraction is used through the rotation ranges of ±3° to ±5°.

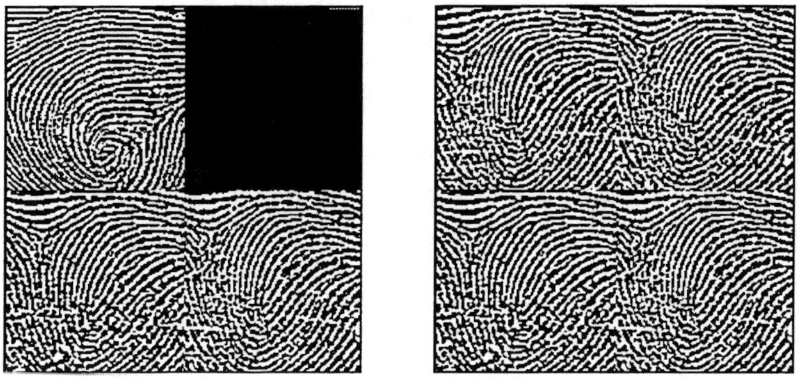

a. b.

Figure 10. BJTC input planes using multiple reference prints. In both cases the print to be identified is in the lower left. In (a), the two references are from different fingerprints. In (b) the three references are 0° and ±5° rotations of the same print.

4.5. Tradeoffs in using multiple prints in the input plane

This study has demonstrated that a fingerprint can be very effectively displayed using a 128x128 pixel area in the input plane. Earlier studies[5,6] have experimentally demonstrated good correlation performance using half of this input area. Since most SLMs currently used for signal processing have active areas of 256x256 pixels or larger, it makes sense to display more than two fingerprints at a time at the input to the JTC system. There are two ways that multiple reference prints can be used to gain processing advantage. The input print can be screened against several reference prints at once[5,6]. This is useful when the system must identify which fingerprint in a database is being matched, or if the verification system is designed to respond to more than one of the user's fingers. This is a useful feature since otherwise if a system user injures his identification finger, the system may be unable to verify identity. The other way that multiple references can be used is to display multiple rotations of the reference[6]. This expands the rotation range over which the system performs well. Both of these approaches are investigated here.

Figure 10 (a) and (b) show the input plane as used for displaying a scene print with two and three references. The print to be verified is displayed on the lower left in both cases. In 9(a) the reference prints are different, and in 9(b) they are 0° and ±5° rotations of the same print. The statistical results for two and three different reference prints are shown in Figure 11 (a) and (b), and the results for two and three rotations of the reference print are in Figures 12 (a) and (b). The two references for Figure 12(a) are rotated by ±2.5°. In the case where multiple rotations of the print are used, the

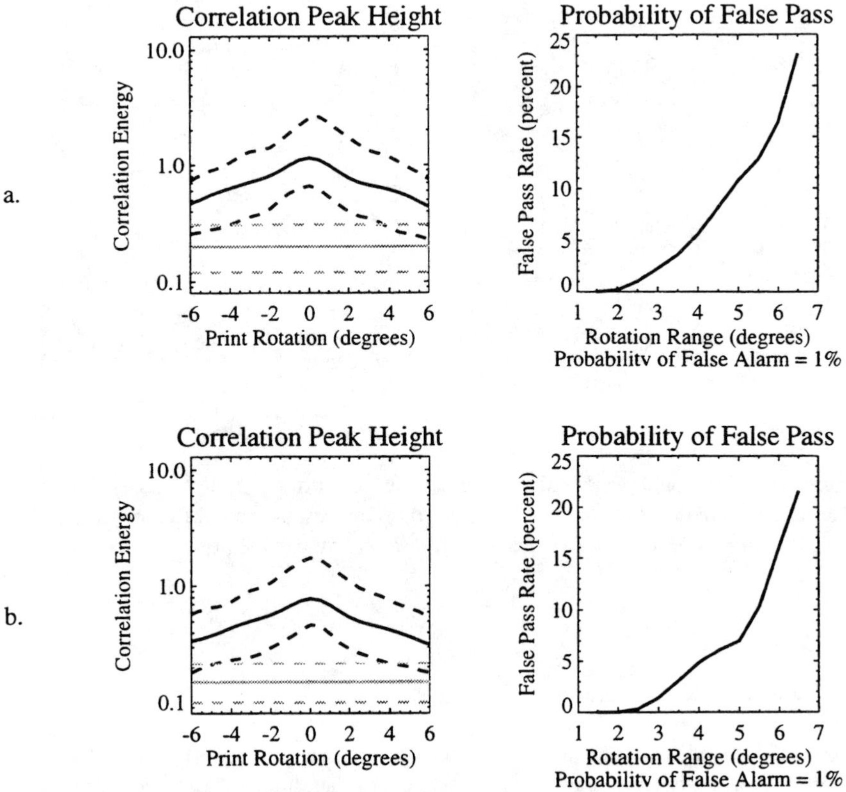

Figure 11. Correlation peak height and resulting probability of false pass for the BJTC based fingerprint recognition system with multiple reference prints for different fingers displayed in the input plane. The graphs in (a) were obtained using two references and those in (b) used three references.

correlation peak is determined by the largest response at any of the expected peak locations. Only the correct location was used to determine the correlation peak response for three different prints. For the simulations used to determine statistics for the unmatched correlation, the peak response was taken as the largest response at any of the possible expected peak locations in both cases.

Figure 11(a) shows that when two different references are used, the expected correlation peak intensity drops by a factor of two. Fortunately the expected unmatched correlation intensity drops by a similar amount, and the resulting false pass performance is similar to, but slightly degraded from, the results for comparing a single pair of fingerprints. As was the case with the Fourier plane windowing described in the last section, a significant gain in processing efficiency has been bought with a small

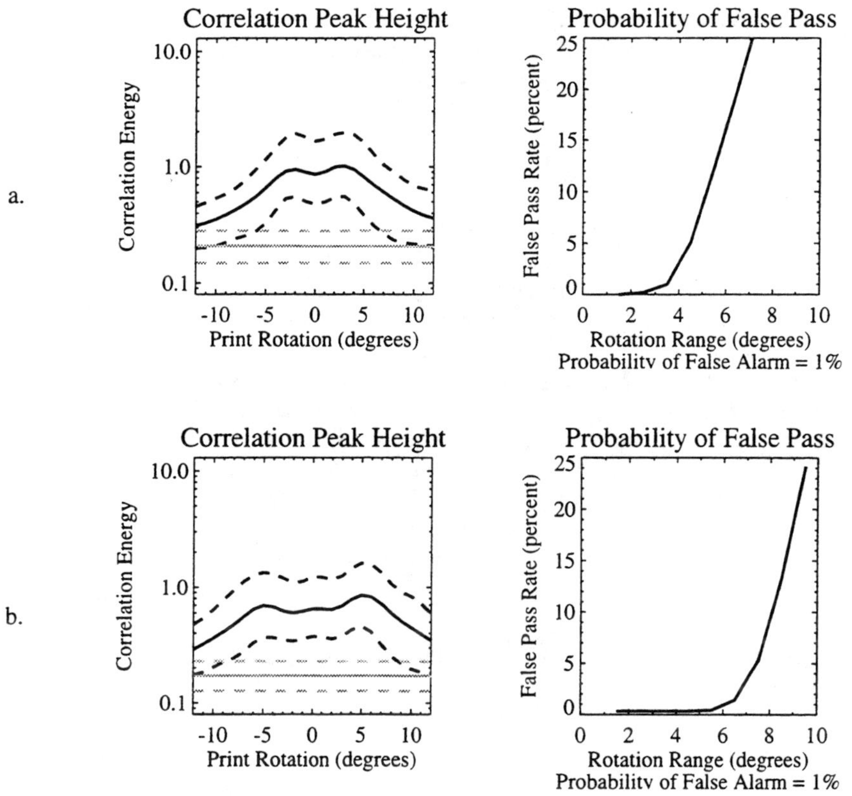

Figure 12. Correlation peak height and resulting probability of false pass for the BJTC based fingerprint recognition system with multiple reference prints for different rotations of the same finger displayed in the input plane. The graphs in (a) were obtained using two reference rotations and those in (b) used three rotations.

degradation in performance. When the number of references is increased to three, there is additional degradation in both the correlation peak intensity and in the system performance. However, overall system performance is still very good over a rotation range of ±2°.

Multiple references can also be used to display more than one rotation of the same print in order to extend the angle over which the system performs well. The results shown in Figure 12(a) show that for two reference prints with rotations of ±2.5°, the acceptance range where the verification system to performs well is extended to ±6°. As was the case with unmatched references, the expected correlation height with multiple references is significantly less than for a single reference. In the case here, the smallest correlation measurement in the exactly on-axis case overlapped with the largest unmatched peak height, resulting in a probability of false pass greater than zero even for small rotation angles. If a CFAR better than 1% was demanded, this system would have a high false pass rate at all rotation ranges.

5. CONCLUSIONS

The Fourier plane binarization techniques of frame subtraction and windowing, both of which produce a large improvement in the correlation performance of a BJTC for a small penalty in additional processing, have been reviewed. A technique for assessing the system performance of a fingerprint verification system under a CFAR constraint has been presented. These techniques have been used to show that a BJTC based fingerprint verification system can be made with very good performance (low probability of false pass). It has been shown that by using the Fourier plane windowing technique, and/or displaying multiple reference prints in the input result in an increase of processing throughput by a factor of two or more and can be realized for a small penalty in system false pass performance. The systems were simulated using a 1% false alarm rate. The single reference verification system had a false pass rate of under 1% for an input rotation tolerance of up to ±3.5°.

The system using multiple references for the purpose of gaining rotation range increased this tolerance to ±6°.

6. REFERENCES

1. B. Javidi, "Nonlinear joint power spectrum based optical correlation", *Appl. Opt.*, **28**, 2358-2367, (1989).

2. B. Javidi, and C. J. Kuo, "Joint transform image correlation using a binary spatial light modulator at the Fourier plane", *Appl. Opt.*, **27**, 663-665, (1988).

3. C. S. Weaver and J. W. Goodman, "A technique for optically convolving two functions", *Appl. Opt.*, 5., 1248-1249, (1966).

4. B. V. K. Vijaya Kumar, C. Chen, and J. D. Brasher, "Correlation Filters assessed in terms of probability of detection and probability of false alarm", *Optical Pattern Recognition IV*, D. P. Casesent, ed., *SPIE Proceedings* Vol. 1959, p. 12-22, Orlando, FL, 13 April 1993.

5. T. J. Grycewicz, "Fingerprint identification with the joint transform correlator using multiple reference fingerprints", *Optical Pattern Recognition VI*, D. P. Casesent and T. H. Chao , ed., *SPIE Proceedings* Vol. 2237, p. 249-254, Orlando, FL, 19 April 1995.

6. T. J. Grycewicz and B. Javidi, "Experimental comparison of binary joint transform correlators used for fingerprint identification", scheduled for publication in *Opt. Eng.* in September,1996.

7. B. Javidi, and J. L. Horner., "Optical pattern recognition for validation and security verification", *Opt. Eng.*, **33**, 1752-1756, (1994).
8. K. H. Fielding, J. L. Horner, and C. K. Makekau, "Optical fingerprint identification by binary joint transform correlation", *Opt. Eng.*, **30**, 1958-1961, (1991).
9. F. Song *et al.*, "Fingerprint identification by using a joint-Fourier correlator and a liquid crystal light valve", *Optics and Laser Tech.*, **26**, 29-37, (1994).
10. J. Rodolfo, H. Rajbenbach, JP. Huiginard, "Performance of a photorefractive joint transform correlator for fingerprint identification", *Opt. Eng.*, **34**, 1166-1171 (1995).
11. B. Javidi and J. L. Horner, "Single spatial light modulator joint transform correlator", *Applied Optics* **28**, 1027-1029 (1989).
12. B. Javidi, J. Wang, and Q. Tang, "Multiple-object binary joint transform correlation using multiple level threshold crossing", *Appl. Opt.*, **30**, 1231 1241 (1991).
13. T. J. Grycewicz, "Applying time modulation to the optical joint transform correlator", *Opt. Eng.*, **33**, 1813-1820, (1994).
14. M. Schonleber, G. Cedilnik, and H. J. Tiziani, "Joint transform correlator subtracting a modified Fourier spectrum", *Appl. Opt.* **34**, 7532-7537, (1995).
15. R. K. Wang, C. R. Chatwin, L. Shang, "Synthetic discriminant function fringe-adjusted joint transform correlator", *Opt. Eng.* **34**, 2935-2944, (1995).
16. S. K. Rogers, J. D. Cline, M. Kabrisky, and J. P. Mills, "New binarization techniques for joint transform correlation", *Opt. Eng.* **29**, 1088-1093 (1990).
17. W. B. Hahn and D. L. Flannery, "Basic design elements of binary joint transform correlation and selected optimization techniques", *Opt. Eng.* **31**,896-905 (1992).
18. B. Javidi, J. Horner, A. Fazlollahi, and J. Li, "Illumination Invariant pattern recognition with a binary nonliniar joint transform correlator using spatial frequency dependent threshold function", *Photonics for Processors, Neural Networks, and Memories, Proc. SPIE* **2026**, 100-106 (1993).
19. T. J. Grycewicz, "Fourier plane windowing in the binary joint transform correlator for multiple target detection", *Appl. Opt.* **34**, 3933-3941 (1995).
20. J. W. Goodman, *Introduction to Fourier Optics*, McGraw Hill, San Francisco, c. 1968.
21. J. Wang and B. Javidi, "Multiobject detection using the binary joint transform correlator with different types of thresholding methods", *Opt. Eng.* **33**(6), 1793-1805 (1994).
22. C. I. Watson and C. L. Wilson, NIST Special Database 4, Fingerprint Database, National Institute of Standards and Technology, Advanced Systems Division, Image Recognition Group, Gathersburg, MD (1992).

Optical Processing Systems

Encrypted optical memory using double random phase encoding

Bahram Javidi, Guanshen Zhang, and Jian Li

University of Connecticut
Department of Electrical and Systems Engineering
Storrs, CT 06269-3157

ABSTRACT

We describe an encrypted optical memory using double random phase encoding at the input plane and the Fourier plane. This technique allows the images to be stored as independent white complex stationary processes. Experimental results and computer simulations are presented.

1. INTRODUCTION

Optical encryption methods using random phase encoding have been proposed recently.[1,2] In this paper, we describe an encrypted optical memory using double random phase encoding in both the input plane and the Fourier plane. Each stored image is encrypted and can be read out by a unique code or a universal code. In this method, the image to be stored is first multiplied by a random phase function. The Fourier transform of the product of the image and the random phase function is then multiplied by another random phase function in the Fourier domain. Taking the inverse Fourier transform, an encrypted image in the output plane is obtained which is then stored in the holographic optical memory.[3-5] The encrypted memory is a stationary white noise if the two encrypting random phase functions are two independent white sequences uniformly distributed on $[0, 2\pi]$.[2] Compared to the other encoding techniques,[6,7] the double phase encoding technique proposed here converts the images to be stored into white Gaussian stationary random processes which provides the maximum entropy. This makes it very difficult to decrypt the memory without the knowledge of the phase functions used in the encryption.[2,8]

In Section 2, the encrypted optical memory using double phase encoding is analyzed. In Section 3, optical experiments and computer simulations of the proposed encrypted memory are presented. In Appendix, the signal-to-noise ratio of the optical memory is determined in terms of the number of the images stored and the space-bandwidth product of the random codes and the stored images.

2. ENCRYPTED OPTICAL MEMORY

Let $f_i(x, y)$ denote the ith optical image to be stored in the memory. Let $n_i(x, y)$ and $b_i(\alpha, \beta)$ denote two independent white sequences uniformly distributed on $[0, 2\pi]$, respectively. Here x and y are the coordinates of the space domain, and α and β are the coordinates of the Fourier domain. In the encryption process, the random phase function $\exp[jn_i(x, y)]$ is used in the space domain and the random phase function $\exp[jb_i(\alpha, \beta)]$ is used in the Fourier domain. The encrypted memory can be represented as:

$$\varphi(x,y) = \sum_{i=1}^{M} \{f_i(x,y)\exp[jn_i(x,y)]\} * \mu_i(x,y) \tag{1}$$

where M is the total number of data images stored in the memory, $\mu_i(x, y)$ is the inverse Fourier transform of $\exp[jb_i(\alpha, \beta)]$, and "*" denotes the convolution operation. The ith data image can only be decrypted when the corresponding "key" , $\exp[-jb_i(\alpha, \beta)]$, is used for the decryption. If we use $\exp[jb(\alpha, \beta)]$ to encrypt all the images, then $\exp[-jb(\alpha, \beta)]$ can serve as a universal key to decrypt all the encrypted images.

To decrypt the memory, the Fourier transform of the encrypted memory $\varphi(x, y)$ is multiplied by the decoding mask $\exp[-jb_i(\alpha, \beta)]$. This causes the encoding phase function $\exp[jb_i(\alpha, \beta)]$ to be canceled out by the decoding mask $\exp[-jb_i(\alpha, \beta)]$ that serves as the "key" for decryption. Therefore, the original data image can be recovered in the space domain because the stored image is positive and the phase function $\exp[jn_i(\mathrm{x}, \mathrm{y})]$ is removed by an intensity sensitive device, such as a video camera.

When the image is retrieved from an optical memory, the ghost images (the images not decoded) degrade the quality of the recovered image. The degradation depends on the statistical distribution of the ghost images. The double random phase encoded optical memory is capable of transforming all ghost images into a white Gaussian stationary background noise over the output plane which are orthogonal to the image to be retrieved. The number of the images that can be stored and successfully retrieved depends on the requirement of applications. The quality of the recovered image can be measured by some metrics such as signal-to-noise ratio (SNR) which may be defined as the ratio of the energy of the decrypted image to the energy of the noise in the area where the image exists. In the appendix, we derive the SNR for the random phase encoded memory defined by Eq. (1). The SNR is proportional to the space-bandwidth product of the random codes and inversely proportional to the number of the images stored in the memory. In practice, the limit of the number of the images that may be stored in one memory depends on the requirement of SNR.

3. OPTICAL EXPERIMENT AND COMPUTER SIMULATIONS

To demonstrate the encryption technique for optical memory, we use the optical set-up shown in Fig. 1. An argon-ion laser with $\lambda = 515$ nm is used as the coherent light source. Two 4-f structures are cascaded. The first structure performs the encryption and the second structure performs the decryption. The collimated beam is split into object and reference beams by a beam splitter BS for hologram recording. The Fourier transform lenses L_1, L_2, and L_3 have a focal length of 400 mm, and L_4 has a focal length of 250 mm. The shorter focal length of L_4 matches the decrypted image size to a CCD camera. The optical paths for the object beam and the reference beam from the beam splitter to P_3 are adjusted to be equal, and the angle α between the two beams is about 7.4°. This arrangement results in fringe spacing 4.0 μm. The holographic optical memory used in the experiments is considered as a planar hologram.

In the memory recording process, the data image to be encrypted is placed at P_1 and a random phase mask M_1 is attached to it. The Fourier transform of the product of the image and the random phase function is obtained at the Fourier plane P_2 by lens L_1. It is multiplied by another random phase mask M_2 that is placed at P_2. Taking another Fourier transform by lens L_2, the encrypted image data is obtained at P_3. By placing an optical recording media at P_3, a holographic optical memory is obtained. By repeating this proc-

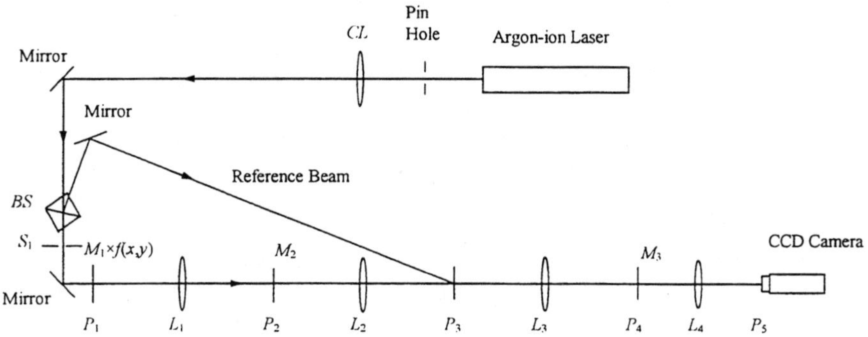

Figure 1. Optical set-up used to demonstrate the encrypted optical memory: *CL*, collimate lens; *M*'s, mirrors; *BS*, beam splitter; *L*'s, Fourier transform lenses; S_1, shutter.

ess, multiple images can be recorded on the same memory with different random phase masks that are independent from one another. Therefore, a specific data image can be decrypted from the memory by the corresponding key while other data remain encrypted.

To decrypt the image data, the shutter S_1 is closed and the optical memory is placed at P_3. The reference beam now acts as a uniform light beam illuminating the holographic optical memory. The Fourier transform of the encrypted image is obtained at P_4 by lens L_3, where it is multiplied by the phase mask M_3 that is the complex conjugate of the random phase function M_2. Taking another Fourier transform by lens L_4 we obtain the original (decrypted) image data at the output plane P_5 by the CCD camera, which removes the random phase modulation M_1. Here $M_3 = M_2{}^*$ serves as the "key" for decryption. It is obtained by producing a complex conjugate mask of M_2. Without this key, the encrypted image can not be recovered. When a multiple-image memory is placed at P_3, the desired image can be decrypted with the corresponding phase mask M_3.

The decoding process requires good alignment. If the recorded images are real the alignment of the phase code is only required at the Fourier plane since the phase modulation in the space domain can be removed by an intensity sensitive device. For each image, different code (key) is used for recording. As a result, the corresponding keys must be used for each of these images. In practice, we can use an electrically addressed spatial light modulator as a "phase mask" in the Fourier plane. Therefore, the alignment needs to be done only once at the system set up time.

In the optical experiment, the images of the characters "E" and "Q" are used as the data images shown in Fig. 2(a) and they are encrypted sequentially in the same memory using different random phase masks. The random phase masks M_{1E} and M_{2E} are used to encrypt the image of "E", and M_{1Q} and M_{2Q} are used to encrypt "Q". The four masks are independent white sequences, 32 × 32 pixels (7.2 mm × 7.2 mm in size) two dimensional arrays. The phase masks are made by bleaching[9] the gray scale photographic films of uniformly distributed white sequences. The phase of each pixel is independent of other pixels and approximately distributed uniformly on [0, 2π]. To obtain a random phase mask, a two-dimensional gray scale matrix of a random array uniformly distributed in [0, 255] is first displayed on a monitor and then a photo picture is taken. To get relatively accurate phases in [0, 2π], a three-level scale of 0, 128, and 255 is displayed on the side of the random array when the picture is taken. After bleaching, the phases corresponding to these three levels are measured by interferometry to ensure that they have the phases of 0, π, and 2π. Calibration has to be made until the desired phases are obtained by adjusting the condition of mask making process. The image of "E" is first encrypted with M_{1E} and M_{2E}, then "Q" is encrypted with M_{1Q} and M_{2Q}. The encrypted images are recorded on a KODAK high speed holographic film SO-253 at P_3 with the use of the reference beam. The encrypted holographic memory is shown in Fig. 2(b). The reason that the encrypted memory is not quite a white noise is the limited space-bandwidth product of the optical system due to the finite size and the limited number of pixels of the random phase masks used in the Fourier domain. This means that images can only be encrypted up to certain frequencies, resulting in limited bandwidth encrypted output images (white noise has an infinite bandwidth). The decrypted data images obtained in P_5 are shown in Fig. 2(c). They are obtained by using M_{3E} and M_{3Q}, respectively. We can see that the images have been recovered, though there is some clutter noise on the recovered images which comes from the incorrectly decoded images. This is due to the limited system bandwidth and limited space-bandwidth product of the codes. If a wrong phase mask M_3 is used to decrypt the images, only noise is obtained. Fig. 2(d) shows the result of applying a wrong mask to the optical memory in which the images "E" and "Q" are encrypted.

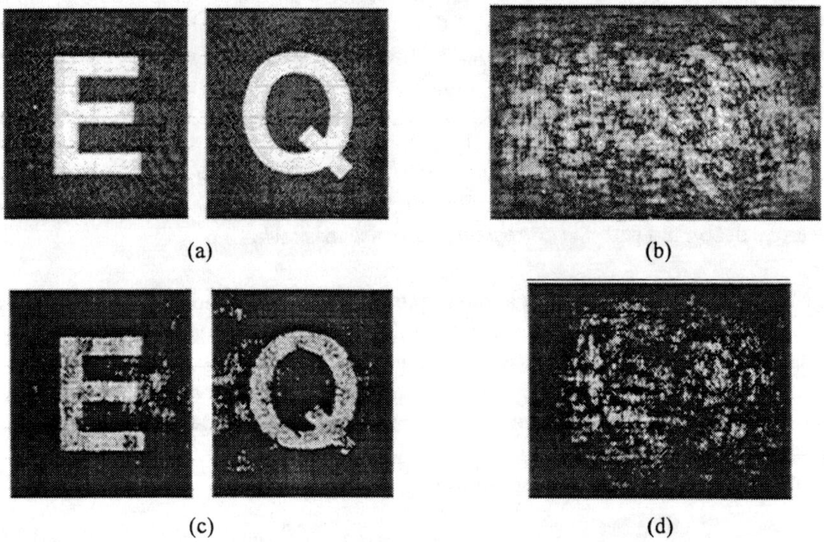

Figure 2. Experimental results of the encryption and decryption. (a) Images of "E" and "Q" used for encrypted memory. (b) Encrypted holographic optical memory with "E" and "Q" stored. Different masks are used for the encryption. (c) Decryption of "E" and "Q" from the memory in (b). The authorized codes are used for the decryption. (d) The decryption result with a wrong code.

The space-bandwidth product affects the whiteness of the background noise which is from the undecrypted images while the system bandwidth directly affects the quality of the decrypted image. Figure 3 shows a computer simulation for decrypting a memory where the image "E" is stored using 256 × 256 random phase codes. The image "E" is recovered when different system bandwidth is used. The result shows that the quality of the recovering can be improved by increasing the system bandwidth.

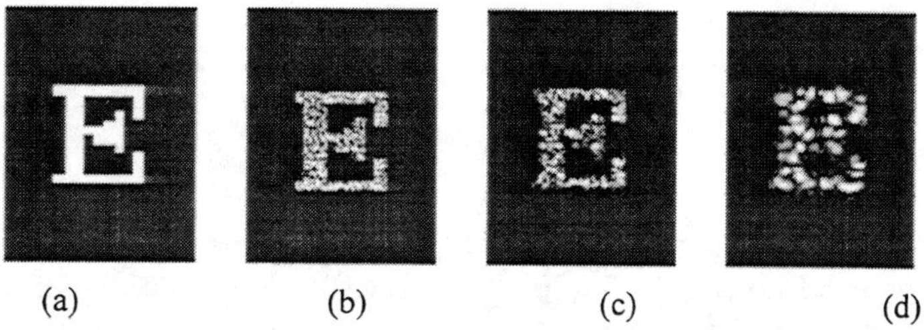

Figure 3. Computer simulation for decrypting a memory where the image "E" is stored using 256 × 256 random phase codes. The image "E" is recovered when different system bandwidth is used. (a) Original image of "E". (b) Decrypted image of "E" with the system bandwidth of 200 × 200 pixels. (c) Decrypted image of "E" with the system bandwidth of 100 × 100 pixels. (d) Decrypted image of "E" with the system bandwidth of 50 × 50 pixels.

3. CONCLUSION

We have described an encrypted optical memory using double random phase encoding technique in the input plane and the Fourier plane. The double random phase encoding allows the images to be stored as independent complex white stationary processes. The output signal-to-noise ratio is shown to be proportional to the space-bandwidth product of the random codes and inversely proportional to the number of the stored images. Tests are provided to demonstrate how images can be encrypted into the optical memory and how they can be reconstructed by a code.

APPENDIX

In this Appendix, we determine the output signal-to-noise ratio (SNR) for the random phase encrypted optical memory.

Let us assume that there are M images $f_i(\xi)$ ($i = 1, 2, ..., M$) stored in an optical memory. Two random phase masks $\exp[jn_i(\xi)]$ and $\exp[jb_i(\nu)]$ are used to encode the ith image in the space domain and the Fourier domain, respectively. ξ and ν are the coordinates of the space domain and the Fourier domain, respectively. The images and the phase masks are sampled to have N pixels.† The encrypted memory can be represented as:

$$\varphi(\xi) = \sum_{i=1}^{M} \{f_i(\xi)\exp[jn_i(\xi)]\} * \mu_i(\xi) \qquad \text{(A1)}$$

where $\mu_i(\xi)$ is the inverse Fourier transform of $\exp[jb_i(\nu)]$.

† One dimensional discrete notations are used for simplicity.

To retrieve the kth image $f_k(\xi)$, the phase mask $\exp[-jb_k(\nu)]$ is used at the Fourier plane. The signal at the output plane is

$$
\begin{aligned}
f_k'(\xi) &= \varphi(\xi) \otimes \mu_k(\xi) \\
&= f_k(\xi)\exp[jn_k(\xi)] + \sum_{\substack{i=1 \\ i\neq k}}^{M} \{f_i(\xi)\exp[jn_i(\xi)]*[\mu_i(\xi)\otimes \mu_k(\xi)]\} \qquad \text{(A2)} \\
&= f_k(\xi)\exp[jn_k(\xi)] + \sum_{\substack{i=1 \\ i\neq k}}^{M} \{f_i(\xi)\exp[jn_i(\xi)]*\mu_i'(\xi)\}
\end{aligned}
$$

where $\otimes$ denotes the correlation operation and $\mu'_i(\xi)$ is the correlation of the two independent random processes $\mu_i(\xi)$ and $\mu_k(\xi)$. The first term is the product of the kth original image $f_k(\xi)$ and a random phase function $\exp[jn_k(\xi)]$ that can be removed by an intensity sensitive device if the image is real and positive. The summation term is the noise term. Each random process $\mu'_i(\xi)$ in the summation is a white stationary process because the random processes $\mu_i(\xi)$ and $\mu_k(\xi)$ are white independent stationary processes. As a result, each term in the summation is a white stationary process and independent of one another, and the noise in Eq. (A2) is s stationary white process. $\mu'_i(\xi)$ is the Fourier transform of a phase-only function. Therefore the power spectral density of the noise term in Eq. (A2) is[2]

$$
\begin{aligned}
P &= \frac{1}{N}\sum_{\substack{i=1 \\ i\neq k}}^{M}\sum_{\xi=0}^{N-1}|f_i(\xi)|^2 \\
&= \frac{1}{N}(E_{\text{total}} - E_k)
\end{aligned}
\qquad \text{(A3)}
$$

where E_k is the energy of the restored image given by:

$$E_k = \sum_{\xi=0}^{N-1}|f_k(\xi)|^2 .$$

and the total energy of the M stored images is:

$$E_{\text{total}} = \sum_{i=1}^{M}\sum_{\xi=0}^{N-1}|f_i(\xi)|^2 .$$

To measure the noise corruption of the retrieved image, we define the SNR for the kth retrieved image $f_k(\xi)$ as

$$\text{SNR}_k \equiv \frac{\text{Energy of the restored image } E_k}{\text{Energy of the noise overlpping the stored image } f_k(\xi)} \qquad \text{(A4)}$$

The energy of the noise overlapping the retrieved image can be represented by $N_kP = \frac{N_k}{N}(E_{\text{total}} - E_k)$, where N_k is the area (total number of the pixels) of the kth image. Thus the signal-to-noise ratio is

$$\text{SNR}_k = \frac{NE_k}{N_k(E_{\text{total}} - E_k)} . \qquad \text{(A5)}$$

When the images encrypted have the same energy, the SNR can be simplified as

$$\mathrm{SNR}_k = \frac{N}{N_k(M-1)} \; . \tag{A6}$$

From this equation we find that SNR is proportional to the space-bandwidth product of the random codes and is inversely proportional to the number of the images stored in the memory. The number of the images that may be stored in the memory depends on the available output SNR.

REFERENCES

1. B. Javidi and J. L. Horner, "Optical pattern recognition for validation and security verification," Opt. Eng. **33**, 1752-1756 (1994).
2. Ph. Refregier and B. Javidi, "Optical image encryption based on input plane and Fourier plane random encoding," Opt. Lett., **20**(7), 767-769 (1995).
3. D. Psaltis, "Optical memory disks in optical information processing," Appl. Opt. **29**, 2038-2057 (1990).
4. M. A. Neifeld, D. Psaltis, "Programmable image associative memory using an optical disk and a photorefractive crystal," Appl. Opt. **32**, 4398-4409 (1993).
5. J. W. Goodman, *Introduction to Fourier Optics*, McGraw-Hill, New York (1967).
6. E. N. Leith and J. Upatneiks, "Holograms: their properties and uses," SPIE Journal **4**, 3-6 (1966).
7. J. F. Heanue, M. C. Bashaw, and L. Hesselink, "Encrypted holographic data storage based on orthognal-phase-code multiplexing," Appl. Opt. **34**(26), 6012-6015 (1995).
8. A. Papoulis, *Probability, Random Variable, and Stochastic Processes,* 2nd edition, McGraw-Hill, New York (1984).
9. R. J. Collier, B. Burckhardt, and L. H. Lin, in *Optical Holography*, Chapter 10, (1971).

OPTICAL NEURAL NETWORKS FOR CLASSIFICATION INTO ARBITRARY CLASSES

H. H. Arsenault, S. Chang, P. Gagné and O. Gualdron*
COPL, Université Laval, Québec, PQ, Canada GIK 7P4
*Universidad de Santander, Bucamaranga, Columbia

Keywords: Optical processing, invariant pattern recognition, image processing

ABSTRACT

Some important concepts of neural networks are similarity, generalization, invariance and training. Some neural networks are supposed to be able to classify objects according to hidden similarities. All of those concepts are put into question by the consideration first put forward by Watenabe that from a purely logical point of view, similarity is a purely arbitrary concept. It can be shown that this implies that the notion of invariance is also arbitrary , that so-called hidden similarities and generalization cannot exist without some external criteria. Such criteria are either implicit in the training algorithms or must be imposed explicitly. This imposes severe limitations on what neural networks can accomplish. However there are some positive implications; neural networks can be designed to classify objects into arbitrary classes. Applications to optical neural networks and examples will be presented.

INTRODUCTION

Watanabe showed[1] that from a purely logical point of view, similarity is a purely arbitrary concept. He called it the *theorem of the ugly duckling*. This seems to fly in the face of some alleged capabilities of neural networks, such as the ability to find hidden similarities and generalization. Not much attention has been paid to this theorem (which becomes rather obvious after one has understood it properly). But the arbitrariness of similarity has some important consequences.

The ugly duckling theorem simply uses the fact that decisions on the degree of similarity between two objects depends on the weights put upon the on the properties used to classify the objects; for example, consider a two balls and two cubes, one of each green and one of each red. If all the weight is put on the color, clearly the object most similar to the red cube is the red ball, but if all the weight is put on the shape, then clearly the red cube is most similar to the green cube. Of course sometimes there are reasons to put more weight on some properties than on others, and that is why the theorem's statement is qualified by the words "from a purely logical point of view".

What are the implications of this theorem? If similarity is a purely arbitrary concept, then it means that from a purely logical point of view, there are no fundamental "hidden" similarities between objects, and if that is so, it is not possible for neural networks to find them. This means that neural networks that purport to find hidden similarities are really finding similarities programmed (perhaps unknowingly) into the algorithm, like a person discovering his own footprints as he walks around a deserted island. It also puts into question concepts such as "generalization", a term that has been thrown about rather loosely in recent times. Indeed, the concept of generalization depends heavily on the concept of similarity; In most cases I have seen in pattern recognition, generalization usually means that a filter is not too sharp in its recognition ability, and that the target for which it has been designed can still be recognized after more or less deformation. I believe that calling this "generalization" is an abuse of language: generalization should refer to the ability to infer, it should refer to a greater ability, not to a lesser one.

In most digital pattern recognition implementations, neural or otherwise, the measured values are usually pixel values, because that is how the images are stored in computers. In the simplest cases, all pixel values are given the same weights; for example, in the classical matched filter, all the pixels of the prototype are given the same relative weight, and the degree of similarity is measured by the inner product of the prototype with the pixels of the unknown target. In practice it is the Fourier transforms that are used, but of course the inner product of two objects is equal to the inner product of their Fourier transforms. In more refined pattern recognition techniques, the pixels are not given equal weights, and the pixel values are used to define some character of the object that is used to classify it.

In neural networks, the input values are usually the pixel values of the inputs; but sometimes the input values are the results of some pre-processing of those pixel values; for example we shall later describe a case where the input values to a neural network are optically pre-processed in order to reduce the images from two dimensions to one, and to introduce some invariance to the orientation of the input. Note that the arbitrariness of similarity is independent of whether the processing is linear or not.

I have formulated an important corollary to Watanabe's theorem: if similarity is a purely arbitrary concept, then so is invariance! Indeed, invariance means that after some transformation, the new object is still similar to the previous one; but as we have seen, "similar" is a slippery concept. For transformations such as translations, rotations and scale changes, it is possible to define (with some care) properties that will have the same measurements after the transformation. For example, lengths are preserved during translations and rotations, but not during scale changes, where only relative lengths are preserved. Under noise, the concept of similarity becomes more arbitrary: how much noise does it take before two objects are not similar any more? Clearly that is an arbitrary choice: in practice, all pixels are usually given equal weights and the similarity decision is made by thresholding.

We shall begin by describing how to classify objects into arbitrary classes. Then we review some recent progress in invariant pattern recognition. The invariant parameters that are of interest include changes of position, of orientation, of scale, signal-dependent noise, and disjoint noise. Over past years, fairly effective methods have been developed to deal with cases where a single invariance is required. For most methods that deal with two invariant parameters, one of them is usually shift invariance, which is guaranteed by systems such as optical correlators that use linear filters. More success has been obtained in dealing with changes in orientation than with changes of scale; that is because rotation is a linear operation whereas a change of scale is not. We review some of the methods recently developed in our laboratories. Not all of them are relevant to the question of classifying objects into arbitrary classes, but all of them deal with one kind of invariance or another.

MEASUREMENT OF SIMILARITY

One question that arises is the following: if similarity is arbitrary, why is it easier to classify objects into certain classes than into certain others? I'm glad you asked: it depends on how you measure the similarity. For example, is similarity is measured by counting the number of pixels having the same values, clearly it will be easier to put objects into classes where the measure of similarity is the number of equal pixels. This means that 1) for any classification method, some thought must be put into how the similarity will be measured; 2) since images are stored into computers as pixel values, it may be useful to change the pixel values into some other parameters before carrying out the classification (or the training, in the case of neural networks). We discuss some ways to do this in the next section.

Inner product

Many similarity measures are carried out by weighed products. If the j-th features of each member of the training set or prototypes X_k consists of a weight vector with elements x_{jk}, an unknown object with weights y_i is classified by comparing its weights with those of the training set. This can be done by means of the inner product

$$w = X^t y \tag{1}$$

where **X** is a matrix whose rows contain the weight vectors $\boldsymbol{x}_k$. The vector $\boldsymbol{w}$ contains the inner product of the unknown vector with all of the weight vectors, and the largest element of the vector $\boldsymbol{w}$ corresponds to the prototype whose position in feature space is closest to that of the unknown object, since the product of two vectors is maximum when they are parallel. This measure is important because it corresponds to operations carried out both by neural nets and by matched filters.

Why is this product a good measure of similarity? Because for binary objects (0,1), it is equal to the number of pixels that are equal to one in both objects involved in the product. This measure is not always good, and can lead to pathological behavior of the network.

Distance functions

An alternate approach and one that is frequently used is pattern classification by distance functions[2]. There are a number of distance functions, the simplest of which is the Euclidean distance between the prototype and the unknown vector

$$D_i = \|x_i - y\|. \tag{2}$$

With the so-called minimum-distance method, the unknown object is attributed to the class containing the prototype having the minimum distance to the unknown object.

There are other useful measures of similarity based on distance. One is the Mahalanobis distance

$$D = (x-y)\, C^{-1}\, (x-y), \tag{3}$$

where **C** is the covariance matrix of a pattern population; this measure, based on a Bayes classifier, minimizes the classification error under certain conditions which assume that the population is randomly distributed with a normal distribution, which is not a very realistic assumption for most practical cases of interest, but which has the advantage of fitting into a nice theoretical mold.

The important point for the purposes of this discussion is that all these measures of similarity modify the weights of the object features in some way in order to obtain a measure of similarity. The features used are usually pixel values, but other features can be used as well.

Non-numeric measures

Similarity concepts may be applied to nonnumeric features. Such features are usually compared by attributing numeric labels to them. The labels can then be manipulated by various techniques such as the ID3 Approach[3], which is a procedure for pattern recognition and classification for synthesizing an efficient discrimination tree for classifying patterns with nonnumeric features. The features are examined in sequence by means of an information-theoretical approach to find a combination of feature values that suffice to determine class membership and that has minimum entropy. This assumes that the features are randomly distributed with a known or assumed probability distribution, so the presence of two objects in the same class means only that this grouping minimizes the number of steps required for classification, and not necessarily that the two objects have many properties in common.

There are other nonnumeric measures of similarity[4], but the comments of the previous paragraph apply to them as well, that is to say that objects are put in the same class in order to optimize some classification scheme. Such methods are useful when objects have many nonnumeric features, but it is debatable whether objects that are put in the same class would be considered similar by one who was not concerned with the time required to classify large numbers of such objects.

Matched filter

The matched filter was originally developed for radar applications, and later applied to images. It has the property that under certain statistical conditions, it is the correlation filter that maximizes the output signal-to-noise ratio. It has played an important role in Optics because in addition to being shift invariant, it can be implemented easily in an optical correlator[5]. The matched filter carries out the correlation operation between an unknown object $f(x,y)$ and a prototype $h(x,y)$

$$c(x,y)=\int_{-\infty}^{\infty}\int_{-\infty}^{\infty} f(\xi,\eta)h(\xi+x,\eta+y)d\xi d\eta \tag{4}$$

Note that the correlation value at the origin c(0,0) is equal to

$$c(0,0)=\int_{-\infty}^{\infty}\int_{-\infty}^{\infty} f(\xi,\eta)h(\xi,\eta)d\xi d\eta \tag{5}$$

If the pixel values of the object $f(x,y)$ and of the prototype $h(x,y)$ are arranged lexicographically into vectors y and x respectively, we see that the correlation at the origin is equal to the scalar product $c = x{\cdot}y$ between the object and the prototype, so the measurement of similarity for the correlation and for the inner product is the same.

Some improved measures of similarity have been proposed, and it has been shown that they can reduce the pathologies of neural networks[6].

CLASSIFYING TARGETS INTO ARBITRARY CLASSES

One consequence of the ugly duckling theorem is that an ideal pattern recognition system should be able to attribute any arbitrary given degree of similarity to any set of objects. It should be programmable to recognize any two objects as being either identical or as belonging to different sets, depending on what is required. A corollary of this is that an ideal system should be able to yield any arbitrary output for any arbitrary input. Is this achievable, or is it a purely academic exercise?

It is possible to design matched filters that yield arbitrary outputs for any given input[6]. A composite filter is a linear combination of matched filters, whose coefficients are chosen in order to obtain a given correlation value. If the correlation values $C_{ij}(0,0)$ between lexicographically arranged objects X_i are put into the columns of a two-dimensional matrix **C**, and if the required outputs for a linear filter with impulse response $h(x,y)$ are arranged into the vector $\mathbf{y}$, the required coefficients for the filter $h(x,y)$ expressed as a linear combination of the training set $X_i(x,y)$ are found by solving the matrix equation

$$z = \mathbf{C}^t \mathbf{y},$$

which is

$$y = [C_{ji}{}^t C_{ji}]^{-1} C_{ji}{}^t z = C^{\dagger} z, \tag{7}$$

where $C^{\dagger}$ is the pseudoinverse of matrix C.

The composite filter determines the correlation values only for the principal correlation peaks. Unfortunately composite filters are usually associated with high sidelobes, and much effort has been expended to reduce the effects of sidelobes. The more components a composite filter has, the more sensitive to noise it is. In practice, at least for optical implementations, no more that five or six components should be used.

Neural nets

How can arbitrary classification be achieved by means of a neural net? In view of our earlier discussion on the equivalence of correlation and inner products, the solution is rather straightforward, especially in view of the fact that it is well known that the backpropagation training method converges towards the pseudo-inverse solution. This means that a neural network trained by backpropagation will approximate the linear solution to the linear problem discussed above. (The nonlinear function of the network comes in during the classification process, not during the training process).

The arbitrary classification could be achieved by a feedforward network where the relation between the input and the output of the network would be

$$y = [C_{ji}{}^t C_{ji}]^{-1} C_{ji}{}^t\, T\, x, \tag{8}$$

where C is the correlation matrix between the training set, x is an input vector, and T is the interconnection matrix between the intermediate slab of neurons and the output slab, whose columns contain the required output vectors for each stored object.

PROGRESS IN INVARIANT CLASSIFICATION

The above methods, as well as some of the following ones, begin with the pixel values of the images to be classified. A more radical approach is described in the next section, where the image is compressed and transformed by means of an optical system; in this case it is modified images that are compared.

Image compression for invariant pattern recognition

A typical optical or digital neural network uses one hologram or stored image for every image class so that we have to correlate each new input image with every hologram or stored image, which slows down the system. On the other hand, if we can combine all the network's weights into a single image, we can obtain all the activations of the neurons by means of a single correlation. In this paper, we show that we can compress the information of an image by using a ring target that focuses the light pattern onto a small rectangular zone.

An optical setup used to integrate the light energy inside a series of concentric rings is depicted in Fig. 1. A telescope formed by lenses l_1 and l_2 projects an input image onto a ring target. Inside each ring is a diffraction grating that, in conjunction with the final lens l_3 , focuses the light to a given point in the output plane.

It seems obvious that an increase of the number of rings will increase the prototype separation (*i.–e.* decrease their similarity), and this is desirable because it makes classification easier. But if we plot the mean similarity vs. the number of rings as in Fig. 2, we find that there is no increased separation after more than about 10 rings; hence we used a 10–ring target in the optical experiments.

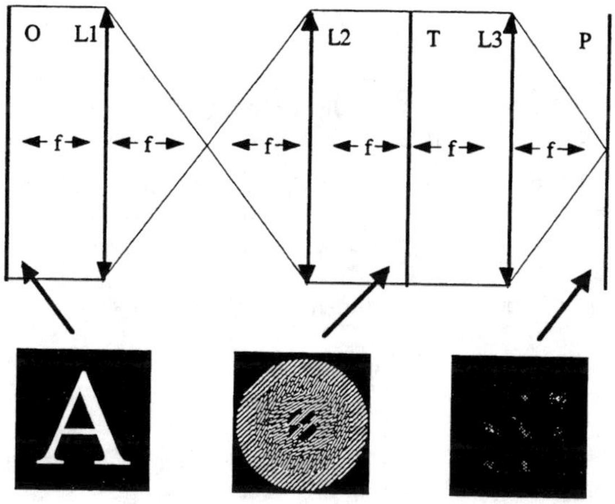

Figure 1: Ring-sampling optical system

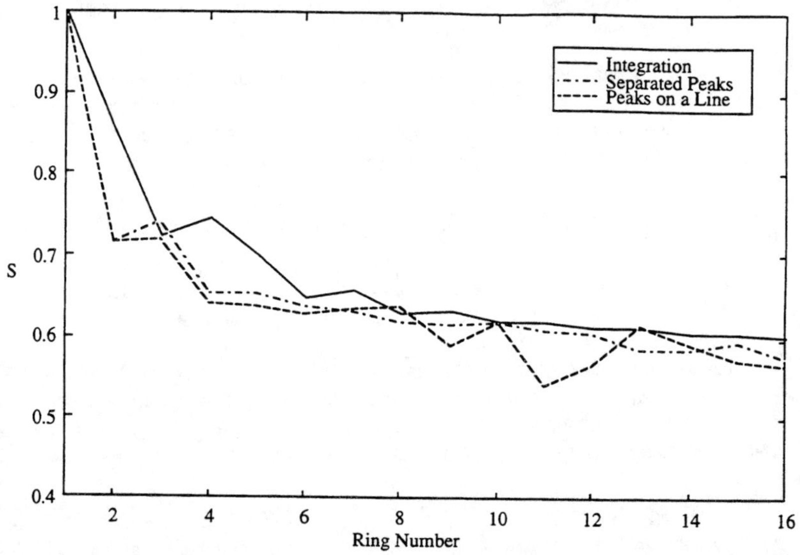

Figure 2: Similarity measure for 9 letters with different numbers of rings.

Learning Rule

A single–layer linear associator is the classifier. There is one neuron for each class to be recognized and its activation corresponds to the inner product of the input image p_l and of the network's weights w_i:

$$o_{il} = \sum_{x,y} p_l(x,y) w_i(x,y) \tag{9}$$

After a training phase with the delta rule, the network will correctly classify the input prototypes, *i.-e.* the difference between t_{il} (the requested target activations) and o_{il} (the actual activations) will be small. The delta rule adjusts neuron activations by increasing (or decreasing) the correlation between the weights and the input prototype. In the present case, we must take into account that the input prototypes can be complex–valued by adding the conjugate of the prototypes in the standard equation. A learning constant ε, $(0<\varepsilon<1)$ must be fine-tuned to avoid divergence.

$$\Delta w_i(x,y) = \varepsilon(t_{il} - o_{il}) p_l^*(x,y) \tag{10}$$

The training is done by iteratively presenting an input prototype and updating the neurons' weights until all activation differences (t_{il}-o_{il}) are smaller than a given value

$$\delta = \max(t_{il} - o_{il}) \tag{11}$$

The Lateral Delta Rule

The scalar product of Eq. (9) is also given by the (0,0) point of the correlation between p_i and w_i

$$k_{il}(x,y) = p_l(x,y) \otimes w_i(x,y) = \sum_{x',y'} p_l(x'-x, y'-y) w_i^*(x',y') \qquad (12)$$

$$k_{il}(0,0) = \sum_{x',y'} p_l(x',y') w_i^*(x',y') = p_l(x,y).w_i^*(x,y) \qquad (13)$$

This equivalence between the scalar product and a correlation makes the implementation of a neural network by an optical correlator possible if we place the weights in a hologram and then read the result in the output plane by measuring the value of the $k_{il}(0,0)$ pixel corresponding to the perfect alignment of the input vector and the weight. But a correlation also generates parasitic activations, corresponding to all the values obtained by translating the weights relative to the input image. Those pixels have no influence in a numerical simulation, but they can distort an optical result enough to cause false classification results.

Figure 3 shows a correlation results with and without the lateral delta rule. Figure 4 shows pictures of fish used for the optical experiment; The compressed and modified vectors for the fish were stacked on a single hologram and Fig. 5 shows the correlation results. These correlation peaks could be used to drive the second stage of an associative memory.

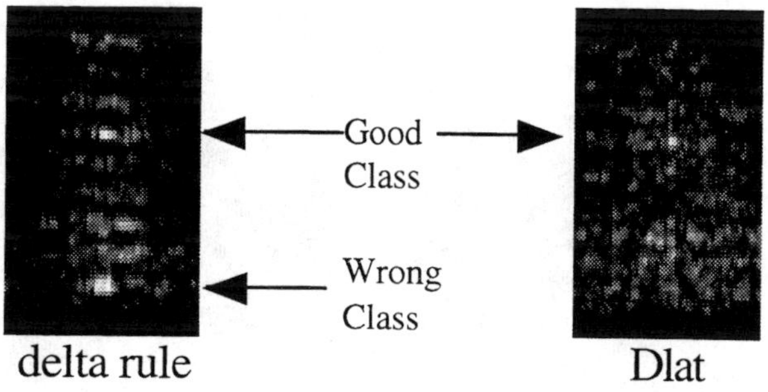

Figure 3: Classification with and without a lateral delta rule

These results show that it is possible to classify input images against multiple classes in parallel with the added benefit of rotation invariance. Although the compression scheme severely compresses the information, the neural network training ensures that enough information is preserved for error-free classification.

Figure 4: Images used for the experiments

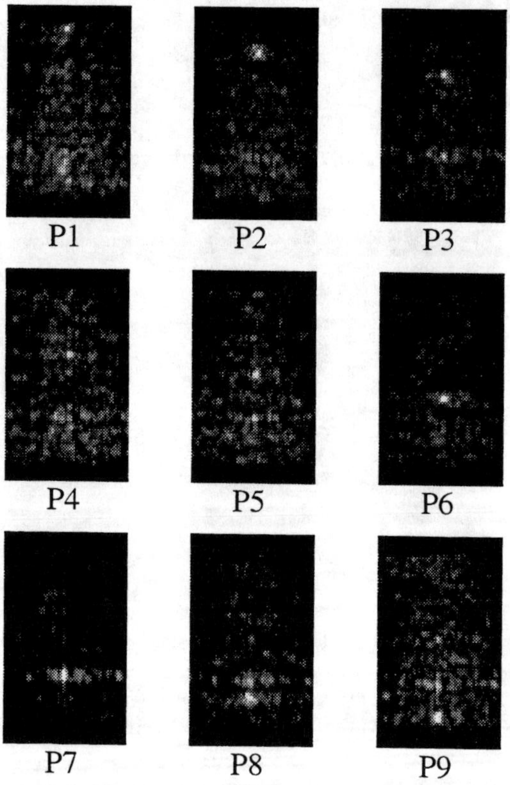

Figure 5: Optical results with the Lateral delta rule

Rotation invariant CHC filter for disjoint noise

A matched filter built from a Circular Harmonic Component (CHC) of the target, ensures full rotation invariance of the intensity pattern in the correlation output[7]. Several filter designs based on CHCs improve the peak sharpness or the discrimination ability of the classical CHF at the cost of a decrease in the SNR. Second-order CHFs are often a good compromise for both an acceptable SNR and peak sharpness of the correlation peak.

A disjoint noise scene $s(r,\theta)$ can be described by the following expression,

$$s(r,\theta) = f(r,\theta) + w(r,\theta)n(r,\theta) \tag{14}$$

where $f(r,\theta)$ is the target and $n(r,\theta)$ is a noise distribution with mean value and variance (m_n, σ^2). The function $w(r,\theta)$ is defined

$$w(r,\theta) = \begin{cases} 0 & \text{where } f(r,\theta) \text{ is defined} \\ 1 & \text{elsewhere} \end{cases} \tag{15}$$

At first sight, the noise problem would seem to arise from the fact that the noise has a non-zero mean, and it could be easily corrected in several ways. One solution could be to block the center of the Fourier plane in the correlator; another would be to use a filter with a zero-mean impulse response. The latter is the case of the classical CHF, because the harmonic angular kernel ensures the zero-mean impulse response of the filter. However, there are strong differences between the response of the classical CHF when it is used with a disjoint noise scene and with an additive noise scene. The high noise resistance of the classic CHF has been proven using a SNR criterion. This SNR criterion is widely used in correlation-based pattern recognition, and involves only the central value of the correlation output $C(x,y)$. Refregier et al. have proposed[8] the Peak to Output Energy Ratio (POE), a more suitable criterion for measuring the resistance of a filter to the presence of non-overlapping noise. They also derived a filter that maximizes the POE criteria for the rotation-variant case[46]. The POE is defined

$$POE = \frac{\left|E\{C(x_0,x_0)\}\right|^2}{E\left\{\overline{\left[C(x,x_0)\right]^2}\right\}} \tag{16}$$

where $C(x,x_0)$ is the correlation output when the target is located at x_0 in the input scene, E{•} means the statistical expected value, and the upper bar means spatial averaging over the whole correlation output. The advantage of the POE metric over the SNR metric comes from the fact that in the former the whole correlation plane amplitude is taken into account.

The optimum CHF for disjoint noise, the rotation invariant filter that maximizes the POE ratio has the form[9]

$$H^{*}(\rho,\phi)=\frac{G_m(\rho)}{B_0(\rho)}e^{im\phi}=\frac{1}{B_0(\rho)}CHC_m\left[F(\rho,\phi)+m_n W(\rho,\phi)\right]. \quad (17)$$

with

$$B_0(\rho)=\int_0^{2\pi}\left[\left[\frac{|W(\omega)|^2}{2\pi}*\left[S_n(\omega)-2\pi.m_n{}^2\delta(\omega)\right]+|F(\omega)+m_n W(\omega)|^2\right]d\phi\right. \quad (18)$$

We have tested the filter for the objects shown in Fig. 6, against correlated disjoint noise. The noise samples were sixteen air photographs of a forest and of two urban scenes (residential and industrial).

Following the work of Yaroslavsky[10], we replace the spectrum of the noise by the mean spectrum of the objects to reject and that of the noise. The filter obtained in this way belongs to the category of sub-optimum filters. This approach has indeed yielded a considerable improvement in the classification accuracy, and we recommend that it be used when it is required to improve discrimination between targets. Although the filters were not designed for discrimination against the other objects, the performance was excellent. In addition to yielding improved performance over filters not designed for disjoint noise, the recognition was rotation invariant.

Figure 6. Targets used in the experiments from left to right and up to bottom, f-18, light, space shuttle, ambulance, Ferrari, formula, police and Porsche.

Scale-adapted wavelet transform

The wavelet transform is a multi-resolution representation of a signal. But unfortunately the WT is a dimension-expanding operation. Theoretically it doubles the number of dimensions of the input signal. Despite this inconvenience, wavelet transforms are widely used.

We have introduced[11] a scale-adapted wavelet transform (SAWT). The scale factor in the SAWT is a function of the input; it adaptively changes its value according to the local structure of the input. The output of a SAWT provides a multiscale representation of the input while keeping the same dimensions as the input.

The wavelet transform is

$$Wf(s_x,s_y,u,v) = \int_{-\infty}^{\infty}\int_{-\infty}^{\infty} f(x,y)\,\frac{1}{\sqrt{s_x s_y}}\,h\left(\frac{x-u}{s_x},\frac{y-v}{s_y}\right)\,dx\,dy \tag{19}$$

Where, s_x, s_y are scale factors in the (x, y) directions, and u, v are shift factors.

The modified wavelet transform is

$$Wf(u,v) = \int_{-\infty}^{\infty}\int f(x,y)\,\frac{1}{s(x,y)}\,h\left(\frac{x-u}{s(x,y)},\frac{y-v}{s(x,y)}\right)\,dx\,dy \tag{20}$$

Not all the values of s are necessary in all cases. The representation is only meaningful if the local structure is broader than the resolution scale. The scale function s(x,y) should be related to the local structure.

There are many ways to design the scale function s(x,y). We use the gradient, because it is a good indicator of the local structure of the input signal. We define s(x,y) as a linear ramp function of the gradient of the input:

$$s(x,y) = \begin{cases} 1 & |\nabla f(x,y)| > D_{max} \\ s_{max} - (s_{max} - 1)\dfrac{|\nabla f(x,y)|}{D_{max}} & \text{otherwise} \end{cases} \tag{21}$$

where, s_{max} and D_{max} are given values.

This transform can be used for multi-resolution edge detection, The SAWT is also useful for pattern classification, because it is a multi-resolution representation. Table 1 shows results for the classification of 8 aircraft. The performance of the SAWT is compared with that of the phase-only filter.

The prototype was aircraft number 3, whose correlation was normalized to one. The results shows that the SAWT considerably decreases the cross-correlations. The improved performance is due to the fact that the phase-only filter preserves only the high-frequency edges of the targets, whereas the SAWT preserves all the edges.

Table 1: Cross-correlations of 8 aircraft

Object	SAWT based	phase-only
#3	1.00	1.00
#1	0.06	0.12
#2	0.04	0.14
#4	0.03	0.09
#5	0.05	0.06
#6	0.02	0.08
#7	0.01	0.04
#8	0.02	0.06

Optimum filters

In recent years, when the parameters controlling the performance of filters have became better understood, researchers have begun to develop filters that optimize one or more parameters that affect the recognition and discrimination ability of matched such as signal to noise ratio, optical efficiency, correlation peak energy, peak to sidelobe ratio, peak to (total) correlation energy ratio and peak sharpness. A performance criterium cannot be increased without a corresponding decrease in some other parameters, so it is not possible to maximize all the parameters at once.

It is possible and necessary to find some compromise between the performance criteria for a given task, because maximizing only one criterium always results in poor performance. Refregier's optimum tradeoff filter[12] allows to find a good compromise between peak energy, signal to noise ratio, and peak to correlation energy. There is a tradeoff between signal to noise ratio and peak to correlation energy for Refregier's optimum tradeoff filter and our Locally nonlinear matched filter[13].

One can choose the parameter values that correspond best to the particular situation under consideration. For example, if the noise is moderate, one can well sacrifice some signal to noise ration in favor of peak sharpness for improved discrimination. The bad news is that such curves cannot be predicted in advance and must be calculated point by point for all the objects under consideration, a very time-consuming task. A worthwhile endeavor would be to devise a way to approximate such curves rapidly.

CONCLUSION

Pattern recognition has reached a point where it is possible to classify objects independently of position, orientation and scale. It is also possible to optimize systems in order to approach a required performance by trading off various performance measures. The ability to process 2-D data in parallel by means of Optics can be used to classify an input object among multiple classes of 2-D objects in real time. The ability to take into account the fact that noise is often disjoint and correlated means that more realistic results can be expected.

The methods described can be implemented either optically or digitally. When the number of data are small, digital processing is usually more convenient, whereas Optics is better adapted to the massively parallel processing of data such as images.

There are other methods developed by us or by others that are not described in this paper due to lack of space. Most noteworthy is a method proposed by Neiberg and Casasent for the classification of targets under out-of-plane rotations, scale and position[14]. We have also studied this method and will report on it at a later date.

REFERENCES

[1] S. Watanabe, "Pattern Recognition, Human and MechanicaL,",John Wiley & Sons (1985).

[2]J. T. Tou and R. C. Gonzalez, *Pattern recognition principles*, Addison-Wesley, Reading, MA (1974).

[3]J. R. Quinlan, *Learning efficient classification procedures and their application to chess end-games*. In R. S. Michalski, J. G. Carbonell and T. M. Mitchell (Eds.), *Machine Learning*, Tioga, Palo Alto, CA.

[4]Y.-H. Pao, *Adaptive Pattern recognition and neural networks*, Addison-Wesley (1989).

[5]VanderLugt, A., *Signal detection by complex filtering*, IEEE Trans. Inform. theory, IT-10, 139 (1964).

[6] H. Arsenault, *Pattern recognition, similarity, neural nets and Optics*, 15th Congress of the International Commission for Optics, *Optics in complex systems*, Garmisch-Partenkirchen, FDR, Aug. 5-10, 1990, Proc. SPIE vol **1319**, pp. 380-384 (1990).

[7] H. H. Arsenault, D. Asselin, S. Chang, L. Leclerc and Y. Sheng, "Recent Results in Rotation Invariant Pattern Recognition," in *Optical Pattern Recognition*, J L Horner and B Javidi, eds., Critical Reviews of Technology Series, Proc. SPIE **CR40**, 75-75 (1992).

[8] Refregier P., Javidi B., and Zhang G., “Minimum mean square error filter for pattern recognition with spatially disjoint signal and scene noise”, Opt. Lett. **18**, 1453 (1993).

[9] O Gualdron and H. H. Arsenault, “Optimum rotation invariant filter for disjoint noise scenes:, Appl. Opt.**35**, 2507-2513 (1996).

[10] L. Yaroslavsky, “Is the phase-only filter and its modifications optimal in terms of the discrimination capability in pattern recognition?”, Appl. Opt. **31**, 1677 (1992).

[11] S. Chang and H. H. Arsenault, “Scale-adapted wavelet transform”, Opt. Lett. **21**, 363-365 (1996).

[12] P. Refregier, "Optical pattern recognition: optimal trade-off circular harmonic filters" Opt. Comm. **86** 113 (1991).

[13] O Gualdron and H. H. Arsenault, “Optimum rotation invariant filter for disjoint noise scenes:, Appl. Opt.**35**, 2507-2513 (1996).

[14] L. Neiberg and D. Casasent, “Feature space trajectory classifier neural network”, Proc. SPIE **2353**, 276-292 (1994).

Phase-only SLMs for optical pattern recognition and security systems

Yunlong Sheng, Luiz Gonçalves Neto*, Danny Roberge**, Lixin Shen**
and Giles Paul Hus

Université Laval, Centre d'Optique, Photonique et Laser,
Département de Physique, Québec, Canada G1K 7P4. sheng@phy.ulaval.ca

* Present address: Sao Paulo University, Laboratory of integrated systems
Department of Electrical Engineering, C.P. 8174, Sao Paulo-SP 01065-970, Brazil

** Present address: MYTEC Technologies Inc.
10 Gateway Blvd. Suite 430, Don Mills, Ontario M3C 3A1, Canada

ABSTRACT

Original experimental techniques for obtaining the phase-only modulation and the complex amplitude modulation with the commercial twisted nematic liquid crystal televisions (LCTV) are described. The techniques are based on the Jones matrix theoretical analysis. We use the programmable phase-mostly LCTV to optically implement the phase-only matched filters, phase-only circular harmonic filters, phase-only synthetic discriminant functions and the adaptive composite wavelet matched filters for pattern recognition. We also implement the phase-only computer generated holograms, speckle-free kinoforms and optical kinoform encryption using random phase encoding for security applications.

1. INTRODUCTION

A spatial light modulator (SLM), that provides full range complex amplitude modulation is an ideal component for the programmable holography and real-time optical pattern recognition using the holographic filters. Such a SLM is a real time wave front manipulator. However, a SLM is addressed by a single electrical or optical signal. It is impossible to control independently and simultaneously the phase and amplitude at each pixel of the SLM by a single signal. Thus, the continuous complex modulation can be obtained only with a combination of at least two SLMs[1-4], or of two sub-pixels in a micro-pixel, which is optically under-resolved[5]. The system complexity is then increased.

Most SLM's provide modulations on the optical intensity or polarization. The SLMs that provide continuous phase-modulation therefore are of interest. A nematic liquid crystal device (LCD) can provide the continuous phase modulation[6]. Amako et. al. fabricated a homogeneously aligned nematic LCD for phase-only modulation with the molecular orientations aligned parallel to the panel faces and to each others without twist[2]. This approach requires special fabrication of the phase LCD, which still has coupled amplitude modulation[2] of more than 10%. In the twisted nematic LCD the phase and amplitude modulations are inherently coupled. Based on the Berreman's theory on dynamics of the twisted nematic LC cell, Konforti et. al. showed that the phase-only modulation can be obtained when the applied voltage is below an optical threshold and the incident light is polarized parallel to the front molecular director of the cell[7]. The similar conditions may be obtained for the phase-mostly modulation using the Jones matrix theory for the twisted nematic LC cell developed by Lu and Saleh[8,9], as will described in Sec.3A. Pezzaniti et. al. determine the eigenpolarization states experimentally by the Mueller matrices measurement with the imaging polarimeter. The elliptical eigenpolarization state input can be phase-only modulated. This regime has no limitation on the bias voltage[10]. However, theoretical explanation to this regime is yet to be developed.

We obtain the phase-only modulation with commercial liquid crystal television (LCTV), which has advantage of reliability and low-cost. Especially, when one found that the monochrome panels of the Epson Crystal Image video projector provide the full 2π range phase retardation, several laboratories were interested in investigating this LCTV for optical pattern recognition and switchable hologram applications[11-22]. In Sec. 2 and 3 we describe the mechanism of the phase/amplitude coupled modulations in the twisted nematic LC cell, based on the Jones matric calculus introduced by Lu and Saleh[8,9], and the experimental techniques developed by us to obtain the optimal phase-mostly and amplitude-mostly modulations in the commercial LCTV.

The phase-mostly LCTV is useful for encoding the real-time phase-only filters (POF) in the optical correlator for pattern recognition. It also can be used for programmable holography, optical security systems and diffractive optical elements testing systems[23]. We build optical correlators with the programmable phase-mostly and complex amplitude modulation SLMs. The phase only matched filter[24] is a kinoform in the Fourier plane of an optical correlator. With a continuous phase modulation SLM the POF yields the correlation output on the optical axis, so that it has 100% theoretical light efficiency[25] and utilizes the full available space bandwidth product (SBWP) of the SLM. The POF amplifies high frequency components and produces then a sharp auto-correlation peak with good discrimination capability. In Sect. 6 we show implementations of a variety of POF's. All the implementations demonstrate the advantages of the POF's for requiring low SBWP, having high light efficiency and good discrimination capability. The low resolution is a basic problem of the electrically addressed SLM. However, for encoding the POF's the LCTV of 200 x 200 pixels already produces satisfactory results. The drawback is the speed of operation at the TV rate, which may be not fast enough for automatic target recognition applications, but could be fast enough for security applications.

The twisted nematic LCTV provides phase and amplitude modulations which are inherently coupled. The optimal filters for the coupled-mode SLMs have been proposed[26,27,28]. In those algorithms the filter value at each pixel is chosen to be on the operating curve of the SLM and to maximize one of the selected metrics: correlation peak intensity, SNR and others. A laboratory result has been shown[29] that used a LCTV in a high-contrast mode having a full range of amplitude modulation and about only $2\pi/3$ range phase modulation. In this case, the optimized filters have better performance than that of the POF. However, the phase shift range of $2\pi/3$ introduces severe phase mismatching resulting a filter which is far from the original POF. Our approach is to experimentally optimize the phase-mostly LCTV by rotating the polarizer and analyzer and using the *in situ* characterization of the LCTV to minimize the coupled amplitude modulation. Indeed for a LCTV having the full range of 2π phase modulation and a 6% variation of the amplitude the benefit from the optimization would be small, since when the SLM is perfectly phase-only, the optimal maximum correlation filter is simply the POF[25]. In Sect. 4 we develop a theoretical analysis on the diffraction of a phase-mostly SLM, showing that the POF response can be obtained from the phase-mostly filter with a slightly reduced SNR.

The POF is sub-optimal in the sense that its SNR is lower than that of the matched spatial filter and its discrimination capability is lower than that of the inverse matched filter. A number of optimal filters and optimal trade-off filters have been proposed[30-34]. The recent synthetic discriminant function (SDF) design relaxes the constraint that the filter is a linear combination of training images to provide much more degrees of freedom for optimization and trade-off of a variety of performance criteria[33]. The primary criteria are the correlation peak sharpness, the robustness to noise and the distortion tolerance. All those filters are complex-valued filters. Their implementation would require a complex amplitude modulation SLM using at least two SLM's. Therefore, the system complexity would increase and the light efficiency would decrease. However, it will be interesting to implement the optimal trade-off filters using the programmable complex amplitude modulation SLM. In this paper we use optical masks, such as the central stop and the wavelet band-pass filter in the filter plane, to block some frequency components in order to improve SNR of the POF. The effect is similar as that of the ternary filters for improving the performance of the binary phase-only filter. We describe also the adaptive composite wavelet matched filter, which provides the invariance to continuous scale variation and is implemented with the complex amplitude SLMs.

We build also optical system with the phase-mostly SLM's for encryption of kinoform using the random phase encoding. Optical phase masks have been proposed for security applications[34,35]. Indeed, a two-dimensional phase mask consists of a very large number of pixels. At each pixel the phase retardation is invisible by ordinary light intensity-sensitive detectors. It is difficult to determine the content of the random phase mask and to reproduce it without knowing the code in the mask. One approach[35] proposed recently by Refrigier and Javidi is encoding a primary image to stationary white noise using two random phase masks in the input plane and Fourier plane respectively. This method can be used to protect high security images. It is impossible to recover the encrypted images without knowing the key random phase mask, which has been used in the encoding. Reconstruction from the encrypted image can be implemented in an optical system using the complex amplitude SLM. However, the use of the random phase masks broadens the Fourier spectrum and thus introduces the speckle noise that can render the optically reconstructed images undistinguishable. We introduce in Section 7 a new approach for encryption of optical kinoforms of the primary images by the random phase encoding. The bandwidths of the phase mask and the computer generated kinoforms are controlled to obtain the speckle-free reconstructed images. The system consisting of two phase-mostly SLMs will be described and optical results will be shown.

2. MODULATIONS IN THE TWISTED NEMATIC LIQUID CRYSTAL DEVICE

A twisted nematic LC cell can be considered as a stack of N birefringent uniaxial crystal slices, whose extraordinary axes are parallel to the slice and are rotated helically from the front to the back of the LC cell through a twist angle α. The light propagates along the z-axis normal to the LC cell. When an electrical voltage is applied across the LC cell along the z-axis, the LC molecules are tilted to the electric field. Assume that[8] the twist angle α is linearly distributed along the z-axis and that the tilt of LC molecules is independent of the position z. The complex transmittance of the LC cell with a polarizer in the front and an analyzer in the rear of the LC cell can be described using Jones matrix calculus[8,9]. The intensity transmittance is

$$T = \left[\frac{\alpha}{\gamma}\sin\gamma\sin(\alpha+(\psi_1-\psi_2)) + \cos\gamma\cos(\alpha+(\psi_1-\psi_2))\right]^2 + \left[\frac{\beta}{\gamma}\sin\gamma\cos(\alpha-(\psi_1+\psi_2)+2\psi_D)\right]^2 \tag{1}$$

The phase shift is

$$\delta = \beta + \tan^{-1}\frac{\frac{\beta}{\gamma}\sin\gamma\cos(\alpha-(\psi_1+\psi_2)+2\psi_D)}{\frac{\alpha}{\gamma}\sin\gamma\sin(\alpha+(\psi_1-\psi_2)) + \cos\gamma\cos(\alpha+(\psi_1-\psi_2))} \tag{2}$$

The complex transmittance is a function of the angles ψ_1 and ψ_2 of the polarizer and analyzer with respect to the molecular director ψ_D in the front slice of the LC cell, and of the applied voltage. The birefringence parameter β is defined as

$$\beta = \frac{\pi d}{\lambda}(n_e(\theta) - n_o) \qquad \gamma = \sqrt{\alpha^2 + \beta^2} \tag{3}$$

where d is the LC cell thickness, λ is the wavelength, n_o is the ordinary and $n_e(\theta)$ is the effective extraordinary indices of refraction as a function of the tilt angle θ. The value of β is related to the applied voltage. When the applied voltage increases, the tilt θ increases and $n_e(\theta)$ decreases, resulting in a decrease of β. The relations between the applied voltage V_{rms}, the tilt angle θ, the effective index of refraction $n_e(\theta)$ and β are non-linear. However, Lu and Saleh[8,9] found a monotonic relation between β and V_{rms}, which is approximately linear for a large range of β.

Eqs.(1) and (2) show the inherently coupled phase and amplitude modulations of a twist nematic LC cell as functions of the ψ_1, ψ_2 and β. For a LCTV the maximum value of β, β_{max}, occurs when the video signal is zero and the brightness bias voltage is minimum. Then, $\beta < \beta_{max}$ decreases with the increase of the applied voltage. When β tends to zero the LC molecular tilt tends to 90° the preceding theory no longer holds.

The twist angle α, the orientation of the front LC molecule director ψ_D and the value of β_{max} are unknown parameters of device, determined by the manufacturer. Those parameters change from one LCTV to another and are different for the red, blue and green panels in the same LCTV video projector. Those device parameters may be determined experimentally with the LCTV operating in a particular regime where the analyzer is parallel to the polarizer $\psi_1 = \psi_2$. The detail of the techniques is described in Refs.19 and 22.

3. PHASE-MOSTLY AND AMPLITUDE MODULATION REGIMES

A. Theoretical regimes

Two special regimes in Eqs.(1) and (2) are of interest:

(1) Phase-mostly modulation: when the polarizer is parallel to the front LC molecular director $\psi_1 = \psi_D$ and the analyzer is parallel to the rear LC molecular director $\psi_2 = \psi_D + \alpha$, Eq.(1) becomes

$$T = 1 - \frac{\alpha^2}{\gamma^2}\sin^2\gamma = 1 - \left(\frac{\alpha}{\pi}\right)^2 sinc^2\left(\frac{\gamma}{\pi}\right) \qquad (4)$$

The intensity transmittance T has a set of maxima at $\gamma = \pm\pi, \pm 2\pi,...$ In the region $\gamma > \pi$ or $\beta > (\pi^2 - \alpha^2)^{1/2}$ the variation of T is small. This is a phase-mostly regime. Eq.(2) becomes

$$\delta = \beta + \tan^{-1}\left(\frac{\beta}{\gamma}\tan\gamma\right) \qquad (5)$$

The phase shift is approximately linear to β when $\beta > (\pi^2 - \alpha^2)^{1/2}$ as shown in Fig.1.

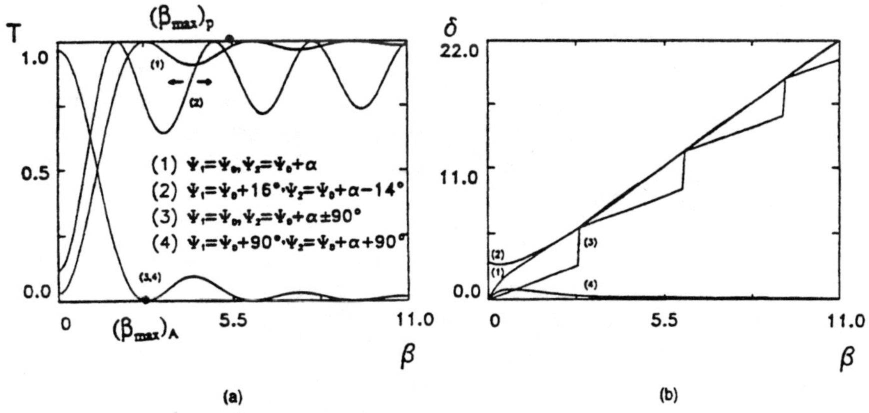

Fig.1. Intensity transmittance T and phase shift δ of the twisted nematic LCD as a function of β for four configurations of polarizer ψ_1 and analyzer ψ_2, α is the twist angle, ψ_D is the orientation of the front LC molecules.

Thus, for the phase-mostly modulation the polarizer should be parallel to the front LC molecular director. The bias voltage should be minimum, such that $(\beta_{max})_p = \beta_{max}$, in order to maximize the phase shift range and to avoid the region of $\beta < (\pi^2 - \alpha^2)^{1/2}$, where a strong drop of amplitude occurs. These two conditions are similar to that for the phase-only modulation in the Berreman's mode[7].

(2) Amplitude-mostly modulation: when the polarizer is parallel to the front LC molecular director and the analyzer is perpendicular to the rear LC molecular director, $\psi_1 = \psi_D$ and $\psi_2 = \psi_D + \alpha \pm \pi/2$, the intensity transmittance T is the complement of that in the phase-mostly mode:

$$T = \left(\frac{\alpha}{\pi}\right)^2 sinc^2\left(\frac{\gamma}{\pi}\right) \tag{6}$$

A high contrast amplitude modulation can be obtained when the bias is adjusted such that $(\beta_{max})_A$ is a first zero of sinc(γ) with $\gamma = \pi$ or $(\beta_{max})_A = (\pi^2 - \alpha^2)^2$. When the video signal is applied, the LCTV works in the region of the first lobe of sinc(γ) with $\gamma \leq \pi$ or $\beta \leq (\pi^2 - \alpha^2)^{1/2}$ as shown in Fig.1. The coupled phase modulation described in Eq.(2) becomes $\delta = \beta$.

B. Experimental optimization of the LCTV modulation

The Jones matrix calculus for the LCTV developed by Lu and Saleh is in good agreement with the experimental measurement for the determination of the device parameters[19,22], α, ψ_D and β_{max}. However, we need still experimental techniques to optimize the LCTV modulations. Two original methods are developed. One is the use of the simple wedge shear plate interferometer to measure the phase modulation of the SLM *in situ*. The other is rotating the polarizer and analyzer and adjusting the bias voltage to manipulate the LCTV operating curve.

The phase shift by the SLM is usually measured using the Mach-Zehnder interferometer. The methods using two slits or Ranchi grating interference fringes is recently introduced[21]. Compared with the Mach-Zehnder interferometry, the wedged shear plate interferometer is extremely simple, easy to align. Compared with the two slits methods, the wedged shear plate interferometer provides the direct measurement of the phase shift with very simple fringe pattern processing, and thereby permitting the *in situ* characterization of the SLMs.

The shear plate is a thick optical flat. Its front and rear surfaces are wedged along the vertical direction to introduce a graded optical path difference between the two surface reflections. A collimated He-Ne laser beam passing through the polarizer, LCTV and analyzer is incident on the wedged shear plate at an horizontal angle θ about 45 deg. as shown in Fig.2a. The interference between the beams reflected from the front and rear surfaces of the wedged plate produces horizontal linear fringes. When the LCTV displays side by side two vertical bands of different gray levels. The two beams of different gray levels can be overlapped in the middle region, as shown in a horizontal section of the wedged plate shown in Fig.2b, because of the lateral shift in the horizontal plane of the beams reflected from the rear surface with respect the beams reflected from the front surface. The interference pattern of the wedged shear plate interferometer therefore consists of three regions. On the two sides are the interferences

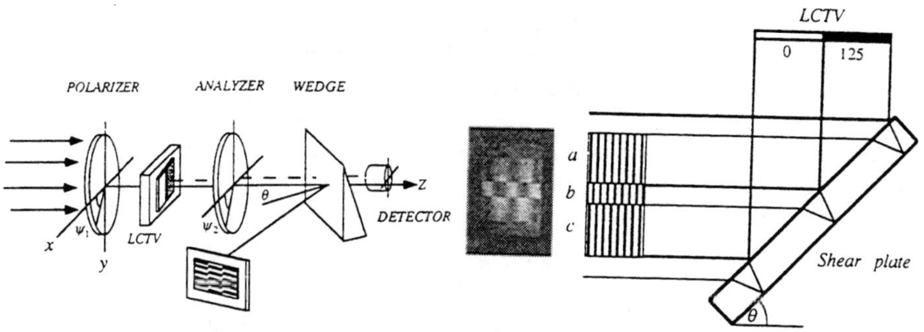

(a) (b)

Fig.2. (a) Wedged shear plate interferometer measuring the phase modulation of LCTV; (b) Interference in the wedge shear plate interferometer. Regions a and c: interference of two wavefronts with the same gray levels. Region b: interference of two wavefronts of different gray levels.

between two beams of the same gray level. In the middle is the interference fringes between two beams of different gray levels. The vertical shift of the fringes in the middle region with respect to the fringes on the two sides is proportional to the difference of phases between the two beams of different gray levels. We obtain then the phase modulation within a narrow vertical region in the middle of the LCTV as a function of the gray level. The intensity modulation is simply measured in one transmitted beam by a photodetector placed behind the wedged shear plate.

The purpose of rotating the polarizer and analyzer and adjusting the bias voltage, under the control of the wedged shear plate interferometer, is to minimize the coupled amplitude modulation in the phase-mostly regime and to maximize the contrast in the amplitude-mostly regime.

For the phase-mostly modulation we set the bias to minimum and display two vertical bars of the gray levels of 0 and 255 side by side on the LCTV. Starting from the theoretical values $\psi_1 = \psi_D$ and $\psi_2 = \psi_D + \alpha$, we rotate the polarizer and analyzer independently until the two gray level beams have an equal intensity after passing through the polarizer-LCTV-analyzer sandwich. During the adjustment we survey and keep the 2π phase shift range unchanged. We found that when $\psi_1 = \psi_D + 16^\circ$ and $\psi_2 = \psi_D + \alpha - 14^\circ$, the amplitude transmittances for the beams of gray levels 0 and 255 were equal. The theoretical explanation is that the rotation of polarizer and analyzer can shift the curve T - β along the β axis with respect to the fixed point $(\beta_{max})_P = \beta_0$, such that $T(\beta_0) = T(\beta_1)$, as shown in Fig.1, where β_1 corresponds to the maximum video signal of the grey level of 255.

For the amplitude-mostly regime we first set the gray level to zero and turned the bias to minimum and look for a null of intensity by independently rotating the polarizer and analyzer. Then, we increased the bias voltage, so that $\beta_0 < \beta_{max}$, and observed the intensity transmittance increasing and decreasing again to a new minimum. The last null of intensity is the starting point of the amplitude modulation, that corresponds to $\beta_0 = (\beta_{max})_A$, as shown in Fig.1. This starting point permits the maximum contrast of the amplitude modulation.

Figure 3 shows the modulation characteristics of the amplitude-mostly and phase-mostly LCTV's, optimized by using the preceding experimental techniques. The phase-mostly LCTV provides continuous 2π range phase modulation coupled with less than 6%

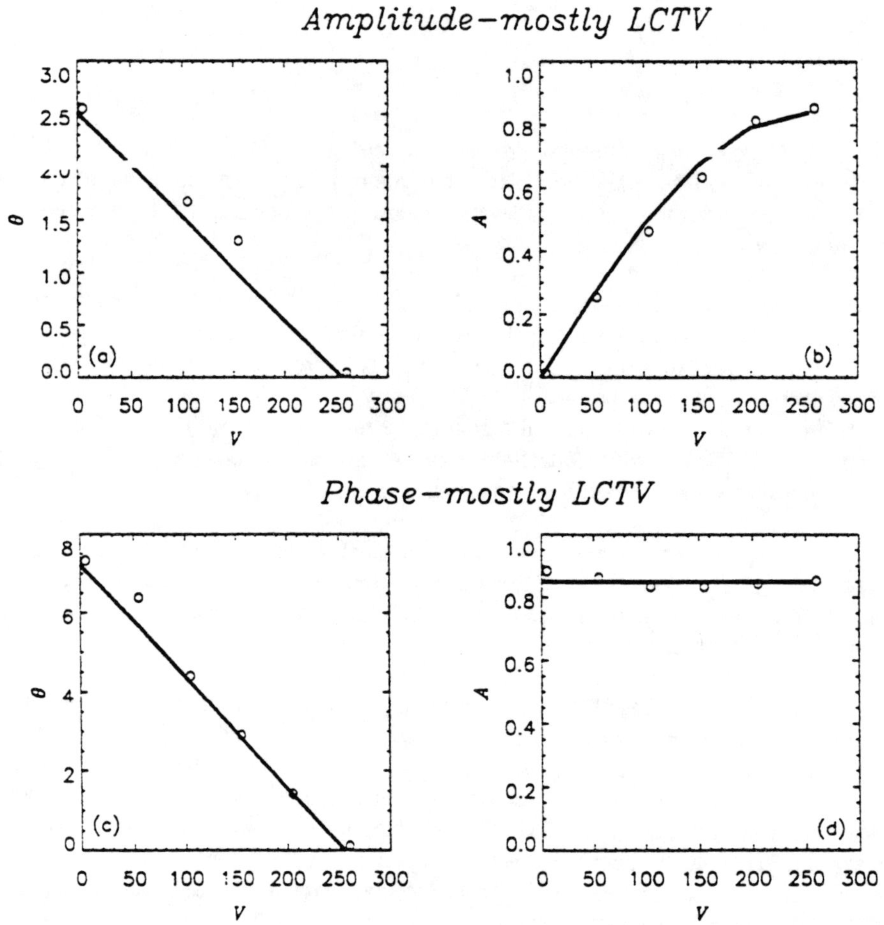

Fig.3. Measured phase (θ) and amplitude (A) modulations as a function of video signal V for amplitude and phase mostly LCTV's.

amplitude variation. The amplitude-mostly LCTV provides high contrast amplitude modulation with a maximum coupled phase modulation of about π. The amplitude-mostly LCTV is the red panel and the phase-mostly LCTV is the green panel of the Epson Crystal Image video projector. The panels have about 2.54 x 2.54 cm^2 aperture and 220 x 320 pixels, each measures about 60 x 55 μm with a center-to-center spacing of about 90 x 80 μm.

4. DIFFRACTION OF THE COUPLED-MODE SLM'S

A Diffraction of the phase-mostly LCTV

Assume that a POF, $\exp[j\delta(u,v)]$ with $0 \le \delta \le 2\pi$, is generated in the computer and is fed as a gray level pattern into the phase-mostly LCTV. The LCTV is controlled by a single electronic video signal, which is proportional to $\delta(u,v)$. The LCTV provides $\delta'(u,v) \neq \delta(u,v)$ for the phase and $A(u,v)$ for the amplitude modulations. The optical filter, $H'(u,v) = A(u,v)\exp[j\delta'(u,v)]$ is a periodic function of $\delta(u,v)$ of period 2π. We expand $H'(u,v)$ into the Fourier series with respect to $\delta(u,v)$[37]

$$H'(u,v) = \sum_{m=-\infty}^{\infty} \alpha_m \exp[jm\delta(u,v)] \tag{7}$$

The complex valued Fourier coefficients

$$\alpha_m = \frac{1}{2\pi}\int_0^{2\pi} A(\delta)\exp[j\delta'(\delta)]\exp(-jm\delta)\,d\delta \tag{8}$$

are the Fourier transform of $A(\delta)\exp[jm\delta'(\delta)]$ with respect to the phase $\delta(u,v)$ and are independent of $\delta(u,v)$. The first order harmonic, $\exp[jm\delta(u,v)]$ with $m = 1$, in the harmonic expansion Eq.(7) is the exact POF. Other terms of $m \neq 1$ are associated to the coupled amplitude modulation and the phase modulation errors of the LCTV. The term of $m = 0$ gives a central spot in the impulse response of the filter. The term $m = -1$ gives a inverted POF impulse response. The impulse responses of the higher order terms, $\exp[jm\delta(u,v)]$ with $|m| \ge 2$, correspond to the simple, double, triple and multiple convolutions and correlations among the impulse responses of the POF. Since no spatial frequency carrier is used, all the diffraction orders are superimposed on-axis contributing to the spread noise in the impulse response. Among the noise terms the zero order is the most visible.

If the phase modulation is linear to the electronic signal: $\delta'(u,v) = c\delta(u,v)$, where $0 \le c \le 1$ is the phase mismatching factor Eq.(8) becomes

$$\alpha_m = \frac{1}{2\pi}\int_0^{2\pi} A(\delta)\exp[j(m-c)\delta]\,d\delta \tag{9}$$

If the phase-mostly LCTV has the full range of 2π phase modulation $c = 1$ and $A(\delta) \sim 1$, then $\alpha_1 \sim 1$ and $\alpha_m \sim 0$ for all $m \neq 1$, the phase-mostly filter is very close to the POF.

Fig.4. Reconstructed images from the coupled-mode LCTV. The image of the space shuttle is the impulse response of the phase-mostly filter and the other images are reconstructed from the kinoforms generated by iterative design algorithms.

Figure 4 shows the optically reconstructed images from the holograms displayed on the phase-mostly LCTV: an array of 10 x 10 spots, an binary image of the letter A and a gray level image of Lenna. The phase-mostly holograms were designed with the iterative method using the experimentally measured operation curve of the LCTV in $\psi_1 = \psi_D + 16^\circ$ and $\psi_2 = \psi_D + \alpha - 14^\circ$ configuration as the constraints in the hologram plane.

B. Diffraction of the amplitude-mostly LCTV

Assume that an amplitude image A(x,y) is fed to the amplitude-mostly LCTV in the input plane of an optical correlator. The complex modulation of the LCTV will be $f(x,y) = A(x,y)\exp\{j\delta[A(x,y)]\}$, where $\delta[A(x,y)]$ is the coupled phase modulation. When the image A(x,y) is binary we have

$$f(x,y) = \begin{cases} \exp[j\delta_0] & \text{when} \quad A(x,y) = 1 \\ 0 & \text{when} \quad A(x,y) = 0 \end{cases} \tag{10}$$

where δ_0 is the maximum phase shift coupled to the amplitude A(x,y) = 1. Hence, when the input image is binary the coupled phase modulation in the amplitude-mostly LCTV affects the input image by only multiplying the image with a complex constant and does not affect the correlation output.

5. FULL RANGE COMPLEX MODULATION

A cascade of a phase-mostly LCTV1 and an amplitude mostly LCTV2 can provide full range complex modulation for a parallel beam. The two LCTVs are combined with an afocal imaging system and are aligned for a pixel-to-pixel superimposition under the control of a microscope. After the careful alignment, a spatial filter is placed in the focal plane of the afocal system to eliminate the grid structure in the image of the LCTV1 on the LCTV2, that minimizes the Moire pattern effect of the residual alignment error.

Each LCTV is placed between a polarizer and an analyzer and is adjusted with the methods mentioned in the above. Both LCTVs are the green panels of the Epson Crystal Image™ video projectors, connected to the computer through a frame grabber. Our computer codes permit to address two monochromatic LCTVs simultaneously and independently with a single color video signal.

The complex transmittances of the two cascaded LCTVs are multiplied: as

$$A\exp[j\delta] = A_1 A_2 \exp[j(\delta_1+\delta_2)] \qquad (11)$$

where the amplitude A_1 is coupled with the phase δ_1 from the LCTV1, and the coupled A_2 and δ_2 are from LCTV2. Obviously, if LCTV1 provides 2π range phase modulation, the realizable complex transmittance values can cover a complete circle in the complex transmittance plane δ - A, as shown in Fig.5, whose radius is equal to $(A_2)_{max}(A_1)_{min}$ that can be normalized to unity. Hence, the effect of the coupled amplitude modulation in LCTV1 is to reduce the maximum amplitude transmittance. If the maximum phase shift of LCTV1 is less than 2π, the cascade of LCTV1 and LCTV2 would be not able to cover a complete circle and to achieve the full range complex modulation.

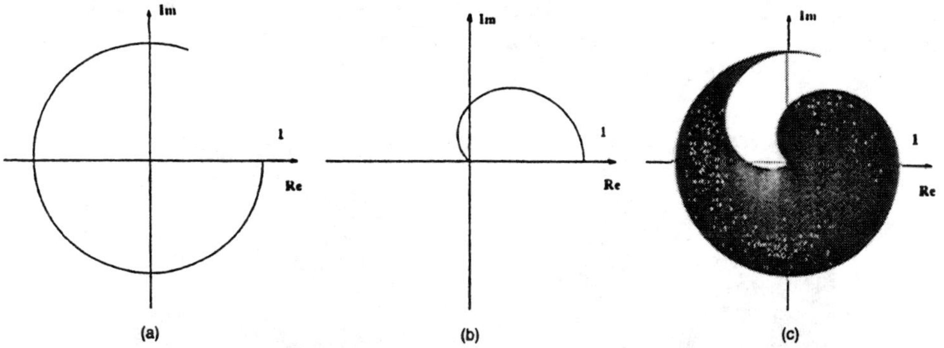

Fig.5. Operating curves for (a) LCTV1 and (b) LCTV2 and (c) the realizable complex amplitude in the complex plane. When LCTV1 provides phase modulation range less than 2π, the realizable complex amplitude does not cover a complete circle.

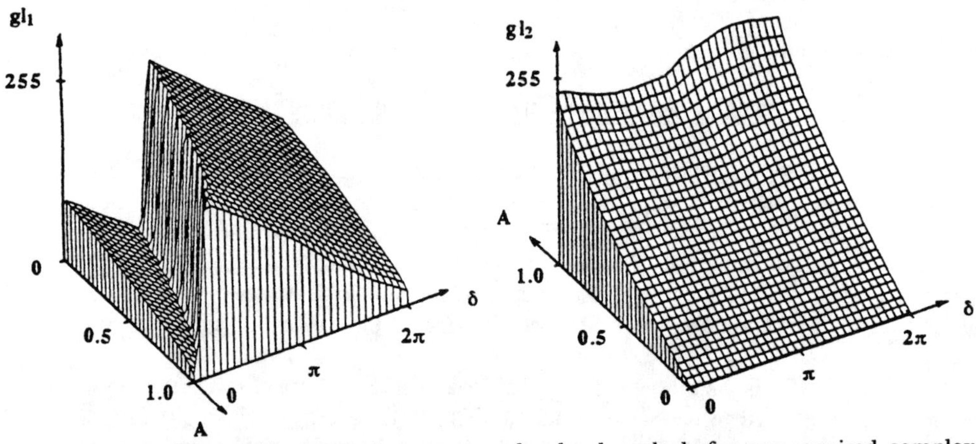

Fig.6. Look-up table generating two gray levels gl_1 and gl_2 for any required complex transmittances.

We encode the complex amplitude SLMs using a look-up table. First, for all combinations of uniformly sampled grey levels of the two LCTVs, gl_1 and gl_2 with $0 \leq gl_1, gl_2 \leq 255$ we calculate the realizable complex transmittances according to the experimental measurement data shown in Fig.3 and the relation (11). Then, we make an inverse mapping to build the look-up table that gives the values of gl_1 and gl_2 as function of the required amplitude A and phase δ, as shown in Fig.6. By the look-up table encoding the coupled modulation of LCTV1 and LCTV2 are combined contributing to the desired complex transmittance. When a pure amplitude modulation of A is required, gl_2 varies from 0 to 255, gl_1 also varies to compensate the coupled phase shift in the LCTV2. When a pure phase modulation is required, gl_1 varies from 0 to 255 and gl_2 also varies to compensate the couple amplitude modulation of the LCTV1. Some complex transmittances in the region $\delta < 0.76\pi$ correspond to $gl_1 = 0$. Those transmittances may be realized solely by the amplitude and the coupled phase modulations of the LCTV2. Since the look-up table is based on the experimental data the effect of the coupled modulations and the nonlinearities of the phase and amplitude modulations of both individual LCTVs are corrected. Figure 7 shows reconstructed images from the Fresnel holograms encoded in the complex amplitude SLM's.

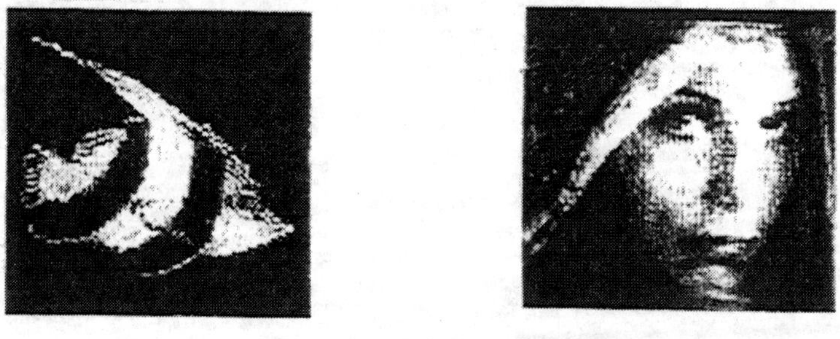

Fig.7. Reconstructed images of switchable complex valued Fresnel holograms.

6. OPTICAL PHASE-ONLY FILTERS USING THE PROGRAMMABLE SLM'S

We build the programmable Vander Lugt type optical correlators using the phase-mostly and the complex amplitude SLMs in the Fourier plane and the amplitude-mostly LCTV in the input plane. The aspect ratio of the LCTV pixels is not equal to unity, resulting in vertical scale different from the horizontal scale in the impulse response of the filters. This fact can be corrected[14] by simply rotating the input LCTV by 90 degrees with respect to the filter LCTV.

A. Phase-only correlations

In the optical correlator using the phase-mostly LCTV to encode the POF the noise owing to diffraction orders of $m \neq 1$ of the phase-mostly LCTV is spread out in the correlation plane. The zero order spot in the phase-mostly filter impulse response yields an image of the input scene. Other high orders in the harmonic expansion of the phase-mostly filter, described in Eq.(7), yield the correlations between the input image and the multiple convolutions and correlations of the POF responses. The SNR in the correlation output depends on the energy distribution among the diffraction orders and on α_m. In the phase-mostly filter the coefficients α_m with $m \neq 1$ are small compared with $\alpha_1 \sim 1$, while the POF term provides narrow correlation peak with high light efficiency. Thus, the output SNR of the phase-mostly filter is usually very high. We use the peak-to-correlation deviation ratio (PCD):

$$PCD = \frac{[I(x,y)]_{max}}{[\sum_{x,y=1}^{N} (I(x,y) - E\{I(x,y)\})^2]^{1/2}} \qquad (12)$$

as a measure of the filter performance, where I(x,y) is the correlation intensity. Since the phase-mostly filter yields a narrow correlation peak, we do not exclude the correlation peak intensity from the standard deviation of I(x,y) in the denominator. The values of PCD's of the phase-mostly filters obtained in the optical experiments was ranged about 30.

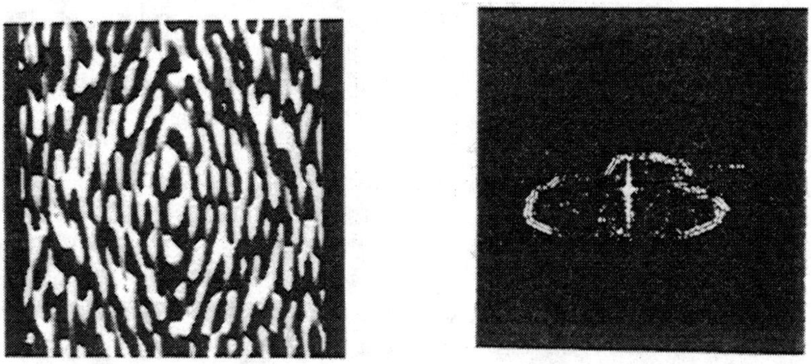

(a) (b)

Fig.8(a) POF generated in computer as a 16 grey levels 200 x 200 size matrix for an images of the tank T-34/76; (b) Impulse response of the POF coded in the phase-mostly LCTV

(a) (b)

Fig.9 (a) Input scene with an image of the tank T-34/76 and an image of the helicopter Kiowa; (b) Optical on-axis correlation output from the phase-mostly filter, shown in Fig.8.

The on-axis continuous POF yields a very bright autocorrelation peak, so that with a 10 mW He-Ne laser source we need to put a neutral optical density 2 to reduce the beam power and to avoid saturation of the CCD camera. Since the POF contains no spatial frequency carrier, the SBWP of 200 x 200 pixels of the LCTV seems sufficient to calculate the correlations of an input scene containing four input images of 64 x 64 size with the POFs of 200 x 200 pixels size. Figure 8 shows the POF and the optical impulse response from the phase-mostly SLM. Figure 9 shows the input scene and the optical on-axis correlation output.

B. Phase-only circular harmonic filters

The full rotation invariant correlation was obtained with the continuous phase-only circular harmonic filter (CHF) encoded on the LCTV. The key issue in this implementation is the use of a bandpass filter. The CHFs in the Fourier plane have amplitude equal to zero about the origin. A central stop on the optical axis is then needed. The stop removes also the circular subsampling error about the origin of the CHF in the Fourier plane. Also, a circular aperture has to be used to block the high frequency components. Otherwise, the invariance of the CHF to the continuous rotation of the input images will be jeopardized. Figure 10 shows the CHF displayed on the LCTV and their optical impulse responses.

One problem of the CHF for pattern recognition is that the CHF contains only one component of the object. There could have an infinity of objects that would yield the same center correlation intensities for a given set of CHF's of multiple orders[38]. The programmable phase-mostly LCTV permits to implement a sequence of CHF's of different orders of the same object. From each correlation output we take the coordinates of the correlation peaks. The distances between the correlation peaks of the CHF's of different orders are used as the rotation and shift invariant features for pattern recognition. Those distance features depend not only on the circular harmonic orders used in the CHF bank, but also on all the circular harmonic orders of the object. This geometrical approach of multiple CHF's has good discrimination capability and the ability against cross-correlation false alarms[39].

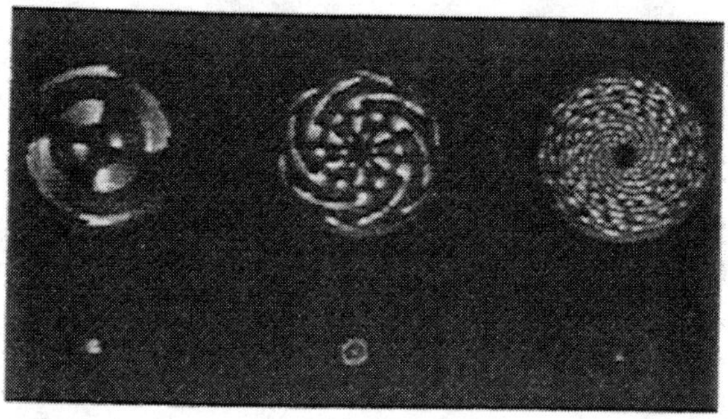

Fig.10 Optical phase-mostly CHF's on the LCTV with a DC block and their optical impulse responses. From left to right: m = 4, 6 and 12.

C. Phase-only composite filters

The composite filter or the synthetic discriminant function (SDF) is a linear combination of a given training set images, that generates desired output for each input image in the training set. The SDF is usually a complex function in the Fourier plane. One of the difficulties with the SDF filters is that only the correlation value at the origin is determined in the filter design. In order to suppress possible sidelobes in the correlation plane, the minimum average correlation energy (MACE) filter or other optimization techniques should be used.

Given a set of training images $\{f_n(x,y)\}$ the conventional phase-only composite filter is defined in the Fourier plane as[41,42,43]

$$H(u,v) = \frac{\sum_{n=1}^{N} \alpha_n F_n(u,v)}{|\sum_{n=1}^{N} \alpha_n F_n(u,v)|} \tag{13}$$

where $F_n(u,v)$ are the Fourier transforms of $f_n(x,y)$. We use a new definition of the phase-only SDF which is the combination of the component POF's, $\phi_n(u,v) = \exp[j\delta_n(u,v)]$, where $\delta_n(u,v)$ is the phase of the Fourier transform of the training images $f_n(x,y)$. This combination is in general not a phase-only function. We set its amplitude equal to unity and define the phase-only SDF as:

$$H(u,v) = \frac{\sum_{n=1}^{N} \alpha_n \phi_n(u,v)}{|\sum_{n=1}^{N} \alpha_n \phi_n(u,v)|} \tag{14}$$

The coefficients α_n in Eq.(14) are determined by the following system of N non-linear equations in the space domain:

$$\left\langle f_k(x,y) \middle| FT^{-1}\left\{ \frac{\sum_{n=1}^{N} \alpha_n \phi_n(u,v)}{\left|\sum_{n=1}^{N} \alpha_n \phi_n(u,v)\right|} \right\} \right\rangle = c_k \qquad (15)$$

where FT^{-1} denotes the inverse Fourier transform and c_k is the desired central correlation output for the input $f_k(x,y)$. This system of non-linear equations is solved with an iterative method using the delta learning rule. The initial solutions is $\alpha_n^0 = c_n$. In the iterations the coefficients α_n are modified with the delta rule as that used in the neural network : the amount of learning is proportional to the error between the actual output and the desired output. Assume that after the i-th iteration the filter is with the coefficients α_1^i, α_2^i, ..., α_N^i and the center correlation is m_n^i for the input image $f_n(x,y)$, at the i+1 th iteration the α_n are given by

$$\alpha_n^{i+1} = \alpha_n^i + \gamma\,[c_n - c_1 \frac{m_n^i}{m_1^i}] \qquad (16)$$

where γ is an overrelaxation constant, which control the learning speed, and $m_n^i(c_1/m_1^i)$ is the normalized output center correlation for the input image f_n. Since only (N-1) coefficients need to be determined, we let α_1 unchanged during the iterations. The iterations stop when

$$\alpha_n^{i+1} \sim \alpha_n^i \qquad \text{and} \qquad \frac{m_n^i}{m_1^i} = \frac{c_n}{c_1} \qquad (17)$$

The N-1 equations with k=2, 3, ..., N in Eqs.(15) are used for determining α_2, α_3,... α_N.

The filter defined in Eq.(14) is the nonlinear combination of the component POFs. It is well known that the POF lets pass all high frequency components of the image, resulting in a narrow autocorrelation peak and virtually no cross correlation peak. The impulse responses of the component POF's $\phi_n(u,v)$ are approximately orthogonal. The denominator $|\Sigma\alpha_n\phi_n(u,v)|$ in Eq.(14) is equal to

$$\left|\sum_{n=1}^{N} \alpha_n \phi_n(u,v)\right| = [\sum_{n=1}^{N} \alpha_n^2 + \sum\sum_{k \neq n} \alpha_n \alpha_k \phi_n(u,v) \phi_k^*(u,v)]^{1/2} \qquad (18)$$

where the inner products $\phi_n(u,v)\phi_k^*(u,v)$ are the values at the center of the cross-correlations between the impulse responses of ϕ_k and ϕ_n with $k \neq n$, which are approximately equal to zero when the training images have no overlapped edges. Hence, the denominator $|\Sigma\alpha_k\phi_k(u,v)|$ is approximately constant. The combination for the phase-only composite filter in Eq.(14) is approximately separable[42]. The impulse responses of the components POF's, $FT^{-1}[\phi_n(u,v)/|\Sigma\alpha_k\phi_k(u,v)|]$ with n = 1, 2, .., N, are approximately orthogonal.

The iteration algorithm for solving Eqs.(15) converges very fast because the initial solution of the unknown coefficients, $\alpha_n^\circ = c_n$, is already not far from the final solutions. The better the condition of the constant denominator $|\Sigma\alpha_n\phi_n(u,v)|$ is satisfied, the faster the iterations converge. When the number of the training images increases the contribution of the cross-correlations between ϕ_n and ϕ_k to the denominator $|\Sigma\alpha_n\phi_n(u,v)|$ becomes non-negligible. As a consequence, the iterations converge more slowly and the performance of the resultant composite filter deteriorates.

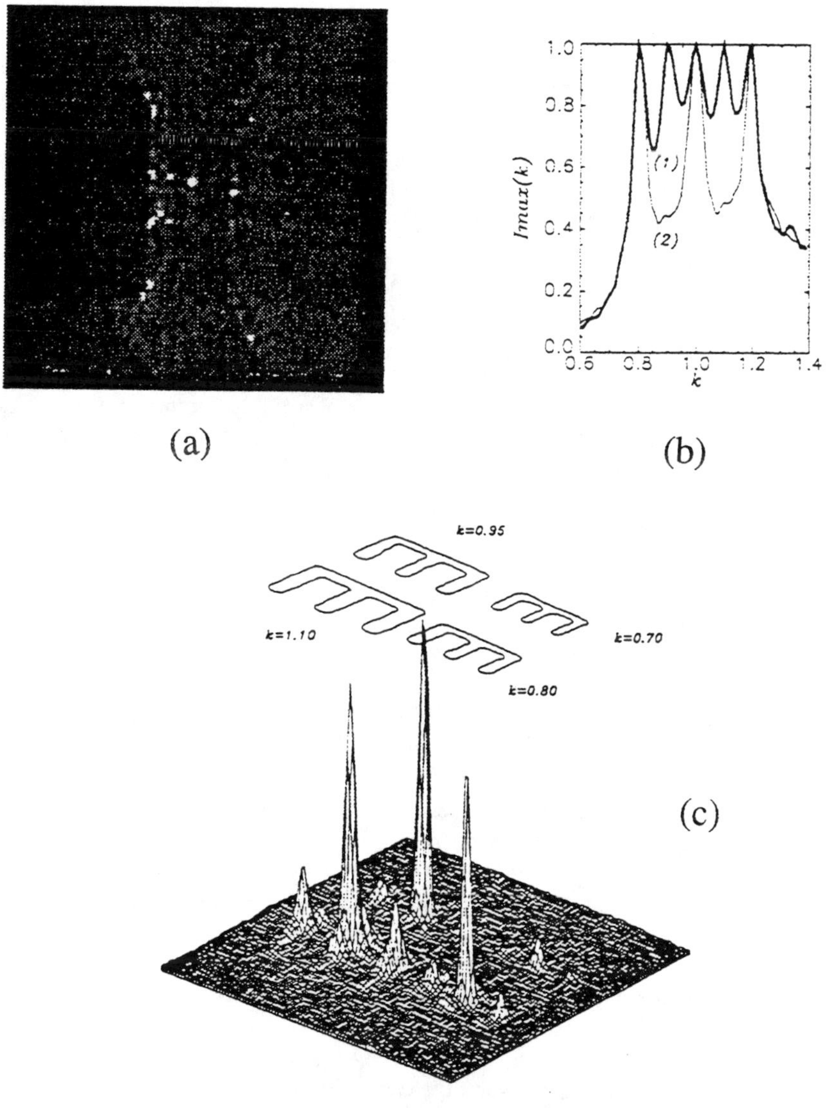

Fig.11 (a) Impulse responses of scale invariant phase-only composite filters; (b) Computer correlation peak intensity as a function of input scale k; (c) 3-D plot of optical correlation intensity. Input scene containing four images are shown on the top of the 3-D plots.

One phase-only composite filter was designed with a training set of five images of the letter E with the scale factors 0.8 - 1.2. The scale interval was of 0.1. The images are of 128 x 128 pixels size. The overrelaxation constant γ was equal to 0.6. The solutions were obtained with only about 10 iterations, two times faster than the iterations for designing the conventional phase-only composite filters defined in Eq.(13). The approximate orthogonality of the component POF's accelerates the iterations to converge. Fig.11a shows the optical impulse responses of the phase-only composite filter, in which the high frequency components of the images are combined to form traces of corner points of the letter E during the scaling.

Curve 1 In Fig.11b shows the computed correlation peak intensity. When the scale k of the input image varies continuously from 0.8 to 1.2 the output peak intensity varies about 20 - 35 %. Curve 2 in Fig.11b shows the cases where the filter was designed with the scale interval of 0.2. In this case, the output peak intensity varies about 60%. Fig.11c shows optical correlation output. The input scene contains four images of different scales as shown on the top of the 3-D plot of correlation output. The image with scale factor k = 0.7 is outside the scale region, where the filter is designed to be invariant, and therefore gives no correlation peak. Other input images generate high intensity correlation peaks, as predicted in the filter design.

In all the cases, the correlations yield narrow peaks at the center without sidelobe. This can be understood by considering our phase-only composite filter as an approximately linear combination of the component POFs. The noise is spread out in the output plane. The peak-to-correlation deviation (PCD) ratio ranges from 25 to 35.

D. Wavelet matched filters

The POF is not optimal in terms of the output SNR. The simple methods for improving the SNR of the POFs are to block some frequency components in the Fourier plane[44] as that does the ternary filter for improving the SNR of the binary phase-only filter. As described in Sec.6B a central DC block and a circular aperture were used to block the low and high frequency components in the correlation of the CHF in order to ensure the continuous rotation invariance of the CHF correlation[39].

The wavelet transform (WT) is a new multiresolution local analysis of signals[45]. For image processing, the WT is efficient for edge detection and image compression[46]. The optical wavelet matched filter (WMF)[47] can provide the correlations of WT's of two images. The filter is the conventional matched spatial filter or the conventional POF multiplied in the Fourier plane by the squared modulus of the wavelet filters. The wavelet filter is the Fourier transform of the wavelet and is a band-pass filter. Thus, the wavelet matched filter can improve the SNR of the POF. The wavelet function and its scale factor should be chosen in the filter design in order to optimize the filter performance[48].

The adaptive composite wavelet matched filter (CWMF)[49,50] can improve the performance of the composite filter. In the example shown in Sec.6C and Fig.11 the phase-only SDF is invariant only to discrete scaling. The adaptive CWMF can be invariant to continuous scaling. The filter yields correlation between the adaptive wavelet transform of the input and that of a composite reference image. The adaptive wavelet transform is used to enhance characteristic points of the images. The continuous scale invariance is achieved by choosing the scale increment of the training images and the scale of the wavelet, in such a way that the correlation peak is invariant to the input image scale variation in the sense of the Sparrow criterion[50].

We built a programmable optical correlator with complex amplitude SLM using three commercial liquid crystal televisions (LCTVs). The input scene is displayed on a LCTV in the amplitude-mostly modulation mode. An afocal imaging system reduces the size of the input, such that the Fourier transform of the input image is matched in size to the filter displayed on the second LCTV. The second LCTV is in the amplitude-mostly mode. The third LCTV is in the phase-mostly mode. An afocal imaging system conjugates the second and the third LCTV's to encode the complex-valued filters.

The adaptive CWMF is synthesized with five training images of the letter E of scale k = 0.8, 0.9, ..., 1.2. The images with k = 1 are of size about 30 pixels. The scale increment of five training images was $\Delta = 3$ pixels. The adaptive WT extracts the characteristic points of the letter E which are simply the corners in the image. Fig.12a shows the correlation peak intensity as a function of the input image scale k. When the input image scale varies $0.8 \leq k \leq 1.2$, the peak intensity varies between 1.0 and 0.9 for the CWMF with $\Delta = 3$, and between 1.0 to 0.4 for $\Delta = 6$. Fig.12b shows the optical correlation outputs of the adaptive CWMF. The three images of the letter E of scale k = 0.80, 0.95, 1.20 are detected. The image of scale 0.95 was not in the training set. It still yielded the correlation peak with the intensity equal to that of the images of k = 0.8 and 1.2. The image of scale k = 0.60 gaves no correlation peak. It is outside the scale range 0.8 - 1.2 for which the adaptive CWMF was designed to be invariant.

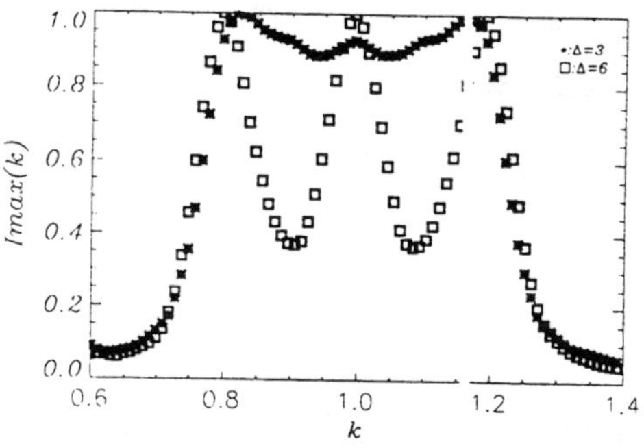

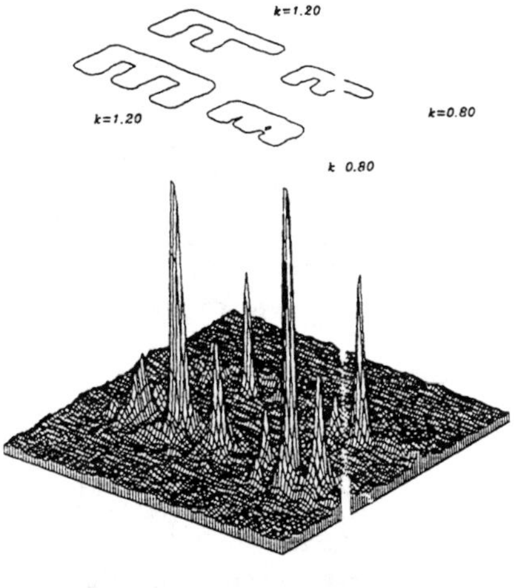

Fig.12 (a) Correlation peak intensities of the adaptive CWMF's for a letter of E; (b) Optical correlation outputs of the CWMF's of a letter of E. Input scene containing four images are shown on the top of the 3-D plots.

7. OPTICAL KINOFORM ENCRYPTION USING RANDOM PHASE ENCODING

Optical phase masks have been proposed for security applications[35,36]. Indeed, a two-dimensional phase mask consists of a very large number of pixels. At each pixel the phase retardation is invisible by ordinary light intensity-sensitive detectors. It is difficult to determine the content of the random phase mask and to reproduce it without knowing the code in the mask.

We propose a new optical kinoform encryption using the random phase encoding. We computer generate a kinoform G(**u**) with $|G(\mathbf{u})| = 1$ of the image f(**x**) to be encrypted. We multiply G(**u**) by a random phase mask (key), $\exp[j2\pi b(\mathbf{u})]$, where b(**u**) is a white noise. The kinoform is a phase hologram. Its content cannot be determined by light intensity detector. Without knowing the key mask, one cannot decode the encrypted kinoform and the image. The phase of the encrypted image is determined by the iterative algorithm in the design of the speckle-free kinoform. It is impossible to recover the encoded image by blind deconvolution, since the image is complex valued. Figure 13 shows the optical system for reconstruction from the encrypted kinoform. An phase-mostly $LCTV_1$ displays the encrypted kinoform, $G(\mathbf{u})\exp[2\pi b(\mathbf{u})]$. The phase mask key $\exp[j2\pi b(\mathbf{u})]$ is introduced to a second phase-mostly $LCTV_2$ that is in the conjugate image plane of the $LCTV_1$. When the two LCTVs are well aligned the random phase introduced on the encrypted kinoform is cancelled by the key, we recover the original image by reconstruction with a Fourier transform.

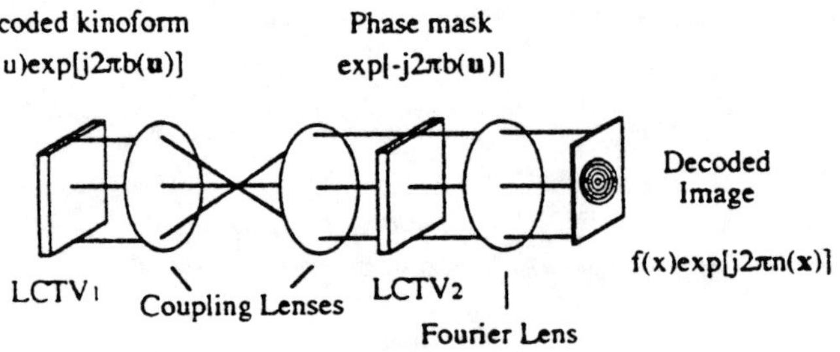

Fig.13 Optical system to reconstruct the image from encrypted kinoforms. The b(**u**) is the random phase mask key;

A. speckle-free kinoform

We multiply the image to be encrypted f(**x**) by a phase function, that yields $g(\mathbf{x}) = f(\mathbf{x})\exp[j2\pi\phi(\mathbf{x})]$, where $\phi(\mathbf{x})$ is a band-limited random phase and is determined by an iterative algorithm for designing the speckle-free kinoform G(**u**) with $|G(\mathbf{u})| = 1$. Figure 14 describes the iterative algorithm. The digital image f_{nm} with $0 \le n,m \le N$ is multiplied by an initial phase function, $\exp(j\phi_{nm})$. The image of N x N pixels size is embedded in an input zero matrix of 2N x 2N pixels size. The input matrix is again embedded into zero matrix of 4N x 4N pixels size before being Fourier transformed. The Fourier domain constraint is to keep the phase of the discrete Fourier transform G_{kl} inside a window of 2N x 2N pixels size, and to force the pixels outside the window of 2N x 2N pixels size

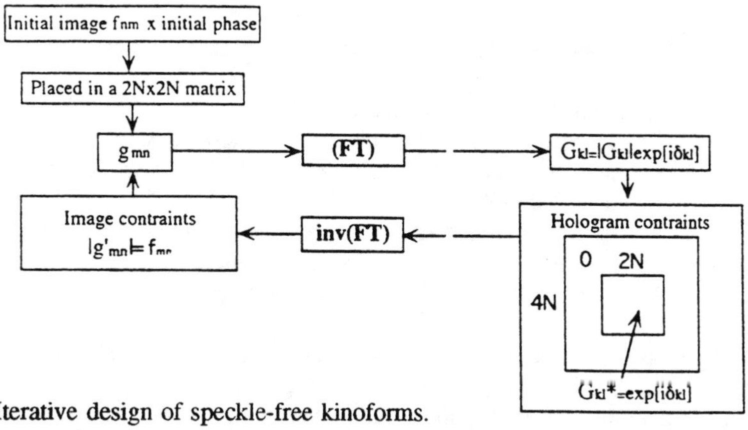

Fig.14 Iterative design of speckle-free kinoforms.

to be zero. This operation reduces the bandwidth of the spectrum $G_{k,l}$ to half. The space domain constraint is $|g'_{nm}| = f_{nm}$. The iteration continuous until the mean square error between f_{nm} and $|g'_{nm}|$ is reduced below a threshold value.

The above iterative design can reduce but not suppress completely the speckle noise in the reconstructed image. In fact, the speckles occur when the phases around a pixel of the image have jumps of π. Those π jumps can be eliminated by the above iteration algorithm. However, there can be isolated zero points[51] in the image plane, that cannot be removed by iteration. An isolated zero point is surrounded by a circle on which the phase ranges from 0 to 2π. Hence, when the iterative algorithm eliminates one π-jump, a new π-jump will occur around this point.

To avoid the speckles from the isolated zeros, we can use the initial phase[52] that is itself a band limited function, containing no 2π phase circle around any point of the image. A spherical phase $\phi_{nm} = -\pi k(n^2 + m^2)$ can be such an initial phase[53], where k is a scale factor which can chosen to change the bandwidth of the phase function $\exp(\phi_{nm})$.

(a) (b)

Fig.15. Optically reconstructed images from the encrypted kinoform (a) without phase mask key ; (b) with the key.

B. Smoothed random phase mask

For encryption of the kinoform we multiply the kinoform with a random phase mask exp[j2πb(**u**)], which can have a large bandwidth. Therefore, there is loss of information in the afocal imaging system, which couples the two LCTV's as shown in Fig.13. We smooth the random phase mask to reduce its bandwidth. In the experiment we generated a random noise b(**u**) in a N/2 x N/2 matrix and then dilate the matrix 4 times. In fact, we embedded b(**u**) into a zero matrix of 2N x 2N and did the Fourier transform followed by the inverse Fourier transform, that resulted in a matrix of 2N x 2N size which contains N/2 x N/2 samples that are equal to the original random noise. In between those samples the matrix values are smoothed by the interpolations.

In the optical reconstruction the smoothed random phase mask exp[-j2πb(**u**)] is introduced into the LCTV2 as the key phase mask. The use of the smoothed random phase reduces also the request for alignment of the two LCTV's. Figure 15 shows the optical results. When there is no key phase mask on the LCTV2, the reconstruction from the encrypted kinoform is just a random pattern. When the key is introduced into the LCTV2 the reconstructed image of the letter A appears, distinguishable and without speckle noise.

8. CONCLUSION

We have developed original experimental techniques for *in situ* SLM characterization with the wedged shear plate interferometer and for optimization of the phase-mostly and amplitude-mostly modulations of the twisted nematic LCTV with rotation of the polarizer and analyzer. We obtained from the commercial LCTV the full range of 2π phase modulation and high contrast amplitude modulation, and also the full range complex amplitude modulation with two cascaded LCTVs. The reconstructed images from the phase-only holograms and complex amplitude holograms show that the modulation characteristics of the LCTVs are well controlled.

We built the Vander Lugt type programmable optical correlators with the phase-mostly LCTV to implement the phase-only filters and also with the complex amplitude modulation cascaded LCTV's to implement wavelet matched filter and adaptive wavelet matched filters. The complex valued filters show improved performance and higher SNR compared with the POFs, but have an increased system complexity and a reduced light efficiency.

We built also optical system using the phase-mostly LCTV for optical kinoform encryption by using the random phase encoding. The bandwidth of the kinoform and phase mask is carefully controlled to avoid the speckle noise and improve the quality of the reconstructed images.

REFERENCES

1. D. A. Gregory, J. C. Kirsch and E. C. Tam, "Full complex modulation using liquid-crystal televisions", Appl. Opt. **31**, 163-165 (1992).
2. J. Amako, H. Miura and T. Sonehara, "Wave-front control using liquid-crystal devices", Appl. Opt. **32**, 4323-4329 (1993).
3. H. Bartelt, "Computer-generated holographic component with optimum light efficiency", Appl. Opt. **23**, 1499-1502 (1984).
4. R. D. Juday and J. M. Florence, "Full complex modulation with two one-parameter SLM's", Proc. Soc. Photo-Opt. Instr. Eng. **vol. 1558**, 499-504 (1991).
5. J. M. Florence and R. D. Juday, "Full complex spatial filtering with a phase mostly DMD" Proc. Soc. Photo-Opt. Instr. Eng. **vol. 1558**, 505-516 (1991).
6. U. Efron, S. T. Wu and T. D. Bates, "Nematic liquid crystals for spatial light modulators: recent studies", J. Opt. Soc. Am. B, **3**, 247-252 (1986).
7. N. Konforti, E. Marom and S. T. Wu, "Phase-only modulation with twisted liquid crystal spatial light modulators" Opt. Letters, **13**, 251-253 (1988).
8. K. Lu and B. E. A. Saleh," Theory and design of the liquid crystal TV as an optical spatial light modulator", Opt. Eng. **29**, 240-245 (1990).
9. K. Lu and B. E. A. Saleh, "Complex amplitude reflectance of the liquid crystal light valve", Appl. Opt. **30**, 2354-2362 (1991).
10. J. L. Pezzaniti and R. A. Chipman, "Phase-only modulation of a twisted nematic liquid-crystal TV by use of the eigenpolarization states", Opt. Letters, **18**, 1567-1569 (1993).
11. D. A. Gregory et al, "Using the hybrid modulation properties of liquid crystal television", Appl. Opt. **30**, 1374-1378 (1991).
12. J. C. Kirsch, D. A. Gregory, M. W. Thie and B. K. Jones, "Modulation characteristics of the Epson liquid crystal television", Opt. Eng. **31**, 5, 963-969 (1992).
13. S. S. Cotariu, S. E. Monoroe Jr. and J. Knopp, "A live input, live filter, liquid crystal correlator" Proc. SPIE, **vol. 1704**, paper 29 (1992).
14. E. C. Tam, "Smart electro-optical zoom lens", Optics Lett. **17**, 5, 369-371 (1992).
15. A. Tanone, Z. Zhang, C. M. Uang, F. T. S. Yu and D. A. Gregory, "Phase modulation depth for a real-time kinoform using a liquid crystal television", Opt. Eng. **32**, 517-521 (1993).
16. G. Paul-Hus and Y. Sheng, "Optical phase-dominant filter using liquid crystal television", Opt. Eng. **32**, 2165-2172 (1993).
17. Sheng and G. Paul-Hus, "Optical on-axis imperfect phase-only correlator using liquid crystal television", Appl. Opt. **32**, 5782-5785 (1993).
18. Z. Zhang, G. Lu and F.T.S. Yu, "Simple method for measuring phase modulation in liquid crystal television", Opt. Eng. **33**, 3018-3022 (1994).
19. C. Soutar and K. Lu, "Determination of the physical properties of an arbitrary twisted-nematic liquid crystal cell", Opt. Eng., **33**, 2704-2712 (1994).
20. L. Gonçalves, D. Roberge and Y. Sheng, "Programmable optical phase-mostly holograms with coupled-mode modulation liquid crystal television", Appl. Opt. **34**, 11, 1944-1950, (1995).
21. A. Bergeron et.al., "Phase calibration and applications of a liquid-crystal spatial light modulator", Appl. Opt. **34**, 5133-5139 (1995).

21. Z. Zhang, G. Lu and F.T.S. Yu, "Simple method for measuring phase modulation in liquid crystal television", Opt. Eng. **33**, 3018-3022 (1994).
22. L. Gonçalves Neto, D. Roberge and Y. Sheng, "Full range continuous complex modulation using two led-mode liquid crystal televisions", Appl. Opt. in press (1996).
23. L. Gonçlves Neto and Y. Sheng, "Speckle-free Fresnel holograms and beam shaping elements", OSA Topical meeting on Diffractive Optics, Boston, OSA Tech. Digest Series Vol.5, 166-169 (1996).
24. J. L. Horner and P. D. Gianino, "Phase-only matched filtering", Appl. Opt. **23**, 6, 812-816 (1984).
25. J. L. Horner, "Light utilization in optical correlators", Appl. Opt. **21**, 4511-4514 (1982).
26. M. W. Fan and J. W. Goodman, "Optimal maximum correlation filter for arbitrarily constrained devices", Appl. Opt. **28**, 3362-3366 (1989).
27. R. D. Juday, "Optimal realizable filters and the minimum Euclidean distance principle", Appl. Opt. **32**, 5100-5111 (1993).
28. R. D. Juday, "Correlation with a spatial light modulator having phase and amplitude cross coupling", Appl. Opt. **28**, 4865-4869 (1989).
29. R. D. Juday, R. S. Barton and S. E. Monroe, " Experimental correlator results with coupled modulators and advanced metrics", Porc. SPIE, vol. **2297**, July (1994)
30. B. V. K. Vijaya Kumar, "Tutorial survey of composite filter designs for optical correlator", Appl. Opt. **31**, 4773-4801 (1992).
31. Ph. Réfrégier, "Optical trade-off filters for noise robustness sharpness of the correlation peak and Horner efficiency", Opt. Letters, **16**, 829-831 (1991).
32. Ph. Réfrégier, B. V. K. Vijaya Kumar and C. Hendrix, "Multicriteria optimal binary amplitude phase-only filters", J. Opt. Soc. Am. A, **9**, 2118-2125 (1992).
33. A. Mahalanobis, B. V. K. Vijaya Kumar S. Song, S. R. F. Sims, and J. F. Epperson, "Unconstrained correlation filters", Appl. Opt., **33**, 3751-3759 (1994).
34. L. P. Yaroslavsky, "Is the phase-only filter and its modifications optimal in terms of the discrimination capability in pattern recognition", Appl. Opt. **31**, 1677-1679 (1992).
35. B. Javidi, J. L. Horner and J. F. Walkup, "An optical pattern recognition system for validation and security verification", Opt. Photo. News, September 13-18 (1994).
36. P. Refregier and B. Javidi, "Optical image encryption based on input plane and Fourier plane random encoding", Opt. Letters, **20**, 767-769 (1995)
37. W. J. Dallas, "Phase-quantization: a compact derivation", Appl. Opt., **10**, 673-674 (1971).
38. P. E. Danielsson, "A comment on rotation-invariant pattern recognition using circular harmonic expansion: Appl. Opt. **28**, 1613-1614 (1989).
39. L. Shen, Y. Sheng and G. Premont, "Theory and optical implementation of the geometrical approach of multiple circular harmonic filters", Appl. Opt. **34**, 4004-4012 (1995).
40. J. L. Horner and P. D. Gianino, "Applying the phase-only filter concept to synthetic discriminant function correlation filters", Appl. Opt. **24**, 851-855 (1985).
41. D. A. Jared and D. J. Ennis, "Inclusion of filter modulation in synthetic-discriminant function construction", Appl. Opt., **28**, 232-239 (1989).
42. M. B. Reid, P. W. Ma, J. D. Downie and E. Ochoa, "Experimental verification of modified synthetic discriminant function filters for rotation invariance", Appl. Opt. **29**, 1209-1214 (1990).
43. J. A. Davis and D. M. Cottrell, "Random mask encoding of multiplexed phase-only

and binary phase-only filters", Opt. Letters, **19**, 496-499 (1994).
44. B. V. K. V. Kumar and E. Pochapsky, "Signal-to-noise ratio considerations in modified matched spatial filters", J. Opt. Soc. Am A **3**, 777-786 (1986)
45. Y. Li, H. Szu, Y. Sheng and J. Caufield, "Wavelet processing and optics", Proceedings of the IEEE, vol.84, No.5, 720-732 (1996).
46. S. Mallat and W. L. Hwang, "Singularity detection and processing with wavelets", IEEE Trans. on information theory, vol.**38**, 617-643 (1992).
47. Y. Sheng, D. Roberge, H. Szu and T. Lu, "Optical wavelet matched filters for shift-invariant pattern recognition", Optics Letters, **18**, 299-301 (1993).
48. D. Roberge and Y. Sheng, "Optical wavelet matched filter", Appl. Opt., **33**, 5287-5293 (1994).
49. D. Roberge and Y. Sheng, "Optical composite wavelet matched filters", Opt. Eng. **33**, 2290-2295, (1994).
50. D. Roberge and Y. Sheng, "Optical adaptive composite wavelet matched filter for scale-invariant pattern recognition", submitted to Appl. Opt. (1995).
51. M. S. Scivier and M. A. Fiddy, "Phase ambiguities and the zeros of multidimensional band-limited functions", J. Opt. Soc. of Am. A, **5**, 693-697 (1985).
52. F. Wyrowski and O. Bryngdahl, "Iterative Fourier-transform algorithm applied to computer holography", J. Opt. Soc. of Am. A, **5**, 1058-1065 (1988).
53. S. Teiwes, H. Schwarzer, H. Aagedal, M. Schmid, T. Beth, F. Reichel and F. Wyrowski, "Speckle-free signal reconstruction with non-periodic diffractive elements on liquid-crystal SLM" OSA toptical meeting, spatial light modulator, tech. digest series, vol.9, 24-27 (1995)

Devices for Optical Processing

Solid Optics: A Review

H. John Caulfield, and Marius P. Schamschula

The Center for Applied Optical Sciences
Department of Physics, Alabama A & M University
P.O. Box 1268, Normal, AL 35762
caulfield@caos.aamu.edu

ABSTRACT

Solid Optics (SO) is 3D optics without air spaces. In some respects it is a 3D version of Integrated Optics (IO), but it shares many features with Conventional Optics (CO) that IO cannot share. After a general comparison of SO, IO, and CO; we note that for many purposes SO is the best choice. We then discuss general approaches, geometries, and figures of merit for SO systems. The final main topic is a review of the two main types of SO system: linear and right-angle.

Keywords: solid optics, compact optical systems

1. INTRODUCTION

Solid Optics (SO) is the 3D counterpart of integrated optics. That, as we will see, is a caricature, but it does serve as a quick, one-sentence definition of what SO is about. Although specific versions of it had been developed earlier, the general concept and its rationale were first discussed in *Solid Optics: the application of precision manufacturing to small, complex optical systems*[1]. Our goal here, with the benefit of an additional nine years of development, is to reassess this field. We will begin with the rationale and follow with illustrative case studies.

2. RATIONALE

What drives us to SO? Why not use Conventional Optics (CO) or Integrated Optics (IO)? Clearly there are circumstances in which CO is the only rational choice, e.g. telescopes, microscopes, and projection TV. Likewise there are cases where IO makes the most sense, especially for optical communications routing, external modulation, etc. SO is not a universal solution, but it is sometimes very attractive. Below, we note some of the issues affecting the choice among CO, IO, and SO.

Another way of saying the same thing is to ask how many spatially resolvable channels (Gabor called them "logons"[2]) the system supports or what is the system throughput. For a system of cross sectional area A and solid angle Ω, the throughput is

$$T = A\Omega.$$

The throughput of a single channel (single mode) is

$$T_1 = \lambda^2,$$

where λ is the wavelength. Thus the number of logons is

$$L = A\Omega/\lambda^2.$$

Clearly IO has a low A and low Ω, so L is small. Most CO and SO systems have much larger A (2D not 1D) and much larger Ω. So T and L are much larger

2.1 Dimensionality

The full use of the interconnection advantages of optics can only be had in 3D[3]. Both IO and IC (Integrated Circuits) restrict access to the plane of the device. Thus we can access an interaction region only from its edges in 2D. 3D optics allows access from the third dimension. This is a tremendous advantage especially in view of the progress in IC. Soon (conceivably quite early in the next century), IC technology will allow considerable electronic functionality within the smallest optically resolvable area. This will lead to smart pixel arrays (smart SLMs) which do not limit the optics in any way. Layers of sources (perhaps from the prior SLM) and smart SLMs can be connected in 3D by optics (lenses, holographic optical elements, etc.) until we reach a final output CCD, CID, flash-addressed RAM, etc. This powerful 3D functionality can not be accomplished (for large arrays) by IC or IO technology. We must use 3D optics.

Note that 3D optics can be either CO or SO. Dimensionality does not impact that choice. Dimensionality does, however, show that unassisted 2D systems (IC or IO) will be less powerful that 3D systems (CO or SO).

Of course, the discussion given above draws distinctions that are too sharp. We include 2D ICs as smart pixels in the 3D CO or SO systems just discussed. These caveats do not diminish the validity of the need for 3D optics. They simply acknowledge that 3D CO or SO is necessary to maximize computational speed.

2.2 Support

When we draw optical components on a piece of paper, they stay where we put them. In real world optics, however, those components can not float freely. In CO we use external mounts — lens mounts, optical rails, benches, etc. In IO and SO, there is no need for such mounts. Alignment is permanent, built-in. The components are mutually self-supporting.

Biology provides good terminology for this. You have an endoskeleton for support of your soft tissue. On the other hand, turtles and crabs have exoskeletons to support their soft tissue. Clearly CO requires an exoskeleton for support. This adds bulk and weight. IO and SO have endoskeletons. No support outside the IO or SO is needed.

2.3 Reflection Losses

Reflection losses occur at interfaces between media. Antireflection (AR) coating is the art form evolved to combat this in CO. Note, however, that the reflection loss relates to the difference in index of refraction across the interface. Thus the problem is maximized when there are many interfaces into and out of air. Air has a near-theoretical-limit low index of refraction. On the other hand, in IO and SO there are no such abrupt changes.

2.4 Component Flexibility

The number and quality of components that can be built into IO is very limited. By contrast, CO and (to a great extent) SO can use very sophisticated components acquired from multiple sources and assembled to form a system. IO components are not assembled. They are fabricated at their desired location.

2.5 Mechanical Robustness and Alignment

Paradoxically, the systems (CO) which allow the most adjustable alignment are also the most prone to inadvertent misalignment, while the systems that allow no *a postiori* adjustment (IO) are the least prone to misalignment. Clearly SO lies somewhere in between CO and IO but much closer to IO. Some small adjustments can be made during assembly, but afterwards it is solid.

A possible annoying feature of a SO system is that it will support microphonics. However any system, even one with an exoskeleton, subjected to mechanical impulses will vibrate. Being one piece, SO and IO will return to their original alignment, whereas CO may not.

2.6 Modularity

The key to low cost and flexibility is modularity. It will appear below that the SO we advocate will be made of the same right triangular and square cross section prisms regardless of function.

2.7 Design Capability

IO optical design is quite different from CO design, because the propagation medium in IO is nowhere near as simple at that of the air assumed in CO. SO is in between those cases. The medium is more complicated than air, e.g. there may be substantial dispersion, but it is less complicated than a waveguide. It is still "free space."

2.8 Susceptibility to Dust, Oil, Scratches, Etc.

In IO and SO, the light passes through no exposed surfaces. The optical path is protected from contamination and scratches.

2.9 Summary

We summarize the eight features listed above in Table 1. For each feature, we assign a grade (A, B, or C) to CO, IO, and SO. A is best, while C is worst.

Clearly, there is no universally best approach. Just as clearly, SO occupies a niche quite different from and often superior to CO and IO. It is a field of optics with distinct advantages over the other two in some cases.

Feature	CO	IO	SO
Dimensionality	A	C	A
Support	C	A	A
Reflection Losses	C	A	A
Component Flexibility	A	C	B
Robustness / Alignment	C	B	A
Modularity	B	C	A
Design Capability	A	C	B

Table 1. Comparison among conventional optics (CO), integrated optics (IO), and solid optics (SO).

3. CLASSIFYING SO SYSTEMS

3.1 On-Axis vs. Off-Axis

The SO system has a central axis, and the light has an optical axis. If they are coaligned, we say the SO system operates on-axis. If reflection or folding is involved, the off-axis SO can avoid the loss entailed by beam splitters. On the other hand, most optical components (lenses, modulators, etc.) work better on-axis than off-axis. This is a vital SO design consideration.

3.2 Multiple Use of Space

If the same space can be used for multiple purposes, the system can be made smaller and lighter. However, a way to separate functionalties (beamsplitters, off-axis operation, etc.) must be introduced.

3.3 Internal vs. External Components

In some SO, the components (sources, lenses, modulators, deflectors, etc.) are attached to and/or built into the outer surfaces. The inside of the SO serves to provide propagation space. In other designs, some components are totally surrounded by SO.

4. PROPAGATION FUNDAMENTALS

Here we want to ignore the components and the system function and inquire only about light propagation. The main issues are volume and form factor as they relate to propagation of light between components' locations. The components themselves will remain unspecified for now.

Volume is an obvious concern. Particularly in SO, weight is proportional to volume. In addition, all systems should occupy as little space as possible for convenience. Finally, it is likely that cost is somewhat proportional to volume. There are two generic ways to reduce volume in SO. First, use high index materials. It is optical path length that counts. Second, make multiple uses of the same space. It is here that there is room for cleverness.

The other critical geometric concern is the shape of the SO system. For a given volume, we prefer a shape compact in all directions. A convenient measurement procedure is as follows: (a) place the object in the minimum sized rectangular parallelepiped box possible and (b) measure the size of the box. By "size" we will use the definition used by American commercial airlines that we call the "airline dimensionality" (AD). By definition,

$$AD = L + W + H,$$

the length plus the width plus the height of the box. Minimizing AD drives the box toward a cube.

For a cube, $V = L^3$ and $AD = 3L$. Combined figures of merit which give equal contributions of 1/2 from V and AD for a cube are easily found in terms of

$$\alpha = AD/(3V^{1/3})$$

($\alpha = 1$ for a cube) and

$$\beta = (AD/3)^3/V$$

(β = 1 for a cube).

We can use α, β, $1/2(\alpha + \beta)$, $\alpha\beta$, $1/\alpha$. $1/\beta$, α/β, β/α, etc. All of these are 1 for a cube. We like to use

$$M = (\alpha + \beta)/2.$$

5. SPECIAL CASES

One of two common special cases is that which uses SO to connect 2D arrays of sources, modulators, circuits, detectors, etc. in some regular fashion. Because these systems are unidirectional and physically cascaded along a single, unbent straight line, we call them "linear systems."

The second common SO processor type is the one that involves 45°-90°-45° cross section prisms (often with mirrors or beamsplitter on one or more faces). The paths often bend by 90°. We call these "right-angle systems." surfaces[4].

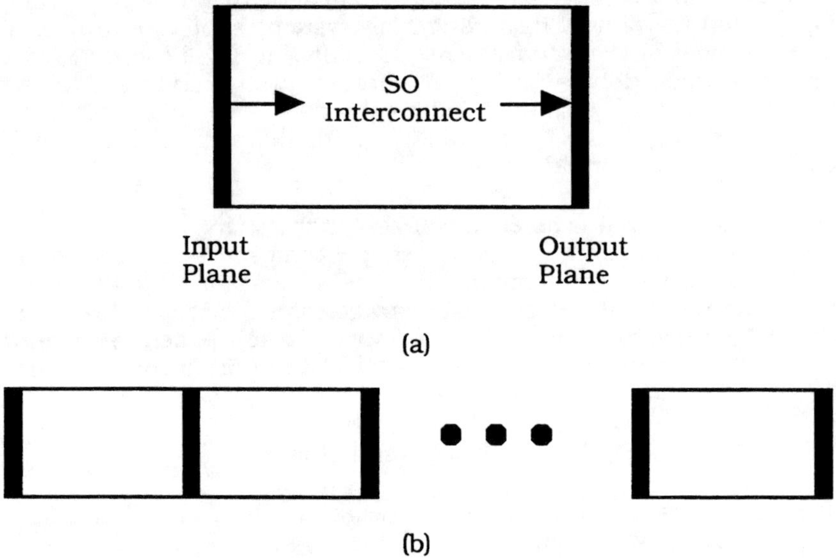

Figure 1. Linear SO systems are composed of modules like (a) singly or linearly cascaded as in (b).

As is evident from the previous section, linear and right-angle systems do not exhaust the possibilities. They do, however, represent a huge fraction of the published papers. Other non-right angle folding geometries often introduce non-normal angles of incidence and thus a source of optical aberration.

5.1 Linear SO Systems

The generic linear SO module is shown in Figure 1. The SO interconnect pattern can range from simple imaging to complex, arbitrary interconnection with or without fanout. Often they use lenslet arrays. A very early discussion of such systems was due to Iga, *et al.*[5] Numerous variations on this theme have occurred involving liquid crystals,[6] and calcite plates.[7] One linear SO system even uses diffusers between planes.[8] These and other systems differ in functionality of the SO interconnections and the I/O layers, not in geometrical concept.

5.2 Right-Angle SO Systems

These systems offer much greater flexibility and much better AD than linear systems. As they often involve considerable folding, they can achieve much lower V as well.

The first right-angle SO system known to us is the RUBIC Cube[9] (a literal cube used for systolic matrix-matrix multiplication). The RUBIC Cube uses a polarization beamsplitter, reflective SLMs, and quarter wave plates to cause the light to traverse the system as shown in Figure 2. Lenses are necessary for highest resolution but are not shown here as they add nothing conceptual to the system.

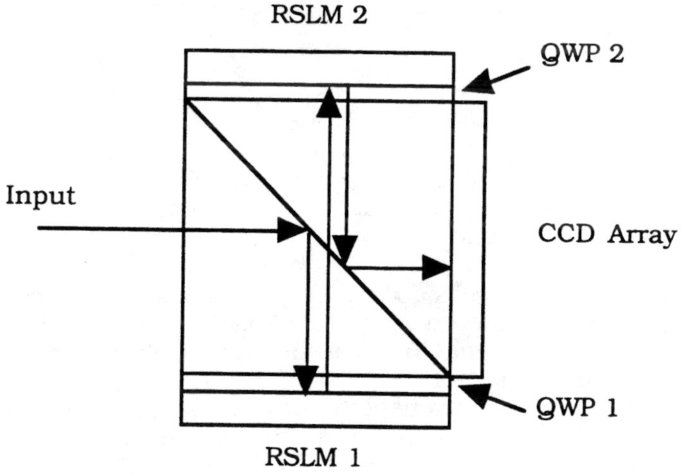

Figure 2. A simplified RUBIC Cube. Matrix A is input on reflective SLM 1 (RSLM 1). The reflected light passes through the quarter wave plate QWP 1 that changes the polarization to the transmissive state. SLM 2 multiplies A components by B components and QWP 2 changes the polarization back to the reflective state. The CCD registers and shifts the products to obtain C AB.

More complex compact systems can be generated by folding along space filling curves such as the Peano and Hilbert curves.[10] An example of such a method of folding the optical axis is shown in Figure 3.

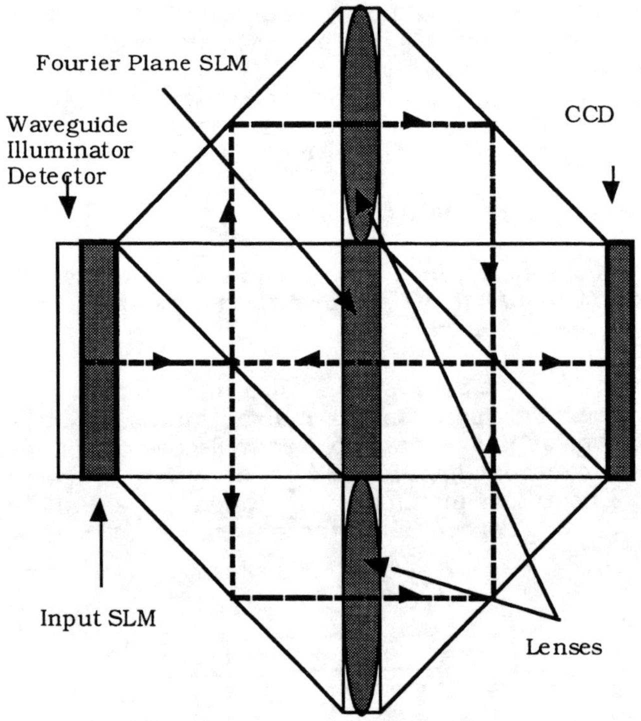

Figure 3. A correlator design based on the self-touching two-dimensional Peano curve. This example is a f/2 system with a collimated input.

Note that in their vertical paths, the rays traverse the same regions twice, saving a length equal to a cube side relative to the functionally-equivalent linear SO system. Now consider a dual rail time-delay or phase shift selector as shown in Figure 4. By switching the polarization before one of the polarizing beamsplitters, we cause the light to go down to the bottom rail. By immediately switching the light back to the transmissive state, we isolate the I/O path indicated with its associated time delay. More complex folded versions are possible,[11] but, by now, they would teach the reader little.

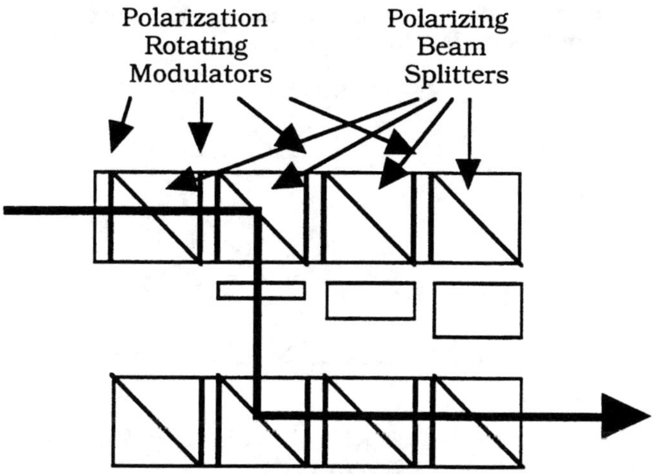

Figure 4. A dual rail optical time delay system. Note that this generic exploded view shows "air" as the low index medium. This could easily be replaced by a solid lower index material.

SO star couplers are easy to design and make[11] as shown in Figure 5. This time, instead of polarizing beamsplitters, we, use polarization-insensitive beam splitter, Such star couplers are lossless in principle (low loss in practice), even (all inputs contribute equally to all outputs), and free of clock skew (to within λ/c or better, in principle). There is also a rectangular cross section space between input and output for a gain medium. All paths through it are equal. Again reflective stars are possible but omitted for brevity.

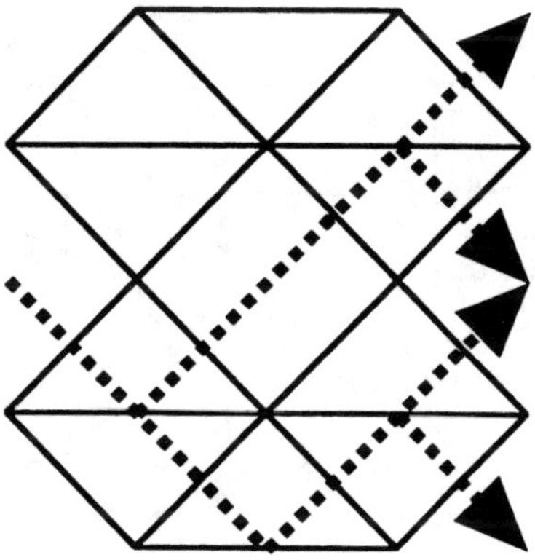

Figure 5. A 4 × 4 perfect star coupler with $T_{ij} = 0.5$ for each of the four beamsplitters.

Essentially the same geometry can be used for a switching network.[11] This time we insert polarizing beamsplitters as shown in

Figure 6. The electro-optic switch (probably KTN Pockels cell is best for low voltage and high speed) selects the paths.

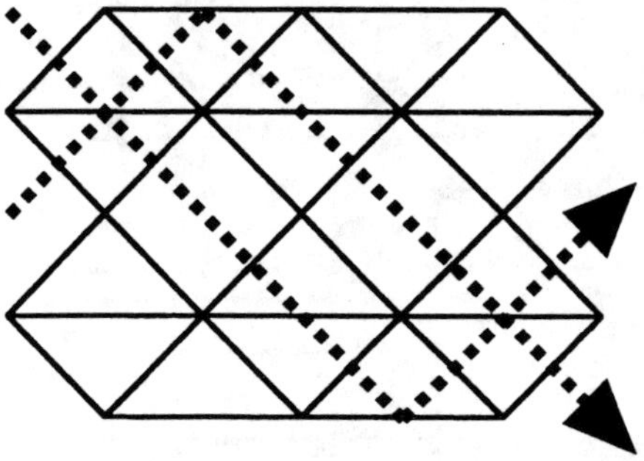

Figure 6. A 4 x 4 switching network with an added layer to permit nonblocking interconnection

5.3 Non-Right Angle SO Systems

By lifting the restriction imposed by right-angle folding, we can design more complex SO geometries[4]. One famous example for such a system is the Litton Compact SO correlator system, shown in Figure 7. Such systems, as in this example, are more difficult to design, since aberrations introduced by non-normal incidence on the reflective mirrors and SLMs.

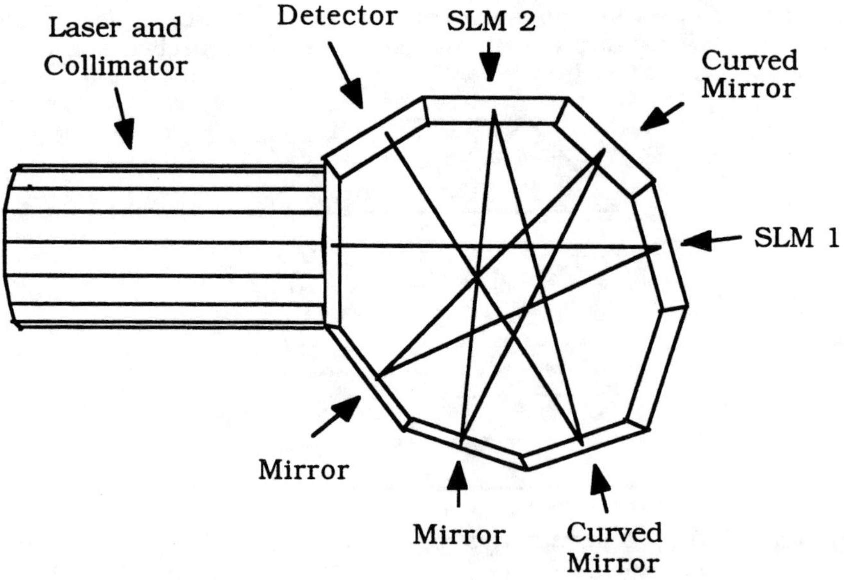

Figure 7. The Litton Compact SO correlator system, sometimes referred to as the "Hockey Puck."

An other standard example of a SO system is the waveguide or "Slab" system developed by Bell Labs. In Figure 8 we show a possible implementation of a correlator using this approach. Note that for reasons of aberrations the incident angles are much smaller than the 45° indicated in the figure.

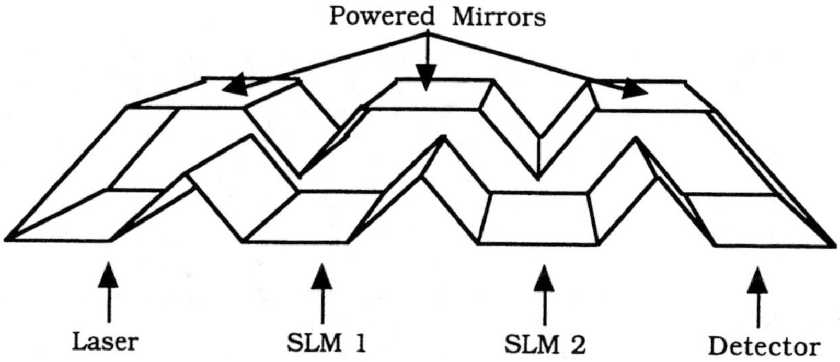

Figure 8. A waveguide correlator, with unused volume cut out.

All of these examples should prepare the reader to understand and appreciate a very general right-angle SO approach based on space-filling curves.

6. THE FUTURE OF SOLID OPTICS

A review such as this often has two useful effects. First, it draws the attention of the review authors to examples they missed. To the authors whose papers we missed, we have two requests: One, forgive us. We did not know or remember. Two, send us your reprints, so we will include you next time. Second, it draws attention to the field thereby causing more people to use the reviewed technology. We sincerely hope this paper will have both effects.

7. ACKNOWLEDGMENT

This work was sponsored by the U.S. Army Space and Strategic Defense Command under Contract No. DASG60-95-2-0001.

8. REFERENCES

1 H. J. Caulfield, "Solid Optics: the application of precision manufacturing to small, complex optical systems," *Proceedings of the SPIE* **748**, 60-62 (1987)

2 D. Gabor, "Theory of communication," *JIEE Part 3* **93**, 429-457 (1946)

3 H. J. Caulfield, "Parallel N^4 weighted optical interconnections," *Appl. Opt.* **26**, 4039-4040 (1987)

4 M. P. Schamschula, P. Reardon, H. J. Caulfield, and C. F. Hester, "Regular geometries for folded optical modules," *Appl. Opt.* **34**, 816-827 (1995)

5 K. Iga, M. Oikawa, S. Misawa, J. Banno, and Y. Kokbun, "Stacked planar optics: An application of the planar microlens," *Appl. Opt.* **21**, 3456-3460 (1982)

6 T. Sakano, K. Noguchi, and T. Matsumoto, "Multiprocessor system using an automatically rearrangable free space multichannel optical switch: COSINE-2," *Appl. Opt.* **32**, 3690-3699 (1993)

7 K. Noguchi, K. Hogar, T. Sakano, and T. Matsumoto, "Rearrangable multichannel free-space optical switch using polarization multiplexing technique," *Electron. Lett.* **26**, 1325-1326 (1990); L. Liu, Y. Yin, H. Peng, W. Xiong, G. Li, B. Wang, N. Wang, L. Shao, F. Liang, and L. Zhao, "Design and fabrication of parallel processor modules by polarization-optical stacked integration," *Int. J. of Optoelectronics* **10**, 39-49 (1995)

8 H. J. Caulfield, H.-I. Jeon, J. Brown, and P. Werbos, "Variable and fixed rank-1 N^4 interconnections," *Appl. Opt.* **29**, 2019-2023 (1990)

9 R. P. Bocker, H. J. Caulfield, and K. Bromley, "Rapid unbiased bipolar incoherent calculator cube," *Appl. Opt.* **22**, 804-807 (1983)

10 M. P. Schamschula, H. J. Caulfield, and A. Brown, "Space-filling Modular Optics," *Opt. Lett.* **19**, 689-691 (1994); M. P. Schamschula, and H. J. Caulfield, "Space-Filling Modular Optics: Expanded Peano and collapsed Hilbert curves," *Opt. Comm.* **111**, 219-224 (1994)

11 J. Qi, M. P. Schamschula, and H. J. Caulfield, "Fast-switching Array-Navigation Optics with Uniform Times (FANOUT)," submitted to the *J. Parallel Distrib. Proc.*

MULTIPLE QUANTUM WELL OPTICAL MODULATORS AND SPATIAL LIGHT MODULATORS

F. C. Jain and S. K. Cheung

Electrical and Systems Engineering Department
University of Connecticut
260 Glenbrook Road, Storrs, CT 06269-2157
Tel. (860) 486-3752, Fax. (860) 486-2447, E-Mail fcj@eng2.uconn.edu

ABSTRACT

This paper describes multiple quantum well (MQW) modulators and spatial light modulators (SLMs) which utilize field dependent changes in absorption, index of refraction, and polarization (birefringence). These optical parameters are changed either by band filling in quantum wells or by modifying the excitonic interaction between electrons and holes. Modulation can be achieved by impressing RF/microwave fields across MQWs in a variety of ways including direct biasing, induction by transporting charges in an adjacent channel (by charge-coupled devices, CCDs, or heterostructure acoustic charge transport, HACT), and by launching a surface acoustic wave (SAW) in a region adjoining MQWs. High contrast Fabry-Perot, birefringent, and HACT spatial light modulators are discussed in detail. One-dimensional Bragg SLMs (which use electrooptic or acoustooptic gratings) and Mach-Zehnder interferometers are also summarized.

I. INTRODUCTION

Spatial light modulators (SLMs) are the key devices in optical data/image processing technology used for the realization of advanced communication and computing systems. Such devices utilize the inherent parallel processing capability of photonic systems to provide an interface with other electronic or optical signal processing elements. Many conferences have been devoted [1-3] to SLMs during the past ten years. Some of the important functions [4,5] performed by SLMs include: optical matrix multiplication and addition (e.g., involving 1,000 by 1,000 elements), incoherent to coherent image conversion, electrical-to-optical conversion, thresholding of analog images, implementation of optical neural networks, and optical cross-bar switches.

A variety of SLM structures, based on various physical effects (used in modulation), have been reported in the literature[6]. These include liquid crystal, magneto-optic, membrane-mirror, and multiple quantum well (MQW) SLMs. Of special mention are the multiple quantum well (MQW) SLMs[6,7-9] which appear at present to be the fastest devices and can be addressed both electrically and optically. It may be mentioned that an optical processor was successfully fabricated using MQW self-electrooptic devices (SEEDs) [10].

MQW modulators can be classified into two broad categories depending on the physical effect utilized: *i*) Band-Filling, and *ii*)Quantum Confined Stark Effect (QCSE). In the case of band-filling, an external bias causes the transfer of carriers into the quantum wells layers which shifts the absorption spectrum to shorter wavelengths (blue shift) [11,12]. In the case of QCSE, the applied field causes a change in the electron/hole energy levels and excitonic interaction which shifts the absorption spectrum towards longer wavelengths (red shift) [13, 14]. Zucker et al. [12] have fabricated MQW Mach-Zehnder interferometeric modulators using both the spectral blue-shift and red-shift. This paper focusses on various types of multiple quantum well (MQW) Stark effect based modulators and spatial light modulators (SLMs).

II. GENERAL MQW BACKGROUND

The manifestation of QCSE in MQW structures has been a topic of intense study since the reporting of excitonic electroabsorption by Chemla et al.[13], T.H. Wood et al.[7], and Miller and coworkers[14]. The confinement of excitons in a quantum well having thickness L_z in the range $a_o/2 < L_z < a_o$ (where a_o is the Bohr radius) is shown[14] to increase their binding energy. This stabilizes the excitons to allow room temperature operation. As a result, the magnitudes of electron-hole oscillator strength M_b, binding energy E_{ex}, and electron/hole energy levels (E_{e1}, E_{hh}, or E_{lh1}) change in the presence of an external field. This shifts the energy of the 2D exciton peak and reduces the magnitude of the absorption coefficient $\alpha(h\upsilon)$. The shift in the absorption spectrum and the reduction in its magnitude result in a corresponding variation in the index of refraction, as dictated by the Kramer-Kronig relation. In addition, the absorption is also dependent on the polarization [15,16] of the incident light. This gives rise to field dependent birefringence. Plots labeled a1-a4 in figure 1(a) show the absorption when E field (polarization) is parallel to the plane of MQWs and curves b1-b4 are for perpendicular polarization in an AlGaAs-GaAs MQW waveguide[15], respectively. The computed [39] birefringence data for the AlGaAs-GaAs MQWs [15] is shown in Fig. 1(b). It is interesting to note that the Stark effect can also be induced optically. Changes in the excitonic absorption under below-gap photo-excitation conditions have been observed by Mysyrowicz et al. [18] and Von Lehman et al. [19]. This is referred to as ac Optical Stark Effect. The optically induced Stark effect was used to measure the excitonic response, which was found to be below 150 femtoseconds. This can be used a measure of upper limit for the frequency of operation of QCSE based devices.

The mathematical formulation of the nonlinear optical parameters is presented below. The imaginary part of the dielectric constant $\epsilon_i(h\omega)$ which includes the excitonic transitions in MQW layers is given by [14,20,21]:

$$\varepsilon_i = \left(\frac{C}{L}\right)\left|\int \Psi_e(x)\, \Psi_h(x)\, dx\right|^2 F(\hbar\omega_{ex} - h\omega) \tag{1}$$

and

$$C = \left[\frac{2\pi e^2}{\varepsilon_o m_o^2 \omega^2}\right] * |M_b|^2 * |\phi_{ex}(0)|^2 * \rho_{ex} \tag{2}$$

where ϕ_{ex} (x) is the exciton wavefunction, $\hbar\omega_{ex}$ is the exciton energy, F is the Gaussian line shape function, ψ_e and ψ_h are electron and hole wavefunctions respectively, L is the width of the quantum well, ϵ_o is free space permittivity, m_o is the electron rest mass, ρ_{ex} is the polarization factor and M_b is the Bloch matrix element [14, 20]. Using the Kramers-Kronig relation, we can express the real part $\epsilon_r(h\omega)$ as:

$$\varepsilon_r(\hbar\omega) = 1 + \left(\frac{2}{\pi}\right) P \int_0^{\infty} \frac{(\omega' \varepsilon_i(\omega')\, d\omega')}{(\omega'^2 - \omega^2)} \tag{3}$$

here, P is the Cauchy principal integral. The complex index of refraction n_c (= n-iκ) is related to the complex dielectric constant. Thus n and κ can be evaluated from ϵ_r and ϵ_i. The extinction coefficient κ is related to the absorption coefficient by

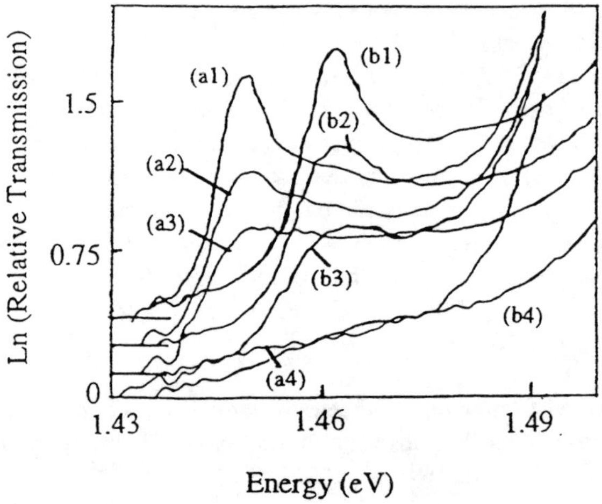

(a)

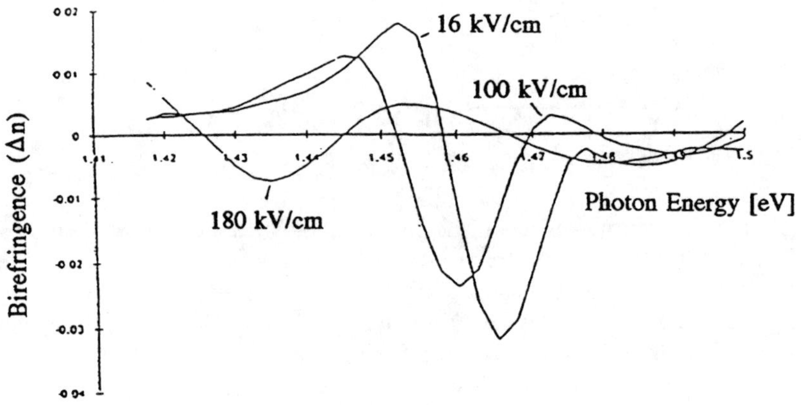

(b)

Fig. 1. (a) Single quantum well absorption spectra obtained with incident polarization [a] parallel and [b] perpendicular to the plane of the layers at effective intensities of (1) 0.8 W/cm^2, (2) 250 W/cm^2, (3) 800 W/cm^2, and (4) 8000 W/cm^2. [See Ref. 15.] (b). Calculated field-dependent birefringence in AlGaAs-GaAs MQWs.

$$\alpha(\hbar\omega) = \frac{4\pi\kappa}{\lambda_o} \tag{4}$$

As the electric field is applied across the MQW layers, the electron and hole wavefunctions (ψ_e and ψ_h) are modified. This results in the variation of α and n, which are commonly referred to as Stark shifts. Generally, the field-induced change in the refractive index, Δn, has a quadratic relation in the low field regime due to the excitonic effects in the MQWs and is given by:

$$\Delta n = -\left(\frac{1}{2}\right) n^3 \left[rE + sE^2\right] \tag{5}$$

where r and s are the linear and quadratic electro-optic coefficients. This is in contrast to the bulk electro-optic effect which is linear and significantly smaller.

Alternately, other nonlinear optical effects such as polarization and birefringence are related to the third order nonlinear optical coefficient D [22]. The nonlinear optical coefficient can be expressed in terms of susceptibility χ [22]. The third order effects which occur both in fibers and in MQWs involving inter subband transitions are expressed using the nonlinear refractive index n^2 as [22]

$$D = n^2 \varepsilon_o E \tag{6}$$

The nonlinear index n^2 is written in terms of susceptibility $\chi^{(3)}$

$$n^2 = n_o^2 + \chi^{(3)} |E|^2 \tag{7}$$

Using an equivalent representation,

$$n = n_o + n_2^2 |E|^2 \tag{8}$$

Equation (7) can be simplified using Eqn. (8) to express the third order nonlinear susceptibility as:

$$\chi^{(3)} = 2n_o n_2^2 \tag{9}$$

Thus, polarization or birefringence changes can be expressed in terms of susceptibility which is field dependent.

Table 1 lists the binding energy of excitons along with quantum well/barrier parameters for various material systems. These can be used to design modulators which operate in various spectral regimes.

III. BASIC MODULATOR STRUCTURES

Intensity modulation can be achieved by varying any one of the following optical parameters: absorption coefficient, index of refraction, and birefringence. Generally, the optical parameters are changed by applying an external RF electric field (voltage). Modulator structures and their system applications differ in terms of the way an electric field is impressed. Electric fields across MQWs can be applied directly (in the form of RF or DC bias) or indirectly (e.g. as in HACT devices which are discussed in the next section). The direct application of electric field is accomplished by using a reverse biased p-i-n or metal (Schottky)-i-n interfaces. Figure 2(a) shows a schematic structure of a multiple quantum well modulator based on electroabsorption. Wood et. al. [7] used the field dependent absorption to obtain optical amplitude modulation of a beam using AlGaAs-GaAs MQWs. The contrast ratio in this type of scheme is low (4:1), whereas, the insertion losses are quite high. With advances in the growth of strained QW layers, Harwit et al. [23] have demonstrated an improved contrast ratio of 8:1 in a coupled-well InGaAs-GaAs MQW structure.

The contrast ratio can be significantly improved by integrating MQW layers within the cavity of a Fabry-Perot (F-P) structure [Fig. 2(b)]. Using a reverse biased p-i-n MQW cavity, we can alter the percent transmission through the F-P structure. This type of modulator is based on electrorefraction. Changes in the index of refraction shift the F-P mode. If the mode wavelength is detuned from the excitonic absorption peak, a lower insertion loss results. Higher contrast can be achieved with lower insertion loss if birefringence is employed as shown in Figure 2(c). In this paper, electrorefractive (change of refractive index with electric field) devices using both asymmetric and symmetric Fabry-Perot structures have been analyzed extensively and discussed in detail in the next section.

High Contrast Fabry-Perot Spatial Light Modulators

Recently, the emphasis has been to design modulators having high contrast ratios. Contrast ratios have been enhanced by manipulating the excitonic behavior by

TABLE 1. STARK EFFECT CALCULATION FOR REFRACTIVE INDEX CHANGE AND ABSORPTION CHANGE UNDER DIFFERENT ELECTRIC FIELD

MQW	$In_xGa_{1-x}As$-GaAs	$In_xGa_{1-x}As$-GaAs	$In_{1-x}Ga_xAs_yP_{1-y}$-InGaAsP	$Zn_xCd_{1-x}Se$-ZnSSe	GaAs-$Al_zGa_{(1-x)}As$	$In_xGa_{1-x}N$-$In_yGa_{1-y}N$
Substrate	GaAs	GaAs	InP	GaAs-	GaAs	Saphire
x	0.12	0.25	0.46	0.72	0.24	0.22
y			0.31			0.06
λ (nm)	948	1064	1064	532	828	
$E_{g,well}$(eV)	1.2359447	1.074076	1.1254	2.282782	1.424	2.909
$E_{g,barrier}$(eV)	1.42	1.42	1.2277	2.9860	1.72123	3.257
E_{hh}(eV)	1.2323217	1.065310	1.1254	2.308224	1.424	2.966
E_{lh}(eV)	1.2879855	1.164269	1.1254	2.393067	1.424	3.325
E_{ge}(eV)	0.05560	0.07723	0.0348	0.0218	0.057	0.014
E_{ghh}(eV)	0.0204	0.024	0.0043	0.0067	0.0148	0.00397
E_{glh}(eV)	0.0368	0.04564	0.017	0.0172	0.0555	0.0137
ΔE_c(eV)	0.1104	0.207554	0.06139	0.3927	0.1783	0.175
ΔE_v(eV)	0.0736	0.13836	0.04092	0.2618	0.1189	0.116 (hh)
L_{real}(Å)	46	46	80	80	50	100
L_e^* (Å)	99	84	126	104	99	122
L_{hh}^* (Å)	64	59	102	102	64	114
L_{lh}^* (Å)	88	79	145	125	91	130
m_e^* (m_0)	0.069	0.069	0.068	0.16	0.067	0.18
m_{hh}^* (m_0)	0.45	0.45	0.84	0.54	0.62	0.73
m_{lh} (m_0)	0.132	0.132	0.105	0.14	0.082	0.17
$m_{hh//}$ (m_0)	0.16	0.16	0.13	0.172	0.105	0.21
$m_{lh//}$ (m_0)	0.28	0.28	0.31	0.315	0.23	0.40
ε(∞)	11.46	11.61	10.232	7.67	10.89	2.832
Δ_{so}(eV)	0.3508	0.3625	0.2017	0.42	0.15	0.017
heavy hole exciton B.E (binding energy) (eV)	0.010	0.0114	0.0121	0.0289842	0.01125	0.0276
c.f.	11.85	11.85	10.44	6.67	10.73	
a_s (Å)	5.6536	5.6536		5.631		0.3221
a_w (Å)	5.7022	5.75495		5.80		0.3277
ε_{xx} ,ε_{yy}	-0.008605	-0.01792663		-0.0191		-0.024
ε_{zz}	0.0079296	0.017851113		0.02365		
C_{11} (10^{11} dyn/cm^2)	11.454	10.9925		7.94		C_{11}/C_{12}=2.76
C_{12} (10^{11} dyn/cm^2)	5.27752	5.1665		4.91		
a (eV)	-2.676	-2.65		-5.04		-3.9
b (eV)	-1.7	-1.7		-1.12		-1.5

1. λ –operating wavelength ;
2. E_{ge},E_{ghh},E_{glh}–electron, heavy hole and light hole ground energy;
3. E_{hh}, E_{lh}–energy gap between conduction band and heavy hole, light hole;
4. ΔE_c, ΔE_v–conduction and valence band offset;
5. L_{real}– real single quantum well size;
6. L_e^*, L_{hh}^*, L_{lh}^* –electron, heavy hole and light hole effective length;
7. m_e^*, m_{hh}^*, m_{lh}^* –electron,heavy hole and light hole effective mass;
8. $m_{hh//}$, $m_{lh//}$ –in-plane heavy hole and light hole effective mass;

ε(∞)–dielectric constant at high frequency;
. Δ_{so}(eV)–orbital splitting energy;
. c.f.: correction factor, where c.f.=$n_{bulk}^2(\lambda)$-1 ;
. a_s –lattice constant of the cladding layer;
. a_w–lattice constant of the active layer;
. ε_{xx} ,ε_{yy}, ε_{zz} - strain tensors;
. C_{11}, C_{12}–bulk elastic constant tensors;
. a–hydrostatic potential;
. b–shear deformation potential;

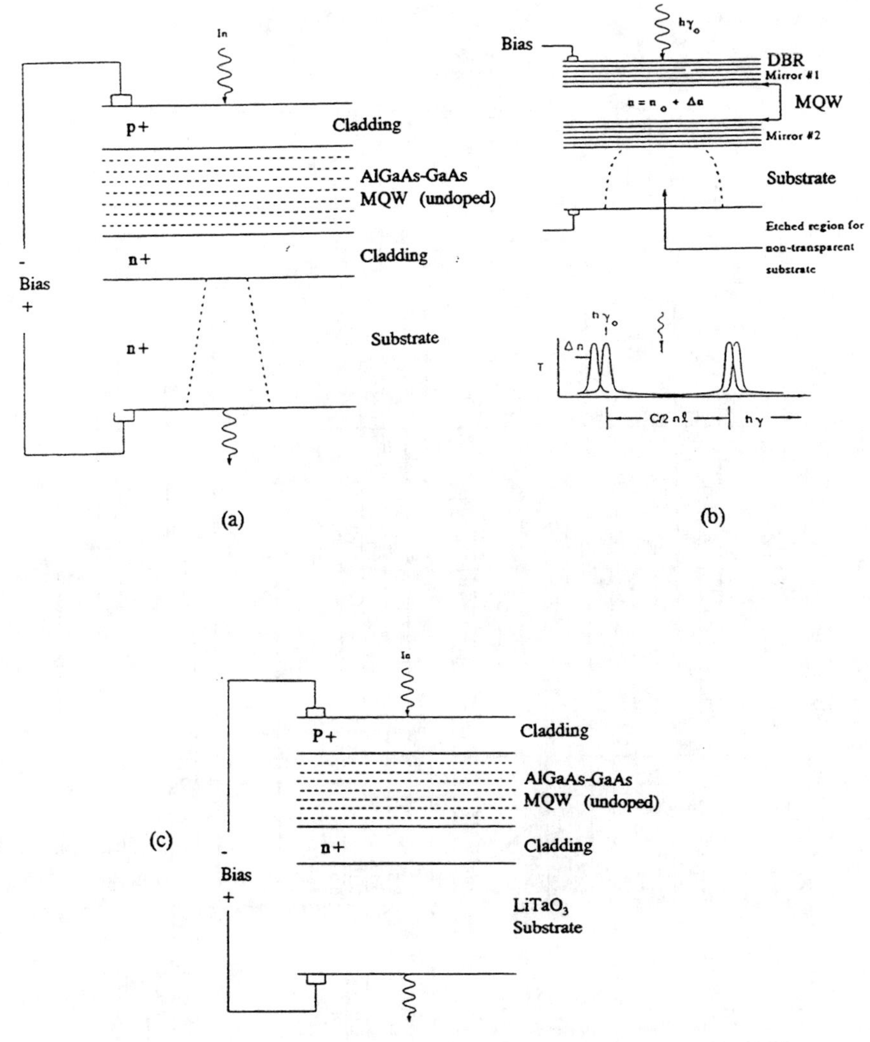

Fig. 2. (a) Schematic of a multiple quantum well modulator using electroabsorption. (b) Schematic of A Fabry-Perot electrorefractive modulator. (c) Strain-induced birefringence in multiple quantum wells transplanted on a $LiTaO_3$ substrate (Ref. Shen et al., OSA 1993).

utilizing coupled-well MQW layers on one hand and by employing a high finesse Fabry-Perot cavity on the other[23,29]. In the case of the MQW Fabry-Perot cavity, the schemes include dielectric $\lambda/4$ mirrors with matched reflectivities, and asymmetrical modulators having variable reflectance. A optical pump is used to vary the reflectivity on the asymmetric scheme [24-28]. The use of a saturable absorber, in which reflectivity can be controlled by an external pump (photoabsorption effect) has attracted significant attention. The on/off contrast ratio as high as 29 dB had been report previously[4]. We reported[29] the fabrication of a F-P modulator which consists of an InGaAs-GaAs MQW cavity sandwiched between AlAs-GaAs quarter-wave mirrors. These devices exhibited a contrast of 1200:1 (30.8 dB) with Stark effect tuning range between 10-12 nm as a function of externally applied voltage.

Figure 3 shows the schematic of a typical tunable F-P cavity optical modulator. It consists of a quarter wave stack of AlAs-GaAs dielectric mirrors (M1 and M2) enclosing an InGaAs-GaAs MQW cavity. The thickness of AlAs and GaAs layers are 805 Å and 697 Å, respectively, in the $\lambda/4$ stack layers. The InGaAs-GaAs MQW layers consists of 50.5 periods, which have about 51 Å of InGaAs and 333 Å GaAs making up one period, with a total thickness about 2.0 μm. Growths were performed in an Intevac (Varian) Molecular Beam Epitaxy (MBE) system equipped with EPI sources.

Figure 4 gives the measured contrast ratio as a function of electric field at 978 nm. Unlike optically pumped asymmetric Fabry-Perot modulators using photoabsorption effect, the Fabry-Perot devices primarily utilize electrorefractive effects to obtain high contrast ratios. The insertion loss of such structure is reduced from 12.5 dB at 978 nm to 3dB by using a MOCVD grown AlGaAs-GaAs MQW system. This requires an appropriate detuning of operating wavelength from the excitonic absorption peak.

Pixel Addressing in SLMs

In this section we describe some specific examples of SLMs, in particular, HACT-SLMs and Birefringent SLMS (BSLMs). HACT-SLMs have the potential of addressing large arrays (1000×1000) of pixels and BSLMs yield high contrast ratios with low insertion losses.

Methods which induce fields across the MQWs indirectly include Heterostructure Acoustic Charge Transport (HACT) addressing, charge-coupled (CCD) addressing and optical addressing. Figure 5(a) depicts the HACT pixel addressing scheme. In this method, fields are induced by signal charges which are packeted and transported by a SAW along the channel (adjacent to the MQW layers). Figure 5(b) shows a charge-coupled imaging scheme on top of the MQW layers. Pixels are addressed when the CCD induces a filed capable of invoking Stark effects in the MQWs. These methods are specially relevant to SLMs having large number of pixels. Figure 5(c) shows an addressing scheme which uses optical READ/WRITE beams operating at wavelengths other than the wavelength causing exciton formation. This method can be used directly (as is the case in optically addressed asymmetric Fabry-Perot modulators) for modulation or pixel addressing in SLMs or conjunction with a photodetector. When a photodetector is used in series with a modulator, the control beam changes the bias across the

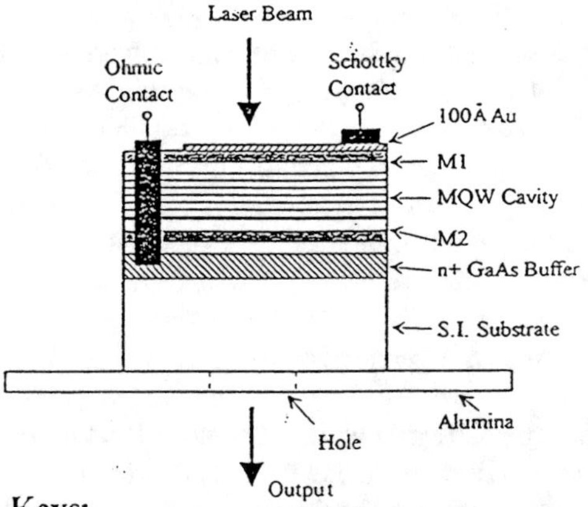

Keys:

M1: 12-period of (69.7nm)-GaAs/(80.5nm)-AlAs layers
M2: 15.5-period of (80.5nm)-AlAs/(69.7nm)-GaAs layers
MQW: 50.5-period of (51Å)-(InGaAs,x=0.18)/(333Å)-GaAs
n+ Buffer: 0.3 μm
Substrate: 502 μm

Fig. 3. Cross-section of a symmetric Fabry-Perot MQW modulator.

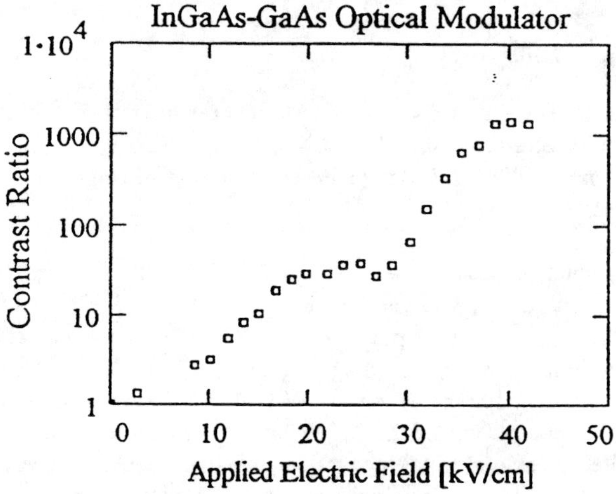

Fig. 4. Contrast ratio as a function of electric field.

(a)

(b)

(c)

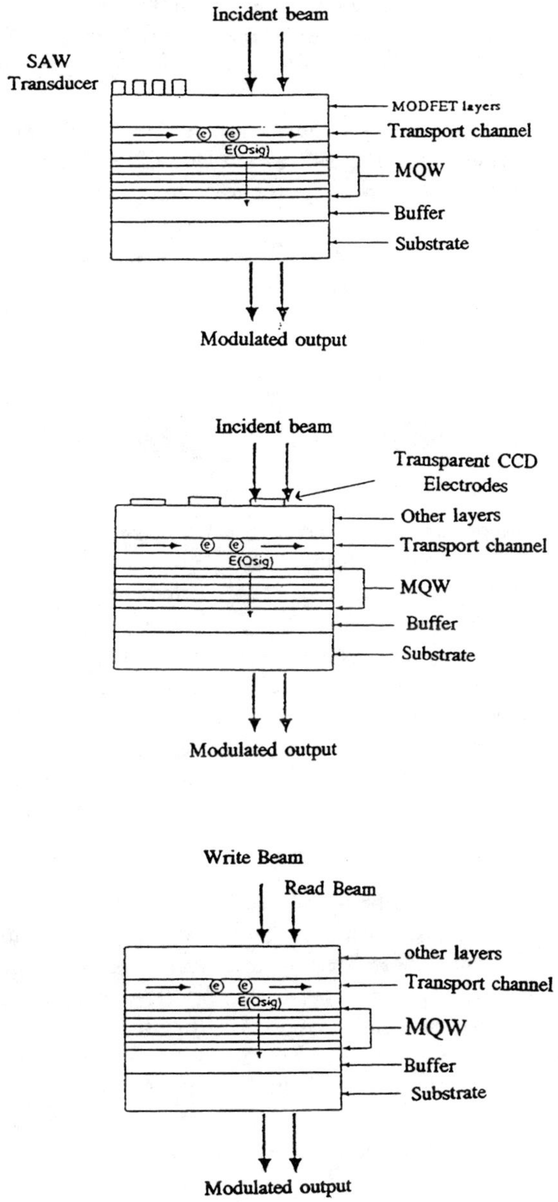

Fig. 5.

Pixel addressing. (a) HACT pixel addressing. Surface acoustic wave (SAW) to transfer signal charge in a HACT channel. (b) Pixel addressing using charge-coupled devices (CCDs). Transfer of a signal charge from one electrode to another by clocking circuitry. (c) Optical pixel addressing. Pixel using WRITE beam to induce charge packet on top of MQW layers.

photodetector and hence the voltage seen by the MQW layers. In asymmetric F-P modulators, the optical pump changes the absorption of the bottom DBR mirror, resulting in an overall change in the transmissivity/reflectivity of the modulator [25].

Optical modulation using quantum well charge-coupled devices (QWCCDs) was demonstrated by Goodhue et al [31]. However, such CCD structures require a complex arrangement of electrodes and clock circuits to address the pixels. We describe an alternative approach [32,33] in which the CCDs are replaced by a HACT type structure. In this device the quantum well channel is similar to that of a modulation doped field effect transistor (MODFET). The signal charge is transported by launching a surface acoustic wave. The electronc application of HACT devices were first proposed by Tanski et al. [34].

SLMs using Heterostructure Acoustic Charge Transport (HACT) Controlled Stark Effect

Fig. 6 describes the basic cross-sectional schematic of the HACT-controlled MQW modulator. The structure can be divided into two parts: (a) one consisting of GaAs(NID)/n-AlGaAs/GaAs(NID)/AlGaAs(NID) heterostructure which establishes a charge transport HACT channel, and (b) multiple quantum well (MQW) layers underneath the channel (which host the excitons produced by the incident laser light). The undoped AlGaAs spacer layer is optional. This is because the carrier mobility and transport in the channel are not critical as in the case of MODFETs. However, the design of NID GaAs caps and n-AlGaAs supply layers is quite critical as described by Tanski et al.[34] and Cullen et al.[35]. The thickness of a GaAs cap and an n-AlGaAs supply layer are such that the channel is depleted with no external bias or signal charge.

The charge packets can be introduced by appropriately biasing the ohmic and Schottky contacts or by optical injection. These charge packets are then transported in the confined channels which are realized by sandwiching the GaAs channel between the n-AlGaAs and NID-AlGaAs layers. The absence or presence of a charge packet, in the region where laser light is incident, controls the magnitude of the electric field in the MQW layers. This in turn affects the excitonic absorption $\alpha(h\upsilon)$ and the modulation depth. The magnitude of the electric field E_1 in the MQW layers can be computed following the procedure used in MODFETs. The channel electron charge, qn_s is known to be related to the electric field in the GaAs channel :

$$qn_s = \epsilon_1 E_1 \tag{10}$$

Here, n_s is the two-dimensional electron gas concentration, q is the magnitude of the electron charge, ϵ_1 is the permittivity of GaAs. Eq. 1 is an approximation which neglects the depletion charge due to ionized impurities. The variation in the electric field E_1 is proportional to the density of signal charge in the channel. The electric field E_1 results in changes in absorption coefficient $\alpha(h\upsilon)$ and index of refraction $n_r(h\upsilon)$ in the GaAs wells. This in turn modulates the transmitted beam. In addition, the

Amplitude/phase modulation - up to 10 μs delays - Bandwidth up to 500 Mhz

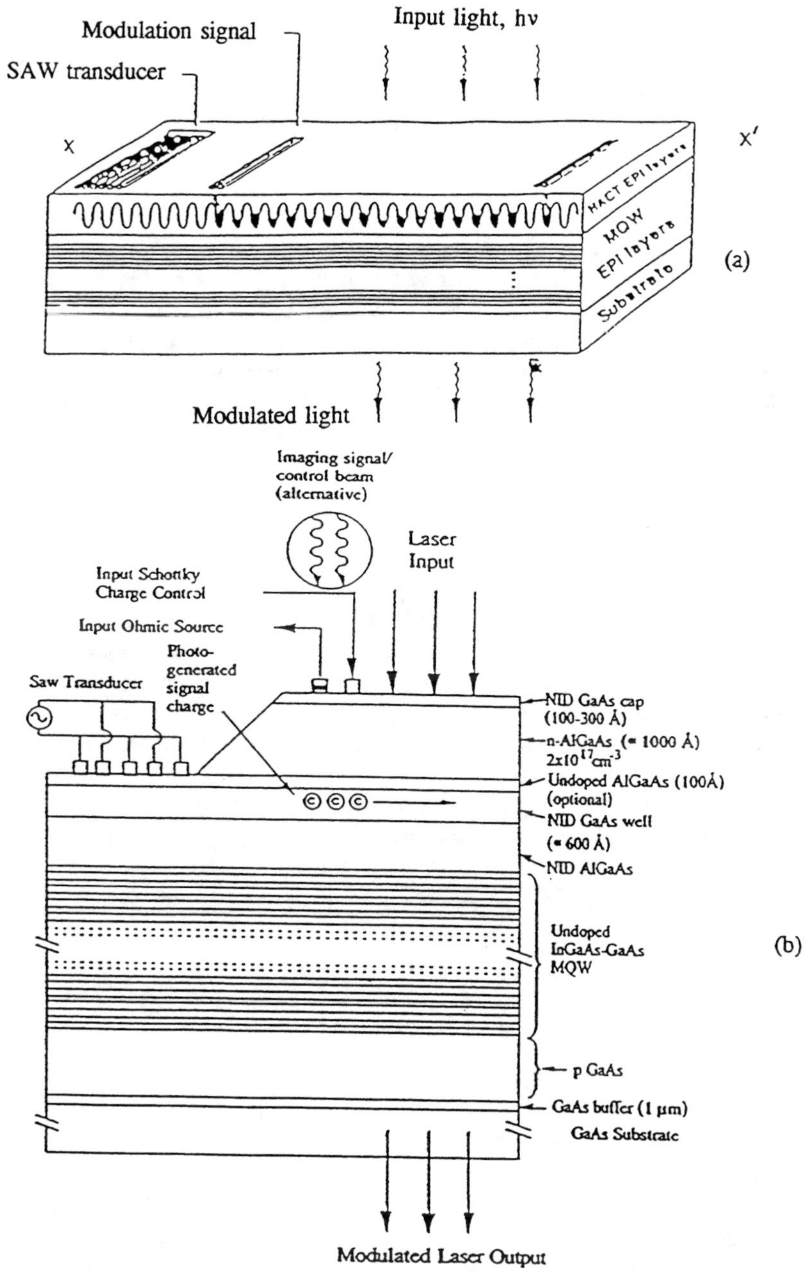

Fig. 6.
Heterostructure Acoustic Charge Transport (HACT) Controlled Multiple Quantum Well optical modulator; (a) 3-D schematic, (b) cross sectional view along x x'.

modulator can be used in reflection mode by having a quarter-wave stack of AlGaAs-GaAs dielectric mirror under the MQW layers[11] .

We have computed the absorption data as a function of n_s (and line charge density n_l) using the MQW model of Lengyel et al.[12]. The line charge packet density qn_l is related to n_s by the SAW wavelength λ_{SAW} :

$$n_l = (\lambda_{SAW}/6)n_s \tag{11}$$

The computation assumes a SAW frequency of 400 MHz ($\lambda_{SAW} = 7.615\mu m$), GaAs well width of 105Å, and AlGaAs barrier width of 95Å. The variation in the density of electrons n_l is shown to cause a significant shift in the absorption coefficient. For example, an electron density of 2.0×10^7 cm^{-1} can result in a $\Delta\alpha$ shift over 1000 cm^{-1}. The maximum value of the line charge density qn_l is determined by the channel capacity and SAW wavelength used in transporting the signal. Peterson et al.[24] have computed charge load capacities of HACT channels, which corroborates our results for n_l [13].

The basic unit of a HACT optical modulator can be adapted to realize 2-D SLMs, as shown in Fig. 7. This figure shows an n-channel HACT addressing SLM operated in the reflection mode. Image charges are injected and packed by the SAW to form pixels on the channels. The channels are separated by either etching of mesas or by ion implantation. The presence of SAW-transported signal charge packets in the channels create a moving image by changing exciton properties in the InGaAs-GaAs quantum well layers underneath the HACT structure. Since the modulated output is proportional to the magnitude of the charge packets and the charge transport efficiency, an image can be formed and read. In addition, the HACT/SLMs have a frame speed, in GaAs, up to ~2870 m/s. The pixel size of the HACT/SLM is determined by the wavelength (or driving frequency) of the SAW. The delay line and frame freezing properties of HACT/SLMs may be exploited for various applications.

HACT Results

Two types of HACT wafers were grown -- (i) MBE 1831 and 1833, which did not have any MQW or dielectric mirror layers, (surface channel HACT (UTC Model 1210) and (ii) MBE 1868 which was fabricated on AlAs-GaAs MQWs (50 by 52 having 20 periods). Samples 1833 and 1868 were measured at 119.8 MHz. While the sample 1833 showed a charge transport efficiency (CTE) of 0.995, sample 1868 (which was grown on MQWs) exhibited a lower CTE (~0.9). The data for the MQW device is shown in Fig. 8 [33b]. The inset shows the HACT structure grown on AlAs/GaAs superlattice MBE#1868. There are about 20 pixels (20 μm each) over which charges are transported before the charge magnitudes degrades to values not causing significant shifts in excitonic properties.

High contrast HACT/SLM [15] fabrication requires the growth HACT layers (for addressing) on MQW Fabry-Perot structures [16]. The insertion losses and sharp

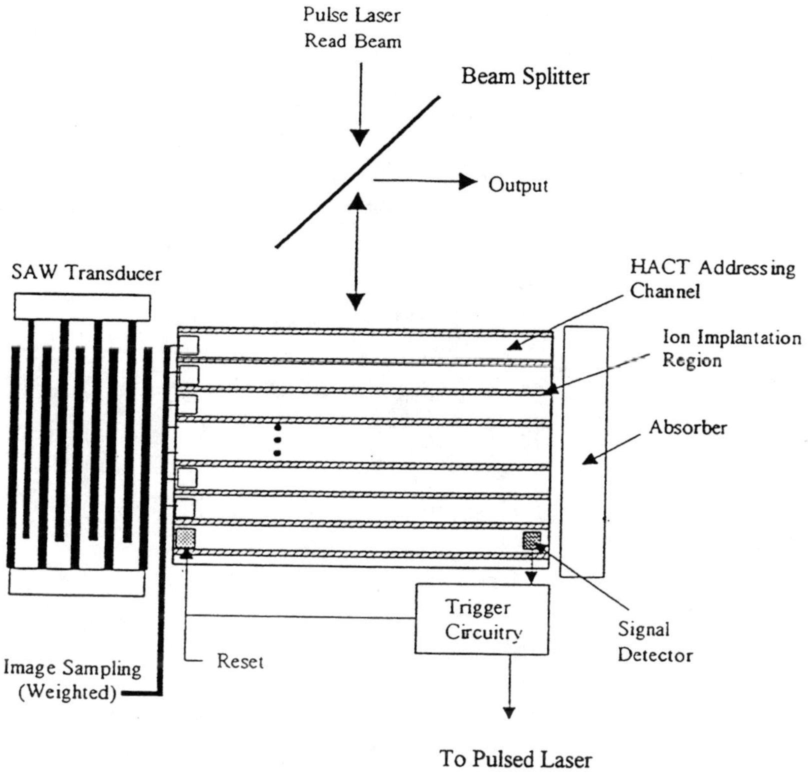

Fig. 7.
A two-dimensional HACT/SLM showing mesa defined rows and a common SAW charge transporter.

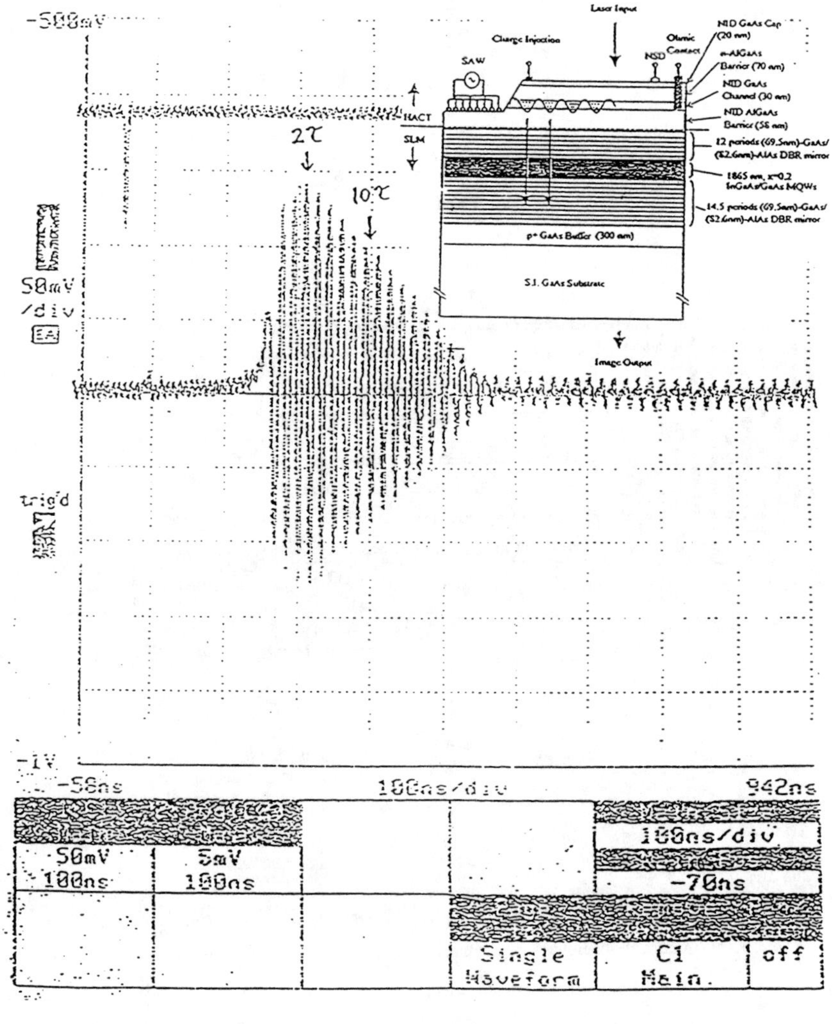

Fig. 8. The charge transport efficiency(CTE) of sample MBE1868. The signal charge decays as it travels about 20τ or 20 pixels.

changes in contrast may be minimized by using a multiple cavity F-P scheme.

High Contrast HACT Spatial Light Modulator Structures

Important issues in the realization of (SLMs), besides modulation, include adequate high on/off contrast ratio and methodology of addressing a large array of pixels ($>10^6$). This section describes a structure which utilizes the high contrast ratio feature of a Fabry-Perot cavity along with Stark effect tuning, and pixel addressing by the heterostructure acoustic charge transport (HACT) scheme.

The basic HACT/SLMs shown in Fig. 6, like other MQW modulators using excitonic Stark effect, offer a modest (10-20:1) contrast ratio [8]. Fig. 9 shows an integrated SLM structure (recently proposed by Jain et al. [26]) having a HACT structure realized on top of a Fabry-Perot cavity which consists of InGaAs/GaAs multiple quantum wells sandwiched between AlAs-GaAs mirrors. The simulated results of the HACT-SLMs are shown in Fig. 10, which demonstrates ~ 160 contrast ratio with 11.7 dB insertion loss. We have reported [28] contrast ratios of 1200:1 in MBE grown Fabry-Perot structures. Based on this work, we believe that high contrast ratio SLMs integrated with HACT addressing scheme can be successfully implemented.

SLM using Birefringence in MQW Structures

Multiple quantum well layers, in systems such as AlGaAs-GaAs, exhibit a very small value of birefringence based on the form anisotropy [37]. However, Wenier et al. have reported [15,30] significant birefringence in MQW waveguides. These observed Δn values vary with wavelength and applied electric field across the AlGaAs-GaAs MQW layers. The enhanced value of Δn has been attributed, by Sonek et al. [16], to the quantum confined Stark effect (QCSE). Birefringence in MQW layers has been utilized in a novel way to realize high performance two-dimensional SLMs. Conceptually, the BSLMs is similar to the liquid crystal light valves LCLVs [38]. In contrast to conventional MQW SLMs described in the preceding sections, the MQW BSLMs operate in conjunction with a polarizer/analyzer combination to convert phase change, due to birefringence, into amplitude modulation. BSLMs are expected to have a faster response than LCLVs [38] and a higher contrast ratio, similar to that of Fabry-Perot SLMs (but without significant insertion loss).

Since MQW layers do not exhibit birefringence when light is normally incident to the plane of quantum wells, MQW BSLMs cannot be used directly as LCLVs. There are at least two ways to get around this problem; (i) use gratings to couple light into the MQW waveguide pixels [39], and (ii) create strain in MQWs [40,41] which result in birefringence for normal incidence. The strain in MQW can be induced by external means. One method is to launch a surface acoustic wave (SAW) which causes strain in the MQW layers [40]. For example, an interdigitated transducer, operating at 100 MHz produces a strain of about 10-4 at moderate operating power levels (about 10 mW for a transducer of 10 Λ aperture; here Λ is the SAW wavelength). Shen et. al. [41] have observed significant contrast in MQW layers by inducing strain using a heated

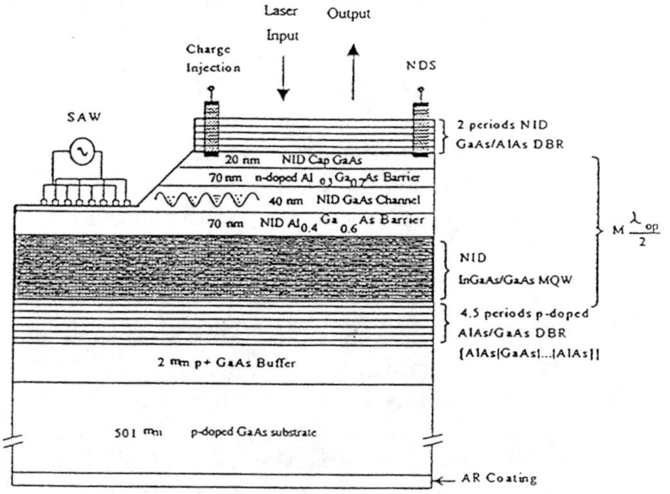

Fig. 9.
A nominal configuration of the HACT-SLM with the HACT embedded inside the Fabry-Perot cavity. The structure contains two DBRs.

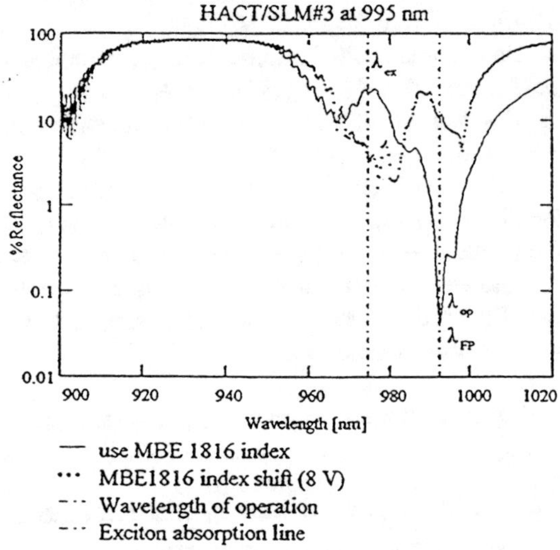

Fig. 10.
The reflective spectra of the HACT-SLM (Fig. 9) operated at 995 nm with a contrast ratio of ~ 160 and 11.7 dB insertion loss.

LiTaO3 substrate (on which MQW layers were transplanted via a lift-off process).

A schematic of induced strain birefringent optical modulator operating in the transmission mode is shown in Fig. 11. The structure includes a p-i(MQW)-n modulator built on a photosensing device. In this structure, the light entering the modulator is perpendicular to its surface. The laser input is absorbed via exciton formation in the undoped MQW region of the modulator. A surface acoustic wave (SAW) transducer is used to induce strain in the multiple quantum layers. The output can be modulated either electrically by modulating the bias, or optically by controlling the intensity of the control beam which in turn governs the voltage drop across the p^+-i-n photodetector. The operation is quite similar to a SEED device except the BSLM uses an i-MQW-region of proper thickness and Δn to yield a phase shift of 90° .

Computations of birefringence (Δn) due to quantum confined Stark effect (QCSE), as a function of wavelength are presented for the AlGaAs-GaAs MQWs in Fig. 1b. The application of electric field causes mixed terms in the index ellipsoid that transform the uniaxial birefringence to biaxial birefringence which results in a phase shift in the MQWs. In the case of an AlGaAs-GaAs birefringent modulator, a thickness of 5.08 μm is estimated for a 90° phase shift (Δn=0.042 at $h\nu$=1.453 eV). The well and barrier thickness in the MQW layers are 94 and 200 Å, respectively. The thickness is expected to be smaller for the coupled-well strained layer InGaAs-GaAs SLMs.

Another type of BSLM employing a waveguide structure is shown in Fig. 12. This structure employs a 45° mirror, similar to the one used in a folded cavity surface emitting laser, and a grating layer for coupling the incident beam into the MQW waveguide. The grating coupling efficiency for TE to TM mode conversion can have a 50% split (with optimum coupling efficiency ~20%) when the input beam is incident at ~31° relative to the direction of grating vector [42-44]. Under modulation, the field-induced birefringence of the MQWs is determined by the control beam, which is incident on a built-in photodetector. The TE and TM modes propagate and undergo a phase shift depending on the pixel length and the magnitude of the field induced birefringence (Δn). The light is totally reflected upward by the etched 45° mirror. The grating dimension needed for effective coupling and the optical path length required for the phase shift dictate a pixel area of ~ 10x60 μm^2. The modulated signal emerges out of an analyzer which is integrated on the surface of the structure. The intensity of the modulated output depends primarily on the birefringence of the MQW layer. The loss due to electroabsorption can be minimized by detuning the operating wavelength from the excitonic peak. These devices can be used to form a two dimensional BSLM array consisting of many pixels as shown in Fig. 13.

IV. ELECTROOPTIC AND ACOUSTOOPTIC MODULATORS

Saw Induced Excitonic Shifts in Multiple Quantum Wells

The introduction of a SAW transducer on top of the MQW layers can produce a new type of device, one in which the electric field is induced by the SAW. The strain produced by the SAW induces an electric field via piezoelectric effect. We have

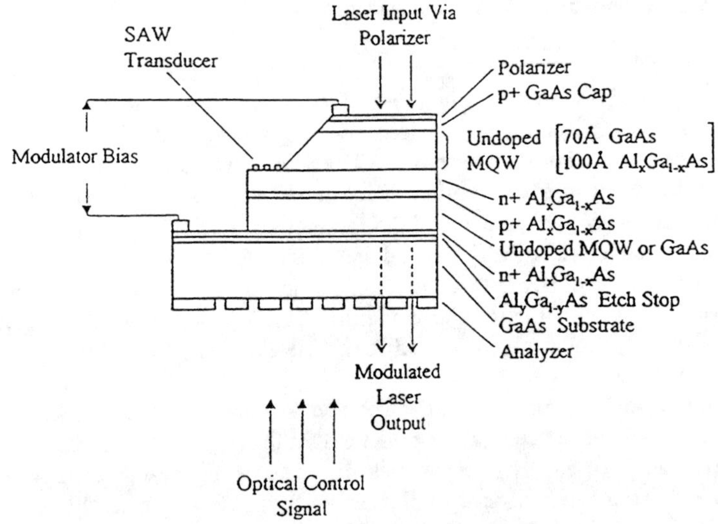

Fig. 11.
Cross-sectional schematic of an induced strain optically addressed birefringent modulator.

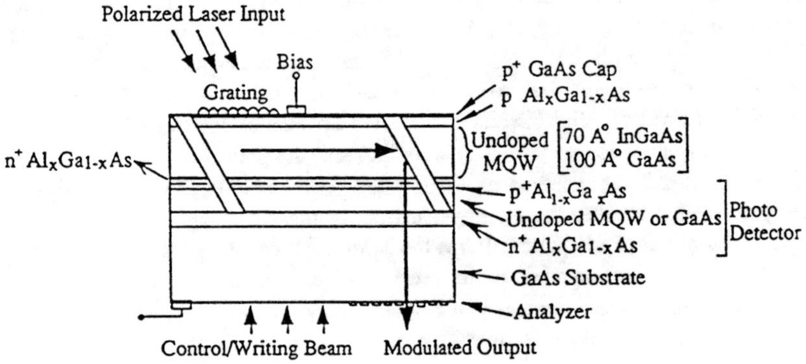

Fig. 12.
Grating coupled modulator using birefringence in MQW waveguide.

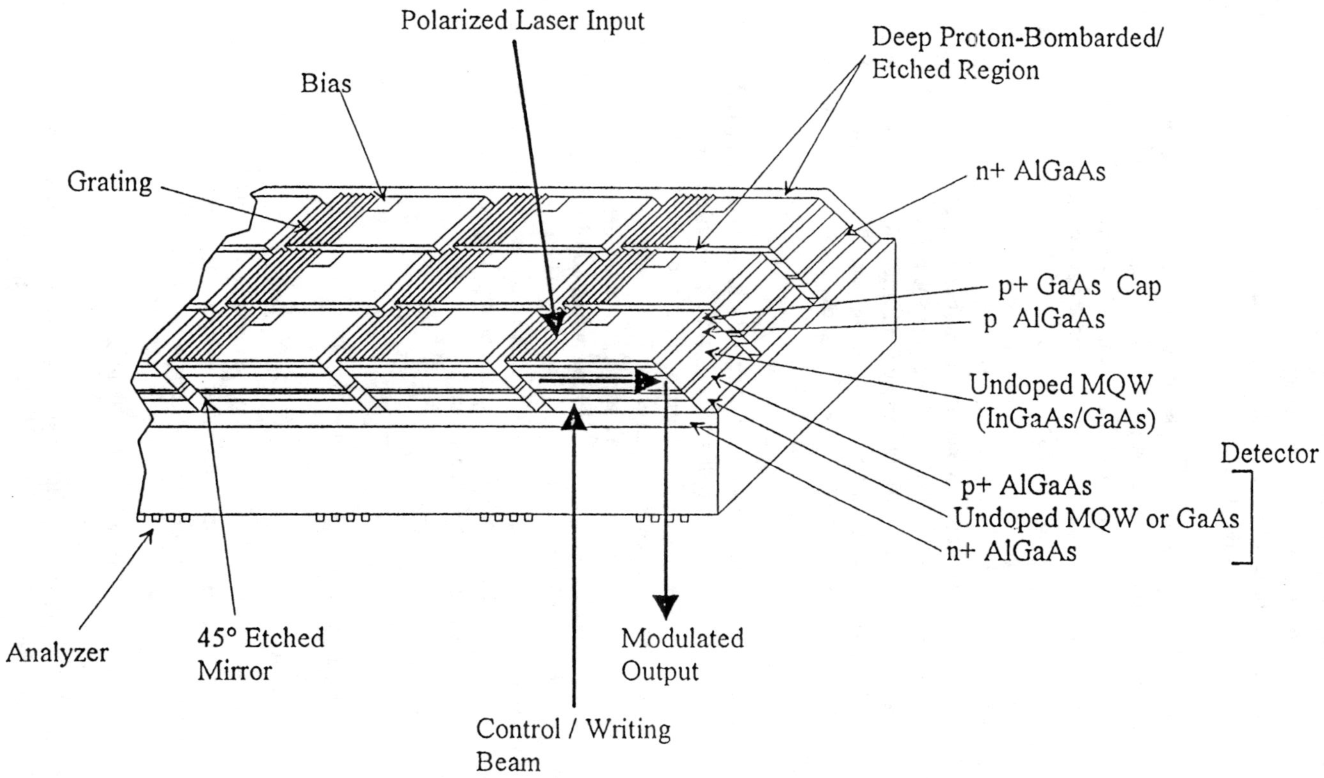

Fig. 13. Three-dimensional view of a grating coupled BSLM.

observed the QCSE by probing the transducer fingers and adjoining region directly with an optical beam. Further investigation of strain effects is in progress. Additionally, we have investigated the possibility of inducing DBRs [33a] using SAW and/or electrooptic gratings for in-plane optical modulators and integrated optical system applications.

The structure of Fig. 14 illustrates the modulation of an optical beam entering from an adjoining waveguide in a direction parallel to the plane of quantum wells. The application of a perpendicular electric field ($E_\perp$) induced by a propagating SAW, results in a shift of the excitonic peak (i.e., QCSE). In this arrangement, SAW is launched using a conventional interdigitated transducer fabricated directly on the active MQW layers which host the excitons. The transducer is positioned so that the SAW is propagating in a direction (x) which is perpendicular to the direction (y) of the incident optical beam. This device has been shown to be the building block of one class of SLMs and optical correlators[10]. We have computed the values of $E_\perp$ and $E_\parallel$ for a transducer on GaAs having a $w/\Lambda=10$ and power P=10mW. Here, w is the aperture of the SAW transducer and Λ is the acoustic wavelength (e.g. $\Lambda=7.16\mu m$ and $2.86\mu m$ at 400 MHz and 1 GHz, respectively). Table II lists the SAW induced electric field magnitudes [Jain and Bhattacharjee, IEEE Photonics Tech. Lett., 1989].

Surface acoustic waves launched at about 878 MHz and having an aperture equal to 50 times that of the SAW wavelength were used to measure the SAW-induced Stark effect shifts. The SAW device is excited by a RF source at the level of 20mW, which produces little change in optical parameters. The power spectrum launched into the SAW modulators is shown in Fig. 15. However, when the incident light was shone in the transducer, it produced a noticeable shift in exciton absorption as shown in Fig. 16. Recently, it was realized that the induced $E_\perp$ field in the image area is smaller than required for obtaining excitonic shifts in large aperture devices. Work is in progress to increase the $E_\perp$ by reducing the aperture while maintaining the same level of RF excitation or by making a unidirectional transducer.

SLMs Based on Bragg Diffraction

The optical modulator of Fig. 14 can be configured as a Bragg modulator/SLM or a switch, as shown in Fig. 17. This device operates as a switch by activating the SAW to direct the beam from an undiffracted position to a diffracted one. The modulation depth (η), when the device is used as a modulator, is given by

$$\eta=(I_o-I)/I_o=\sin^2(\Delta\phi/2) \qquad (12)$$

where I_o and I are the optical beam intensities in the undiffracted direction (zeroth order) in the absence and presence of the SAW, respectively, and $\Delta\phi$ is the phase shift of the transmitted beam. It is expressed in terms of interaction length (l), the free space wavelength (λ_o) of the optical beam, and change in the index of refraction (n):

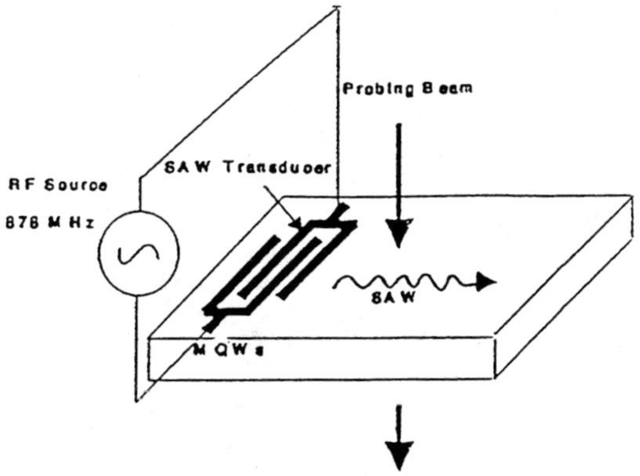

Fig. 14.
The measurement setup for the MQW device powered by a SAW transducer.

TABLE II. Typical SAW-induced perpendicular electric field magnitudes.

Total SAW Power P mW	SAW Frequency f Hz	w/Λ	$E_{\perp} \times 10^4$ V/cm
10	400 MHz	10	.88
20	400 MHz	10	1.244
10	1 GHz	10	2.2
20	1 GHz	10	3.11

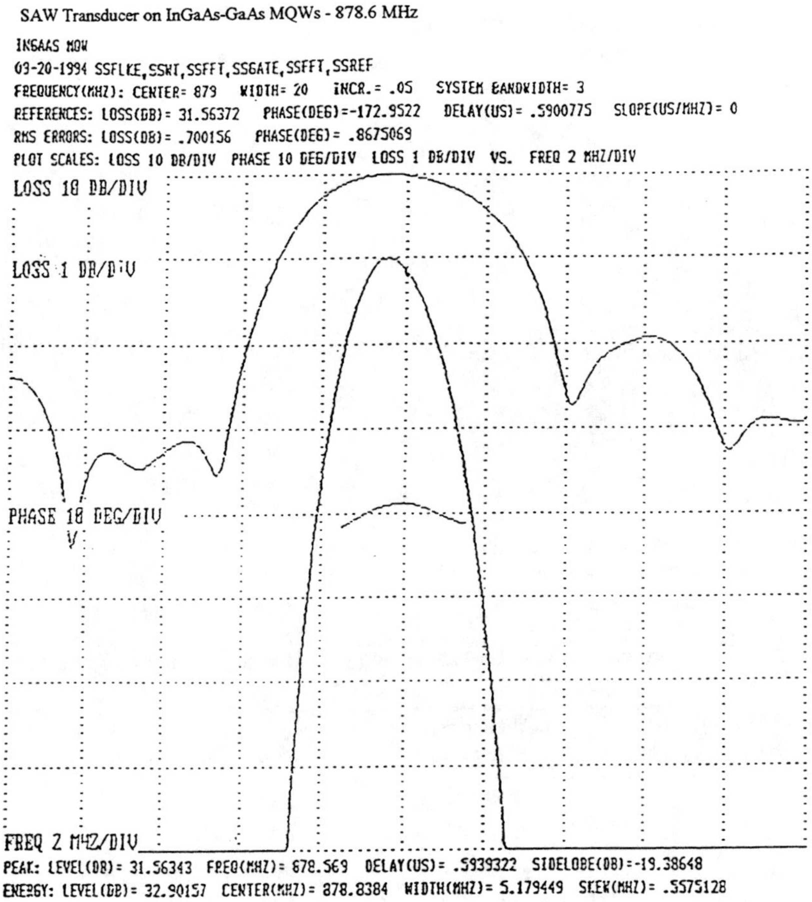

Fig. 15. The launching power spectrum on the SAW transducer.

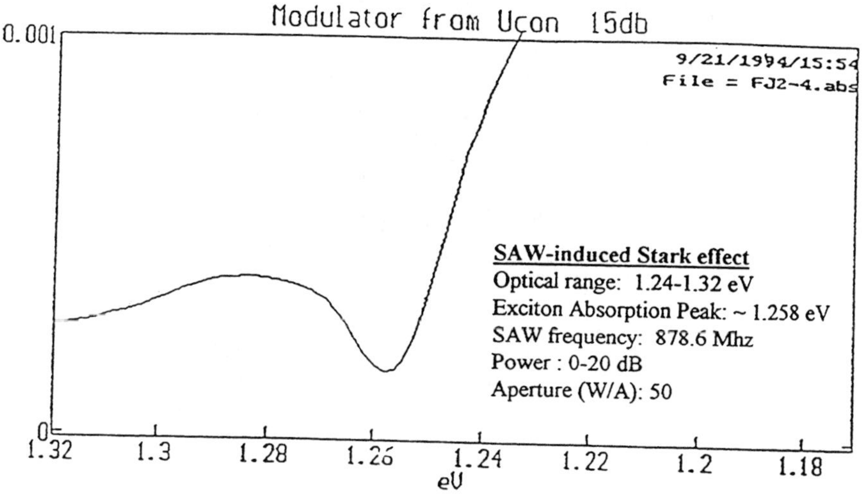

Fig. 16. The Excitonic shift due to the SAW-induced-field in the MQWs.

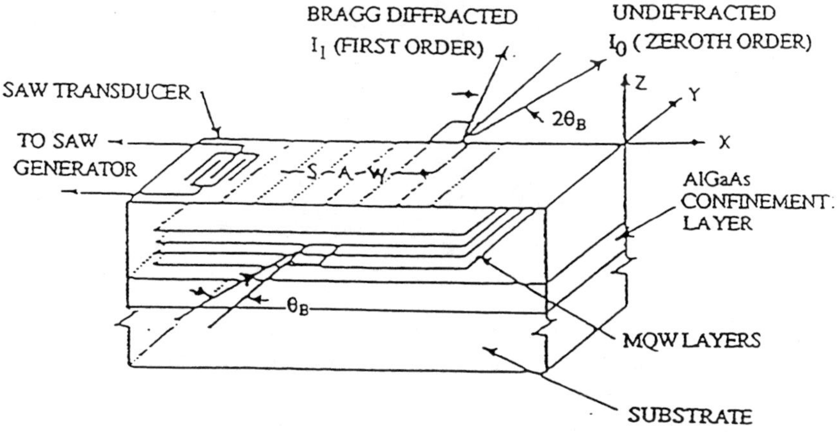

Fig. 17.
Schematic of a side entry optical modulator using SAW-induced quantum Stark effect in MQW layers. (This diagram also shows the Bragg type acousto-optic modulator/switch for an incident light beam at Bragg angle θ_B).

$$\Delta\Phi = 2\pi l \Delta n / \lambda_o \tag{13}$$

$$\Delta\phi = \left(\frac{2\pi}{\lambda_o}\right)\left[(n-1)\delta_o\cos(\omega t - kx) + \int_{-\infty}^{0}\Delta n_{\delta o}dz + \int_{-\infty}^{0}\Delta n_{MQW,1}dz + \int_{-\infty}^{0}\Delta n_{MQW,2}dz\right] \tag{14}$$

where

$$\delta_o = u_z(z=0) \tag{15}$$

$$\Delta n(\omega)\,|_{MQW,\,2} = \left(\frac{c}{\pi}\right)\int_{0}^{-\infty}\left[\frac{\Delta\alpha(\omega')}{(\omega'^2 - \omega^2)}\right]d\omega' \tag{16}$$

$$\Delta\alpha = \alpha(\omega)\,|_{E_\perp=0} - \alpha(\omega)\,|_{E_\perp} \tag{17}$$

The value of modulation depth can be substantially higher if the change in n is significant and/or the interaction length (l) is large. The magnitude of $\Delta\phi$ is quite large since Δn is enhanced by the excitonic effects in MQW structures. The phase shift, in general, can be expressed in terms of various components of Δn. Our objective is to demonstrate that the acousto-optic modulation in MQW layers requires substantially smaller interaction lengths than regular GaAs devices. An important goal is to demonstrate Bragg modulation in the structure of Fig. 18 which utilizes a grating produced electro-optically via the Schottky electrodes [25]. This modulator does not require SAW induced Stark effect. As a result, significantly higher electric fields can be induced inside the MQW layers. Induced electrooptic grating have been proposed

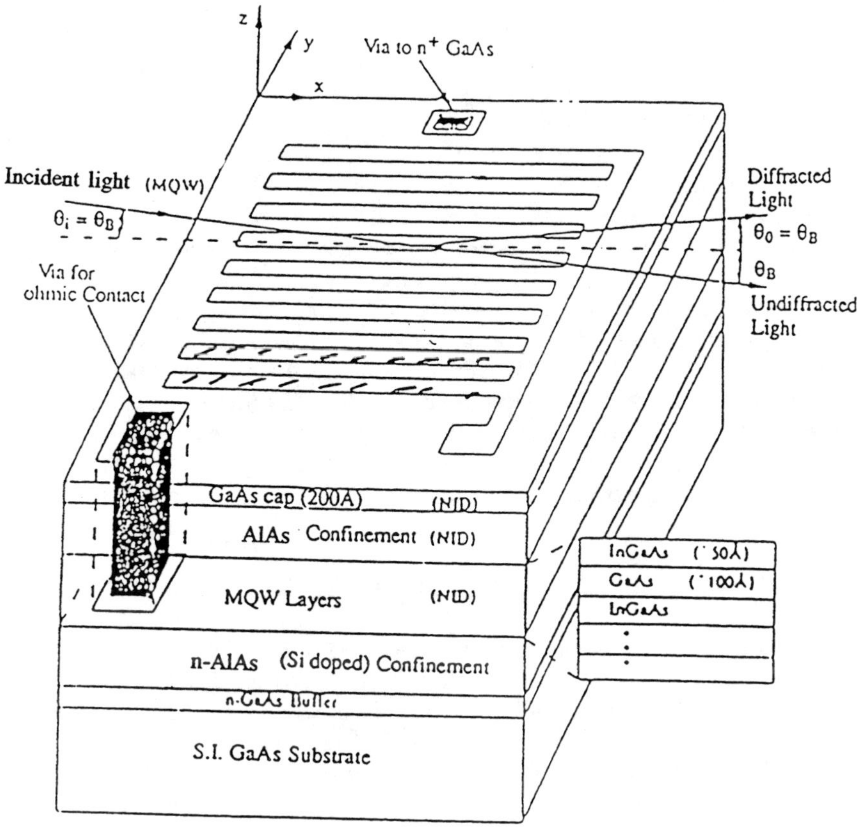

Fig. 18.
Schematic of an electro-optic Schottky Bragg Modulator. [The ray path, shown on top of the comb electrode structure, depicts diffraction taking place in the MQW layers (in the x-y plane) between the AlAs cladding layers.]

to realize tunable filters and other devices [33b].

V. OPTICAL CHARACTERIZATION USING VARIABLE ANGLE SPECTROSCOPIC ELLIPSOMETRY

In order to simulate the performance of MQW modulators, optical characterization was done using a variable angle spectroscopic ellipsometer (VASE). By treating the MQW as a single effective layer, the effective optical constants of the MQW structure can be obtained [46-49]. Figure 19 presents the refractive index and extinction coefficient measured by VASE for an $In_{0.2}Ga_{0.8}As$-GaAs MQW sample. The sample structure consists of 47.5 periods of 46Å-$In_{0.2}Ga_{0.8}As$/96Å-GaAs MQWs with overall thickness of 6770Å. This technique can also be used to obtained the field-induced changes of optical constants in the InGaAs-GaAs MQWs. Figure 20 (a) and (b) depict the field-induced changes of the optical constants, Δn and $\Delta\alpha$, respectively. Figure 21 shows the comparison of the transmission spectra for the MQW Fabry-Perot modulator between experimental and simulated results using the VASE data.

VI. HIGH FREQUENCY OPTICAL MODULATORS

Mach-Zehnder modulators have been fabricated using MQW waveguides [11,12]. Physical effects such as band filling (blue-shift) and quantum confined Stark Effect (red shift) have successfully been employed by Zucker et al. The modulators exhibited a 3dB response frequency of about 5 GHz. Velocity matched electrodes in asymmetric surface normal Fabry-Perot modulators (AFPMs) are expected to improve this response. However, it remains to be seen whether they will exceed the performance of MQW electroabsorption (EA) modulators or Mach-Zehnder LiNbO3 devices.

Electroabsorptive MQW modulators have been reported both in surface-normal as well as waveguide configurations. Barron et al. [50] obtained a 3dB cut off frequency $f_{3dB}=37$ GHz in surface normal AFPM structures. Ramdane et al. [51] have reported 36 GHz operation in waveguide type EA modulators. Several groups have demonstrated integration of EA modulators with lasers. We expect improved dynamic range, insertion loss, and f_{3dB} in electrorefractive (ER) modulators with either velocity matched or surface normal structures.

VII. CONCLUSIONS

HACT SLM structures grown on multiple quantum wells have shown charge transfer in the quantum well transport channel. Preliminary measurement at 119.3 MHz on a HACT device yielded a charge transfer efficiency CTE of 0.9, which is equivalent to a transport distance of 20 pixels. This is significantly lower than obtained for conventional HACT devices [9]. The charge transfer efficiency (CTE) needs to be improved by enhancing the uniformity of HACT (MODFET type) layers grown on the SLM's multiple quantum well structure. In addition, increasing the magnitude of the

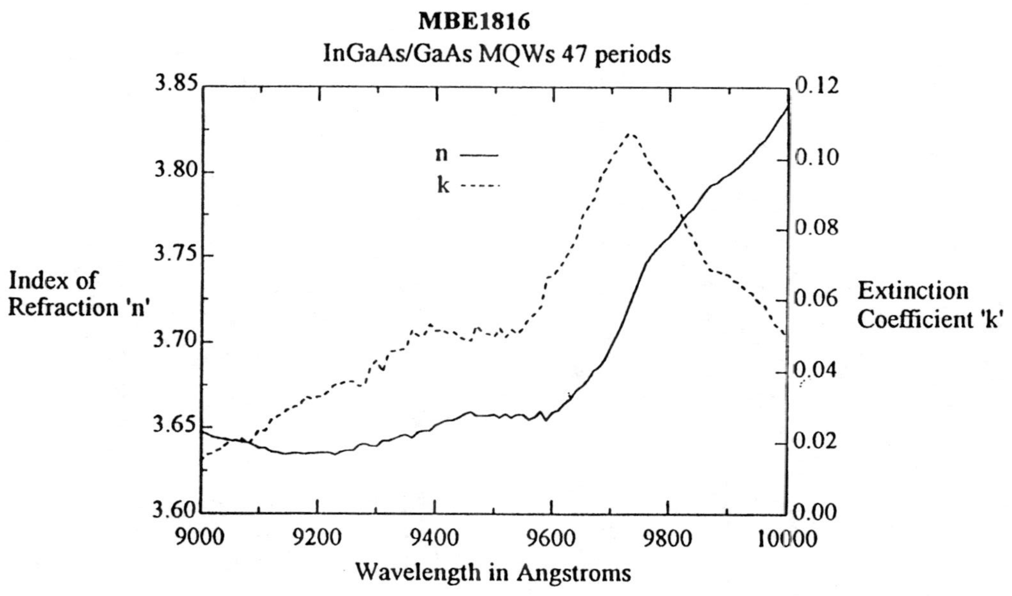

Fig. 19 The refractive index and extinction coefficient of an InGaAs/GaAs MQWs for HACT modulators.

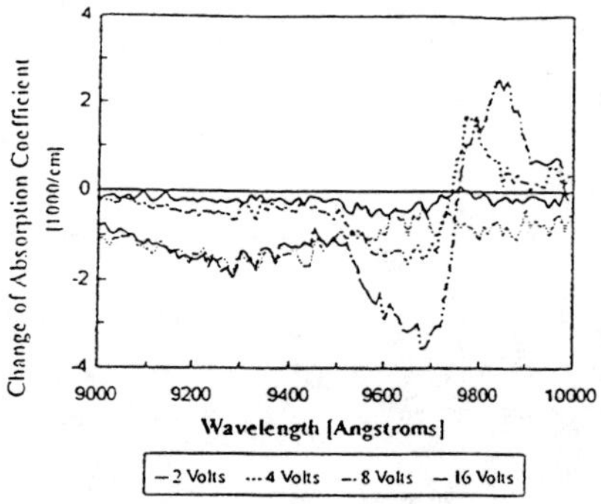

(a)

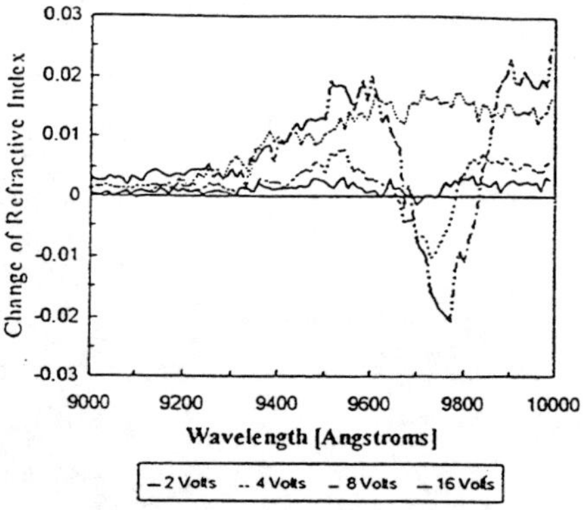

(b)

Fig. 20. The field-induced changes of the optical constants (Δα and Δn) of the InGaAs-GaAs MQW measured by VASE. (a) Δα versus λ; and (b) Δn versus λ.

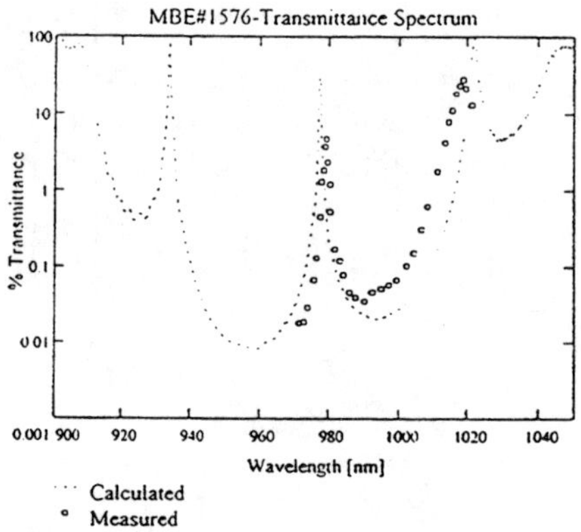

Fig. 21. The comparison of the transmission spectra for the tunable Fabry-Perot optical modulator between the experimental and theoretical results.

injected signal is expected to enhance the line charge density per packet, which should induce sufficient electric field for the modulation of the incident beam. The simulations of HACT/SLM designed at 993-987 nm have yielded a contrast ratio larger than 100. HACT/SLMs greatly reduce the number of interconnects required for pixel addressing (~3x1000 for 1000x1000 pixel array). This, along with their high contrast ratio, make them ideal candidates for high speed signal processing applications.

Based on our successful fabrication of high contrast (1200:1) tunable electrorefractive Fabry-Perot MQW modulators [16], we expect the realization of electrically/optically addressed SLMs using these devices.

Preliminary measurements using a 878 MHz SAW transducer (to induce electric fields in MQWs) showed a slight shift in the excitonic absorption peak (SAW-induced Stark effect) when light was incident in the proximity of the transducer. Increasing SAW power will allow practical modulator structures, in which the incident beam is adequately separated from the transducer, to be realized.

The feasibility of implementing birefringent modulators and SLM structures, using MQW wavegude structures and strain-induced birefringence in MQWs (under normal incidence), has been demonstrated. BSLMs integrated with polarizers/analyzers will result in higher contrast ratios.

The enhanced electrooptic/acoustooptic effect can be utilized to induce Bragg gratings in MQW waveguides, which can be used to realize filters and/or modulators [33b]. In the case of the electrooptic DBR gratings, the tuning of a filter can be achieved by adjusting the magnitude of induced index changes (Δn) in the DBRs. the manipulation of the MQW cavity index (e.g. by electrorefraction) also provides an additional control parameter for modulator/filter tuning. The fabrication of these devices has been attempted but detailed measurements/experimentation are yet to be completed.

During the course of this work, we investigated the direct measurement of optical parameters in MQWs using Variable Angle Spectroscopic Ellipsometry (VASE), a non-destructive technique. Experimental results are in agreement with the theoretical simulations, and have led to design improvements of tunable modulators [52].

ACKNOWLEDGEMENTS

The authors gratefully acknowledge the ONR contract (Dr. William Miceli) and the grant from the Connecticut Department of Economic Development which support this work. This review articles has been made possible by contributions over the years by Drs. K. Bhattacharjee, T. Grudkowski, D. Cullen, R. Sacks, W. Huang, and R. LaComb. In addition, the assistance of Ms. H. Wang, W. Zappone, E. Heller, S. Srinivasan, J. Robinson, and T. Nelson has been valuable in various phases of the manuscript preparation.

REFERENCES

1. Spatial Light Modulators and Applications, Co-sponsored by IEEE/Lasers and Electro-optics Society (LEOS) and Optical Society of America (OSA), September 10-12, 1990, Lake Tahoe, Nevada, Technical Digest Series, vol. 14.
2. Spatial Light Modulators and Applications III, sponsored by SPIE, August 7-8, 1989, San Diego, CA, Proceeding Digest Vol. 1150, U. Efron (ed.).
3. Optical Computing '90, Kobe, Japan, April 1990.
4. J.A. Neff, R.A. Athale, and S.H. Lee, "Two-dimensional spatial light modulators: a tutorial", Proc. IEEE., 78, pp. 826-855, May 1990.
5. P.B. Berra, A. Ghafoor, M. Guzzani, S.J. Marcinkowski, and P.A. Mitkas, "Optics and supercomputing", Proc. IEEE, 77, pp. 1797-1815, December 1989.
6. U. Efron, *Spatial Light Modulator Technology*, Dekker, New York, 1995.
7. T.H. Wood, C.A. Burrus, D.A.B. Miller, D.S. Chemla, T.C. Damen, A.C. Gossard, and W. Wiegmann, "High-speed optical modulation with GaAs/GaAlAs quantum wells in a p-i-n diode structure", Appl. Phys. Lett., vol. 44, pp. 16-18, Jan. 1984.
8. U. Efron, T.Y. Hsu, J.N. Schulman, W.Y. Wu, I. Rouse, and I.J. D'Haenens, "Multiple quantum well-based spatial light modulators", SPIE, Vol. 825, Spatial Light Modulators and Applications II, pp. 8-18, (Proc. of 17-18 August 1987 Conference, San Diego), 1988; T.Y. Hsu and U. Efron, SPIE, vol. 1150, pp. 61-85, (Proc. of 7-8 August 1989 Conference, San Diego, CA).
9. J. Singh, S. Hong, P. Bhattacharya, R. Sahai, C. Lastufka, and H. Sobel, "System requirements and feasibility studies for optical modulators based on GaAs/AlGaAs multi quantum well structures for optical processing", J. Lightwave Tech., 6, pp. 818-831, June 1988.

10(a). A.L. Lentine, H.S. Hinton, D.A.B. Miller, J.E. Henry, J.E. Cunningham and L.M.F. Chirovsky, "Symmetric self-electro-optic effect device: Optical set-reset latch," Appl. Phys. Lett., Vol 52, pp. 1419-1421, Apr. 25, 1988.

10(b). N.C. Craft and M.E. Prise, "Processor Does Light Logic", Laser Focus World, pp 191-200, May 1980; also see: H. Arsenault and Y. Shey, "An Introduction to Optical Computers", SPIE, vol. 778, 1992.

11. J.E.Zucker, K.L. Jones, T.Y. Chang, N. Sauer, B. Tell, K. Brown-Goebeder, M. Wegner, and D.S. Chemla, "Low-loss InGaAsP-InP Multiple Quantum Well Waveguide Interferometer, Electron Lett., 26, pp. 2030, 1990.

12 J.E.Zucker, "High speed Quantum Well Interferometric Modulators for InP based Photonic Integrated Circuits," Microwave and Optical Technology Letters, 6, pp. 6-14, Jan. 1993.

13. D.S. Chemla, T.C. Damen, D.A.B. Miller, A.C. Gossard, and W. Wiegmann, "Electroabsorption by Stark effect on room-temperature excitons in GaAs-GaAlAs multiple quantum well structures", Appl. Phys. Lett., vol. 42, pp. 864-866, May 15, 1983.
14. D.A.B. Miller, D.S. Chemla, T.C. Damen, A.C. Gossard, W. Wiegmann, T.H. Wood, and C.A. Burrus, "Electrical field dependence of optical absorption near the band gap of quantum-well structures", Phys. Rev. B., 32, pp. 1043-1060, July 15,

1985.
15. J.S. Weiner, D.A.B. Miller, D.S. Chemla, T.C. Damen, C.A. Burrus, T.H. Woods, A.C. Gossard, and Wiegmann, "Strong Polarization-sensitive Electroabsorption in GaAs-AlGaAs Quantum Well Waveguides", Appl. Phys. Lett. 47, pp. 1148-1150, 1985; D.A.B. Miller, J. Weiner, and D.S. Chemla, "Electric-Field Dependence of Linear Optical Properties in Quantum Well Structures: Waveguide Electroabsorption and Sum Rules", IEEE J. Q. Electronics, QE-22, pp.1816-1830, 1986.
16 G.J. Sonek, J. M. Ballantyne, Y.J. Chen, G.M. Carter, S.W. Brown, E.S. Koteles, and J.P. Salereno, IEEE J. Quantum Electronics, 22, pp. 1015-1018, July 1986.
17. F. Jain, S. Cheung, G. Drake, C. Chung, and K. Bhattacharjee, "Birefringence Dependent Multiple Quantum Well Light Valves/Modulators and Logic Devices," Conference Proc. on Emerging Optoelectronic Technologies, pp.332-336, (December 18-20, 1991, Banglore, India), Tata McGraw-Hill, 1992.
18. A. Mysyrowicz, D. Hulon, A. Antonetti, A. Migus, W.T. Masselink, and H. Morkoc, "Dressed excitons in a multiple-quantum well structure: evidence for an optical Stark effect with Femtosecond response time", Phys. Rev. Lett., 56, pp. 2748-2751, 1986.
19. A. Von Lehmen, D.S. Chemla, J.E. Zucker, and J.P. Heritage, Optl. Lett., 11, pp. 609, 1986.
20 Y. Kan, N. Nagai, M. Yamanishi, and I. Suemune, "Field Effects on the Refractive Index and Absorption Coefficient in AlGaAs Quantum Well Structures and their Feasibility for Electrooptic Device Applications," IEEE J. of QE, QE-23, pp. 2167-2179, 1987.
21. F.C. Jain, K.K. Bhattacharjee, and T.W. Grudkowski , "Recent Development in Multiple Quantum Well Acousto-Optic Electro-Optic Modulators", Proc. IEEE 1991 Ultrasonics Symposium, Vol. 1, pp 529-536, December 8-11 (Lake Buena Vista, FL).
22. D. Marcuse, *Light Transmission Optics*, 2nd Ed., Van Nostrand Reinhold, New York, 1982.
23. A. Harwit, R. Fernandez, and W.D. Eades, SPIE Proceedings, 1541, 38(1991).
24. R.H. Yan, R. J. Simes and L.A. Coldren, IEEE J. Quantum Electronics, QE-25, 2272 (1989).
25. R. Jin, G. Khitrova, H.M. Gibbs, C. Lowry, and N. Peyghambarian, Appl. Phys. Lett. 59, 3216 (1991).
26. L. Chen, K. Hu, R.M. Kapre, and A. Madhukar, Appl. Phys. Lett. 60, 422 (1992).
27. A. Larsson and J. Maserjian, Appl. Phys. Lett. 59, 1946 (1991).
28. J.F. Hefferman, M.H. Moloney, J. Hegardy, J.S. Roberts and M. Whitehead, Appl. Phys. Lett. 58, 2877 (1991).
29. R.H. Yan, R.J. Simes, L.A. Coldren and A.C. Gossard, Appl. Phys. Lett., 56, 1626(1990).
30. S. Cheung, F. Jain, R. Sacks, D. Cullen, G. Ball, and T. Grudkowski, "High Contrast Fabry-Perot optical Modulator Using Quantum Confined Stark effect Tuning in InGaAs-GaAs Multiple Quantum Well Cavity", Appl. Phys. Lett. 63,

pp.296-298, July 1993.

31. W. D. Goodhue, E.E. Burke, K.B. Nichols, G.M. Metze, and G.D. Johnson, "Quantum Well Charge-coupled Devices for Charge-coupled Device Addressed Multiple Quantum Well Spatial Light Modulators," J. Vac. Sci. Tech., B4, pp. 769-772, 1986.
32. K. K. Bhattacharjee, and F. C. Jain, "Spatial Light Modulators Using Acoustic Charge Transport Controlled Stark Effect in Multiple Quantum Well Structures," Technical Digest, 14, pp. 64-67, (OSA Topical Conference on Spatial Light Modulators, September 10-12, 1990).
33. (a) F. C. Jain and K. K. Bhattacharjee, "Two-Dimensional Spatial Light Modulators Utilizing HACT-Controlled Quantum Confined Stark Effect," SPIE Proc. 1347, pp. 614-622 (San Diego, CA.; July 8-10, 1990).
    (b) F. Jain, S. Cheung, W. Huang, A. DeMaria, R. Sacks, D. Cullen, T. Grudkowski, C. Valerio, and K. Bhattacharjee, "Electro-Optic and Modulators using Strak Effect in Multiple Quantum Well Structures", 1994 Proc. IEEE Ultrasonics Symposium, pp.879-883 (November 1-4, 1994, Cannes, France).
34. W. T. Tanski, S. W. Merritt, R. N. Sacks, D.E. Cullen, E. J. Branciforte, R.D. Carroll, and T.C. Eschrich, "Heterojunction Acoustic Charge Transport Devices on GaAs," Appl. Phys. Lett., 52, pp. 18-20, January 1988.
35. D. Cullen and M. Miller, "First HACT Analog Memory Devices Demonstrated," Proc. IEEE Ultrasonics Symposium, 1, pp. 529-536, December 1991 (Lake Buena, Vista, FL).
36. F. Jain, S. K. Cheung, R. LaComb, T. Grudkowski, R. Sacks, and D. Cullen, "High Contrast Spatial Light Modulator Using Stark Effect Tuning and Heterostructure Acoustic Charge Transport (HACT) Pixel Addressing", OSA Technical Digest Series (Spatial Light Modulators and Applications), Vol. 6, pp. 157-160, March 15-17, 1993 (Palm Springs, CA).
37. J.P. van der Ziel and A.C. Gossard, "Optical Birefringence of Ultrathin $Al_xGa_{1-x}As$-GaAs Multiple Heterostructures", J. Appl. Phys., 49, pp. 2919-2921, May 1978 (also, van der Ziel et al., Appl. Phys. Lett., 28, p. 735, June 1976).
38. T.D. Bearth, W.P. Bleha, and S.Y. Wong, "AC Liquid-Crystal Light Valve", Feb, 1973; J. Grinberg, A. Jacobson, W.P. Bleha, L. Fraas, D.Boswell, and G. Meyer, Opt. Engrg, 14, p.217, 1975.
39. F. C. Jain, S. K. Cheung, and G. Drake, "Spatial Light Modulators, using Birefringence in Multiple Quantum Well Structures", SPIE Vol. 1751, p. 304, Miniature and Micro-Optics (1992).
40. F. Jain and S.Cheung, "Multiple Quantum Well Birefringent Spatial Light Modulator", U.S. Patent, 5,485,014, January 1996.
41. H.Shen, M.Wraback, J. Pamulapati, M.Dutta, J. Kosinski, W. Chang, and P. Newman, and Y. Lu, "A Novel High Contrast Multiple Quantum Well Spatial Light Modulator Based on Polarization Rotation", 1993 OSA Conference on Spatial Light Modulators and Applications, pp.161-164, March 15-17, 1993 (Palm Springs, CA).
42. M.T. Wlodarczyk and S.R. Seshadri, "Analysis of grating couplers in planar waveguide for waves at oblique incident", J. O. Opt. Soc. of Am. A, vol. 2, no. 2. pp. 171-185, Feb. 1985.

43. A. Jacques and D.D. Ostrowsky, "The Grating Coupler: Comparison of Theoretical and Experimental Results", Opt. Comm., vol. 13, no. 1, pp. 74-77, Jan. 1975.
44. R. Ulrich, "Efficiency of Optical-grating Couplers", J. of Opt. Am., vol. 63, no. 11. pp. 1419-1431, Nov. 1973.
45. K. K. Bhattacharjee and F.C. Jain, "Multiple Quantum Well Acousto-optic and Electro-optic Modulators", IEEE 1990 Ultrasonics Symposium, p. 621-624, December 4-7, 1990 (Honolulu, HI).
46. K.G. Merkel, P.G. Snyder, I.A. Woollam, S.A. Alterovitz, and A.K. Rai, "Characterization of Multilayer GaAs/AlGaAs Transistor Structures by Variable Angle Spectroscopic Ellipsometry", Jap. J. of Appl. Phys., vol. 28, no. 6, pp. 1118-1123, June, 1989.
47. J.A. Woollam, P.G. Snyder and K.G. Merkel, "Ellipsometric Characterization of Multilayer Transistor Structures, " Materials, Sci. & Engr., B5, pp. 291-294, 1990.
48. P.G. Snyder, M.C. Rost, G.H. Bu-Abbud, and J.A. Woollam, "Variable angle of incidence spectroscopic ellipsometry: Application to GaAs-$Al_xGa_{1-x}As$ multiple heterostructures," J. Appl. Phys., vol. 60, no. 9, pp. 3293-3302, Nov. 1, 1986.
49. R.M.A. Awesome and N.M. Bashara, Ellipsometry and Polarized Light, Elsevier Science Publishing Co., N.Y., 1989.
50. C.C. Barron, B.J. Thibeault, C.J. Mahon, and L.A. Coldren, "High Contrast Modulation of Asymmetric Fabry-Perot Modulators at 20 GHz", OSA Proc. on Photonic Switching, 16, pp. 104-106, 1993 (J.W. Goodman and R.C. Alferness, eds.).
51. A. Ramdane, Electron. Lett., vol. 30, pp. 1980-1981, 1994.
52. S. Cheung, H. Wang, W. Huang, and F. Jain, "Optical Characterization of InGaAs-GaAs MQWs using Variable Angle Spectroscopic Ellipsometry for Designing Tunable Modulators", to be published in J. Appl. Physics.

Optical Memory and Applications

Nonvolatile holographic storage in photorefractive materials

Demetri Psaltis, Xin An, George Barbastathis,
Ali Adibi, and Ernest Chuang

Department of Electrical Engineering
California Institute of Technology, MS 136–93, Pasadena, CA 91125
e-mail: psaltis@sunoptics.caltech.edu

ABSTRACT

We review methods of nonvolatile storage in photorefractive materials, namely thermal and electrical fixing, two-lambda readout, two photon recording, and periodic refreshing. For each method, we briefly review the physical processes that are responsible for counteracting photorefractive erasure, and present recent experimental results.

Keywords: Holographic storage, photorefractive materials, thermal fixing, electrical fixing, two-lambda method, two-photon method, periodic copying.

1 Introduction

Holographic memories[1] are particularly suitable for applications that demand high data capacity and high transfer rate. Recently, the Holographic Random Access Memory[2,3] (HRAM) and the three-dimensional (3-D) Holographic Disk[4,5] architectures were proposed and demonstrated for high capacity digital storage.[6–8] In this paper we are interested in non-volatile holographic storage, i.e. methods to maintain the information inside a holographic memory system despite various erasure mechanisms that occur in a major class of holographic storage materials, photorefractive crystals.

A typical holographic memory system is shown in Fig. 1. The reference is a plane wave. In the signal arm, a Spatial Light Modulator (SLM) imprints the infor-

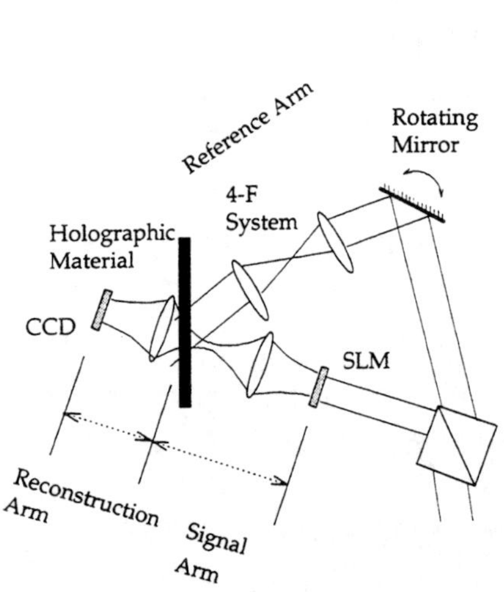

Figure 1: Schematic of an angle-multiplexed holographic memory.

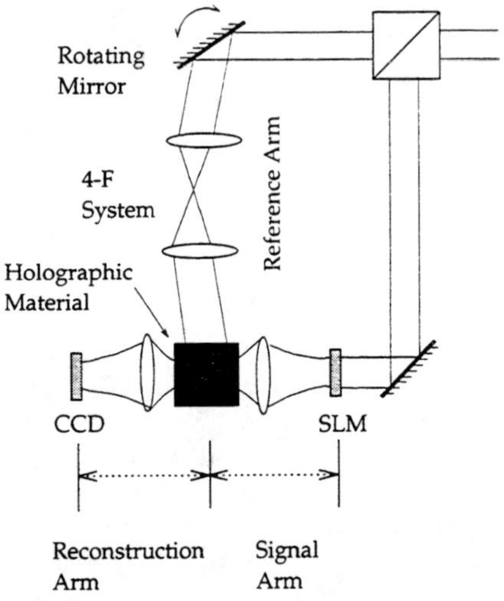

Figure 2: Simplified schematic of the 90° geometry.

mation on the wavefront (usually as amplitude modulation), which is then recorded as a hologram by interfering with the reference. Multiplexing is achieved by changing the angle of incidence of the reference (angle-multiplexing[9,10]) using the mirror and the telescopic 4-F system. The reconstruction is imaged on the CCD camera for detection. This architecture has been used, among others, for holographic disk experiments.[5] A variant, the 90°-geometry, shown in Fig. 2, has been used in high capacity HRAM demonstrations.[7,11–13] Except for angle multiplexing, a number of different techniques have been proposed for holographic storage, such as wavelength,[14,15] phase-code,[16] fractal,[17] peristrophic,[18] and shift[19,20] multiplexing. Usually, it is necessary to combine spatial multiplexing with one or two other techniques (e.g. angle+fractal[12,13] or angle+peristrophic[5]) in order to achieve high capacity. The choice of multiplexing technique is very important in the design of a holographic memory.

The second major decision, equally important, concerns the holographic material. For read-only (or write-once-read-many, WORM) applications, photopolymers like DuPont's HRF-100 are promising.[5,21] On the other hand, rewrittable holographic memories[22] are usually implemented with photorefractive materials.[23–26] Alternatives are photorefractive polymers,[27–29] but they are still in the early development phase in terms of the understanding of the physics of recording, and of practical applications.

Volatility in photorefractive storage is a major concern, in the same fashion as in silicon CMOS memories. Erasure in photorefractive materials is well understood,[30,31] and is closely related to the recording process itself. We will give a brief description here before we start the discussion of non-volatile readout.

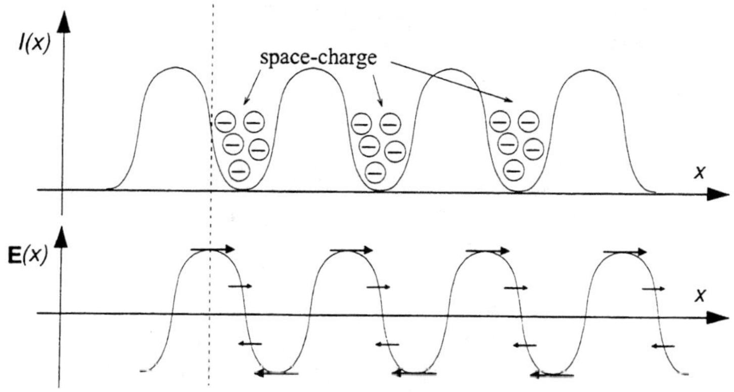

Figure 3: Simple diagram explaining the photorefractive effect in the diffusion-dominated case.

The photorefractive effect is based on photo-induced band-transport of electrons, as shown in Fig. 3. The photorefractive crystal is illuminated by a sinusoidal light pattern $I(x)$. Electrons that happen to lie in energy levels between the valence band and the conduction band (i.e. electrons that belong to deep traps) and were initially located at illuminated regions are excited to the valence band, and move freely to the dark regions, because of diffusion. There they recombine with ions at acceptor sites. (For simplicity we assume that there are no external electric fields or photovoltaic fields, i.e. there is no drift.) When steady-state is reached, excess electrons will have accumulated in the dark regions, generating a space-charge field $\mathbf{E}_{sc}(x)$ out of phase by $\pi/2$ with respect to the illumination. In turn, $\mathbf{E}_{sc}$ locally modulates the refractive index of the crystal via the electro-optic effect, generating a phase hologram. When drift is present, the recording mechanism is very similar, except the phase-shift between $\mathbf{E}_{sc}(x)$ and $I(x)$ is different than $\pi/2$. The quantitative treatment of this process is based on the band-transport equations.[30–33]

The phase hologram would be stable if the space-charge pattern developed during the recording phase could be maintained. Unfortunately this is not true even in the dark, because some of the recombined electrons are thermally excited back to the conduction band, and the space-charge field decays. The dark-storage lifetime depends on the doping species and concentration, and can be as low as a few seconds (typically in $BaTiO_3$) or as high as a few weeks (in lightly doped $LiNbO_3$). When the hologram is illuminated, the thermal excitation of recombined electrons is enhanced by photoexcitation, and the decay is stronger. Therefore, photorefractive holograms are optically erasable. This is convenient for the HRAM application. However, two situations exist when optical erasure is undesirable: hologram readout, and recording of multiple holograms.

To record multiple holograms, one must determine appropriate recording times, i.e. follow an "exposure schedule[25,34,35]", so that, after each exposure, all the stored holograms have equal diffraction efficiency. As a result of the simultaneous recording and erasure of previously recorded holograms, the diffraction efficiency after superimposing M holograms in the same location of a photorefractive medium varies with M as[36]

$$\eta = \left(\frac{M/\#}{M}\right)^2 \tag{1}$$

where the $M/\#$ (M-number) is a system metric, determined by the material constants (refractive index, absorption coefficient, dopant concentrations, electron mobilities), the beam intensities, and the optical system used for the measurement.

Several methods have been devised against erasure in the cases of readout and long-term dark storage in photorefractive materials. Historically, the first attempt was to fix the holograms by generating gratings that are not optically or thermally sensitive. One method is to induce ionic gratings by heating up the crystal,[37] since then the mobility of hydrogen ions increases. Alternatively, ferroelectric gratings are formed when a negative (with respect to the dielectric polarization) electrical pulse is applied,[38] because the ferroelectric domains are locally destabilized.

Instead of fixing the holograms thermally or electrically, another idea is to try to sustain them as long as possible by reading out using a wavelength where the material absorption is low. For example, Fe-doped $LiNbO_3$ has high absorption in the green and blue wavelengths, but is relatively insensitive to red. This two-lambda technique[39] involves trade-offs in the achievable density, because multiple scalar volume holograms cannot be entirely Bragg-matched in a wavelength other than that used for recording,[40,41] except at the expense of allowing Bragg degeneracies.

Holograms can be recorded in Fe-doped $LiNbO_3$ with infrared light by using sensitizing high-intensity green pulses.[42] Commonly, $LiNbO_3$ materials possess shallow traps, i.e. energy levels very close to the conduction band. Electrons from the deep traps are "pushed" to those levels by the sensitization pulses, and can easily be excited to the neighboring conduction band by the low-energy low-absorption infrared light (two-photon process). Since the hologram is recorded and reconstructed in the same long wavelength, the two-lambda readout limitations are eliminated.

Finally, a method that requires neither a complicating fixing process nor multiple wavelengths is periodic copying,[43] as in silicon dynamic random access memories. Holograms that coexist in the holographic memory are periodically read-out, fed-back and re-recorded continuously, each for time long enough to maintain a stable system. This technique requires very robust alignment, since any slight deviation in the direction of the fed-back signal results in the opposite effect of efficiently erasing the stored hologram.

In the following sections we review the methods for non-volatile readout mentioned so far, namely thermal and electrical fixing (section 2), two-lambda (section 3), two-photon (section 4) and periodic refreshing (section 5). We explain the physical mechanisms and limitations underlying each technique, describe the most recent experiments and comment on the potential for practical applications.

2 Fixing of holograms in photorefractive materials

2.1 Thermal fixing

Amodei and Staebler[37,44] found that optically stored holograms in $LiNbO_3$:Fe could be stabilized against readout by heating the crystal during or after the writing process to above $100°C$ and illuminating it at room temperature. They proposed that at high temperatures, where the ionic conductivity dominates the electronic conductivity, the optically induced electronic space charge is compensated by ionic space charge of opposite sign. At room temperature, the ionic conductivity is very small compared to the photoconductivity of the electrons. Therefore, the ionic replica of the holographic gratings remains nearly intact and gives rise to the diffraction and retrieval of the stored information upon illumination. It was first proposed[45] that the mobile ions involved in the thermal fixing process were Si ions, which are common impurities in $LiNbO_3$. Bollmann et. al.[46] suggested that the OH^- ions which enter the crystals during growth were more likely responsible for fixing. The presence of these ions can be detected by their optical absorption at $2.87\mu m$. Vormann et. al.[47] confirmed the importance of OH^- concentration in the fixing process. But they showed that the mobile ions must be positively charged H^+ protons by spatially resolved IR absorption measurement. The migration of protons may happen in various ways, for instance either by jumping or by an exchange of the positions between OH^- and O^{2-} ions. However, the microscopic mechanism is still unclear. Other species such as Li, Nb, O vacancies are also considered likely to be responsible for the fixing.[48] And in his more recent paper, Bollmann[49] assumes interstitial OH^- ion migration and argues that OH^- ions at O^{2-} sites represent positive charges with respect to the unperturbed lattice.

Although there are still some questions remaining in the interpretation of the thermal fixing process, it is generally believed that there exist two processes. Free carriers can be excited by illumination at suitable wavelengths from available donors and trapped at acceptors, leading to charge transport and redistribution. At the same time, protons in the material can move under the effect of an electric field or a concentration gradient if the temperature is high enough. In addition, for more complete treatment, the thermal excitation of electrons should be included.

Thermal fixing can be carried out in one of two ways. The first consists of three steps—recording, fixing, and revealing. During holographic recording, upon illumination by spatially–varying intensity pattern, photo–excited carriers migrate under various transport mechanisms and be trapped at different acceptor sites. This generates a space–charge field which in turn, gives rise to the spatial modulation of the refractive index of the material via the electro–optic effect. During fixing, the crystal is heated to elevated temperatures where the protons become mobile. The field–induced proton migration compensates the spatially-varying electronic charge concentration. In the revealing process, the crystal is cooled down to room temperature. After illumination by uniform light, either coherent or incoherent,

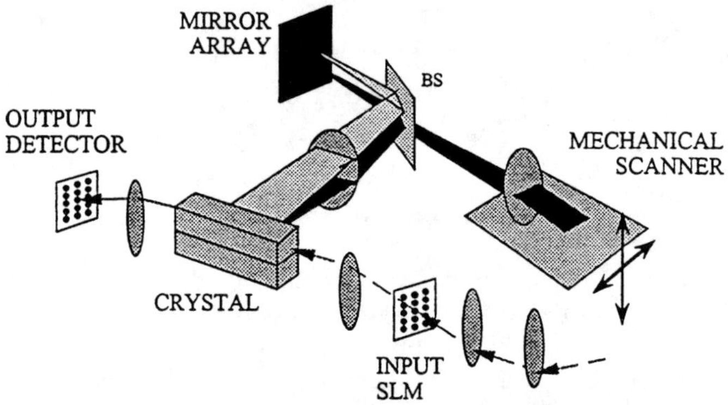

Figure 4: Experimental setup for storage and thermal fixing of 1,000 holograms.

the electronic gratings are partially erased, leaving a net space–charge field which is stable against optical erasure. In the second scheme, holograms are recorded in the crystals at elevated temperatures. Because of the continuous compensation by the mobile protons, the net space–charge field is very small. Since the formed space–charge field inhibits further holographic recording, very high diffraction efficiency can be achieved in this way. However, since the revealing process has to be carried out at lower temperatures, the fixed holograms suffer from thermal contraction.[44]

Fixing efficiency, which is defined as the the ratio of the diffraction efficiencies of the fixed and the original holograms is a very important issue. Experiments using either of the two fixing schemes yield a fixing efficiency ranging from lower than 1% to approximately 50%. Theoretical work has been developed[50,51] to describe the fixing process. They showed that the fixing efficiency depends on the recording geometry, the reduction/oxidation state of the material, the photovoltaic effect, the proton concentration and mobility.

Another important issue is the life time of the fixed holograms. It is determined by the protonic transport under illumination. Amodei et. al. predicted a shelf life as long as 10^5 years at room temperature, while more recent data provided an estimate in the order of years. There is also evidence for electronic compensation in the protonic gratings when the crystal was kept in the dark.[52] It showed a much faster decay of the fixed holograms due to the screening of the protonic grating by electronic compensation.

In an attempt to build a non–volatile large–scale holographic memory system, we thermally fixed 1,000 holograms to determine the fixing efficiency and analyze the fidelity of the fixed holograms. The experiment was done using our 160,000 hologram setup. The system diagram is shown in Fig. 4. The system is capable of storing multiple holograms using angle, fractal, and spatial multiplexing with a segmented mirror array at the center of the 4F system in the reference arm. In our fixing experiment, we used only angle multiplexing in which only the horizontal linear stage moves the focused reference beam along one mirror facet thus scanning the reference beam at the crystal horizontally. In the signal arm, a Spatial Light

Modulator (SLM) consisting of 480×440 pixels was used to imprint the information pattern onto the signal beam. The images to be stored were all binary random–bit patterns. The recording medium—a 0.015% Fe–doped $LiNbO_3$ crystal of dimension 2cm $\times$ 1.5cm $\times$ 4cm was placed at the Fourier–transform plane in the signal arm. The c–axis was cut for the 90° geometry, at 45° to the vertical faces. The absorption constant of the crystal was 0.55 cm^{-1}. To avoid the problems with Fourier–plane holography such as the extremely non–uniform modulation depth, quick buildup of scatter noise etc. a binary random–phase plate was used in the signal arm to diffuse the DC energy. The phase plate consists of 60×50 pixels with 0 and π phase excursions. Because of the finite width of the transition regions between the phase pixels, they appear as dark lines in the image thus causing errors in our data analysis. To avoid errors, the images to be stored also have 60×50 superpixels to match the random–phase plate with each pixel consisting of 8×8 SLM pixels. The reconstructions were monitored by a cooled scientific CCD camera, and grabbed by a personal computer for data analysis. The data analysis had three steps. First a reconstruction was taken as a grey–scale image. The second step is the registration of the reconstructed images to map out the ZERO and ONE regions. In the third step, we calculated the histogram of ZERO and ONE signal and used them as the estimates of the probabilty–density functions of the binary signals. We defined the Signal–to–Noise Ratio (SNR) as,

$$SNR = \frac{\mu_1 - \mu_0}{\sqrt{\sigma_1^2 + \sigma_0^2}} \tag{2}$$

where μ_0 and μ_1 are the means of the ZERO and ONE signals, respectively, and σ_0^2 and σ_1^2 are the corresponding variances.

1,000 holograms were stored with the spacing between two neightboring holograms as 0.003° which was roughly three times wider than the measured angle selectivity. A routine exposure schedule with erasure time constant as 5,200 seconds was used to make the holograms uniform. The average diffraction efficiency was 4×10^{-7}. After the writing, the crystal was placed in an oven preheated to 120°C for 30 minutes. After cooling, the crystal was replaced in the setup and illuminated. We observed that, the uniformity and diffraction efficiency of the holograms were seriously affected by the simple act of removing and replacing the crystal.

We did not observe the evolution of the fixed holograms during the developing phase because there was no way to guarantee the accurate repositioning right after the heating process. Instead, we chose to make measurement after 210 minutes of illumination. During the illumination, the reference beam was scanned through the full angular range continuously in order to avoid fanning. The same registration process was used as in the data analysis to locate a specific hologram for comparison of the diffraction efficiencies and the fidelity before and after fixing. After 210 minutes of illumination, approximately 25% of the original diffraction efficiency was retained by the fixed holograms. In addition to the illumination by the reference beam, we also used UV illumination. After 210 minutes, 27% fixing efficiency was obtained. Fig. 5 shows the reconstructions of the fixed holograms from both experiments. The erasure of the un–fixed holograms with UV illumination is shown at the bottom.

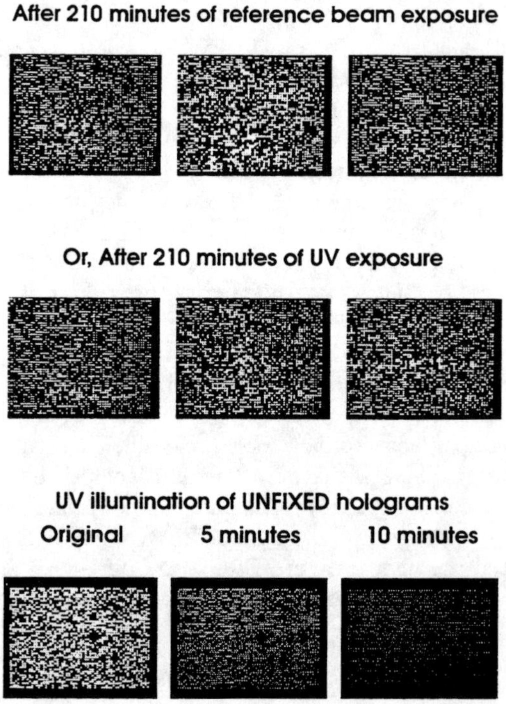

Figure 5: Sample reconstructions from the 1,000 hologram fixing experiment, obtained after illumination with and without fixing.

In order to compare the fidelity of the original and fixed holograms, data analysis was performed on the sample reconstructions. For the same random–bit patterns, there were no visible degradations. The SNR after the fixing dropped slightly. For example, the SNR's for one reconstruction before and after fixing were 5.57 and 5.16. As we pointed out, the reposition accuracy strongly affects the image quality of the reconstructions. Therefore, at least part of the drop could be attributed to the repositioning. Another possibility is the lower diffraction efficiency due to the thermal fixing. This will be a very important factor in the large–scale memory system where the diffraction efficiency is very low at the begining. We measured no errors from both the original and fixed holograms. In order to estimate the bit–error probability, we fit the histogram data with 1st–order χ^2–distribution . For the same recontruction, we achieved 7.03×10^{-7} and 2.86×10^{-6} for the original and the fixed hologram, respectively. The bit–error probability increased by roughly a factor of 2.

2.2 Electrical fixing

Electrical fixing was developed in the early 70's by Micheron and Bismuth as an alternative and hopefully more practical method for non-volatile storage. The method is based on the reversibility of ferroelectric domains in photorefractive crys-

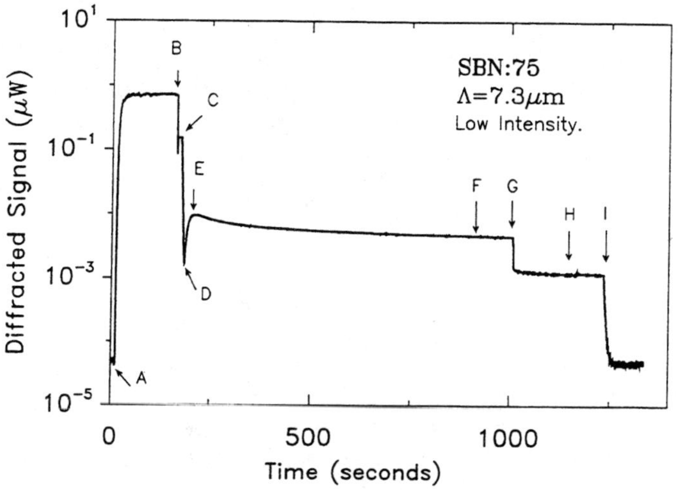

Figure 6: A typical fixing-revealing experiment. A: recording begins; B: writing beams are blocked, and negative voltage pulse is applied; C: optical erasure with non-Bragg-matched beam begins; D: phase reversal in the optical field; E: fixed grating reaches peak value; F: fixed grating reaches steady state; G: positive pulse is applied; H: revealed (compensating) grating; I: optical erasure.

tals. It was first applied in iron-doped barium titanate[38] ($Fe:BaTiO_3$) and then to strontium barium niobate[53] (SBN:x, $Sr_xBa_{1-x}Nb_2O_6$).

Electrical fixing of a single hologram is very simple to implement: during the poling process of the ferroelectric crystal, the optical $\hat{c}$-axis is defined, and is parallel to the spontaneous polarization axis, i.e. the direction in which the ferroelectric domains point. In the simplest version, the holograms are recorded without any applied field; after recording is complete, the hologram is fixed by applying a negative, i.e. antiparallel to the $\hat{c}$-axis electric field $\mathbf{E}_a$. Micheron and Bismuth observed[53] in the SBN specimen used in their experiments the following effects: the field had little effect if $|\mathbf{E}_a| < 700$ V; for 700 V$< |\mathbf{E}_a| <$1000 V the hologram was not optically erasable and the diffraction efficiency η was enhanced; for $|\mathbf{E}_a| >$1000 V, η dropped as $|\mathbf{E}_a|$ increased. By applying a positive electric field to the crystal, the hologram became optically erasable again.

Recently, this fixing process in SBN was revisited and studied extensively,[54–58] with emphasis on the application to practical angularly multiplexed memories. Our current understanding of the fixing process is based on the switching of ferroelectric domains when electric fields are applied.[53,55] We will describe the physics of fixing based on a typical fixing experiment[54] shown in Figure 6.

Consider a photorefractive crystal like SBN, with relatively low coercive field. During holographic recording, the photorefractive space-charge field $\mathbf{E}_{sc}$ builds up and reaches steady state inside the crystal. After recording is complete, the negative pulse $\mathbf{E}_a$ is applied and achieves the following effects: In locations where the total field $|\mathbf{E}_{\mathrm{sc}} + \mathbf{E}_{\mathrm{a}}|$ exceeds the coercive field $\mathbf{E}_{\mathrm{co}}$, the domains are reversed under the electric force applied to them. If $\mathbf{E}_{\mathrm{a}}$ is strong enough, the switching condition is fulfilled only in regions with negative space-charge field. Thus the space-charge grating field triggers the generation of a polarization grating inside the crystal. The additional electric field generated by the reversed electric dipoles screens the photorefractive space charge field resulting in re-distribution of the space-charge (point C in Fig. 6).

Under constant illumination, excess charges that are not trapped by the polarization field are re-distributed. During this time, the electric field experiences a sign reversal (point D in Fig. 6). The total electric field in equilibrium $\mathbf{E}_{\mathrm{tot}}$ (point F in Fig. 6) can be decomposed into two distinct contributions, one due to the polarization grating, and one due to the re-distributed free carriers (electrons). Both components are proportional to the amplitude P_s of the polarization grating.[55]

When a positive pulse is applied to the crystal with the fixed hologram (point G in Fig. 6), the ferroelectric domains are switched back in the direction of the $\hat{\mathbf{c}}$-axis (if the applied field exceeds the coercive field) and thus the ferroelectric domain grating is destroyed. This results in the revealing of the compensating charges and enhanced diffraction efficiency, since screening is cancelled. The revealed hologram is electronic, therefore it is optically erasable (point I in Fig. 6).

The explanation of the fixing effect given so far agrees with the observations of Micheron and Bismuth[53] and Qiao et. al..[54] The fixing efficiency depends strongly on the grating spacing Λ. This is expected, since the efficiency of switching should depend on the relative sizes of the ferroelectric domains and the space-charge field period. The ratio of the efficiency of the compensating grating to the efficiency of the fixed grating increases as $(\Lambda/\Lambda_{\mathrm{D}})^4$, where Λ_{D} is the Debye length. This dependence is predicted theoretically from the charge transport equations, and is also verified experimentally independent of the recording and erasing intensities.[55]

The diffraction efficiency of the fixed hologram in steady state for small grating spacings Λ increases[56] as Λ^p with $p = 1.3 \sim 2.1$ (p decreases with recording/erasing intensity) and peaks at $\Lambda \approx 7\ \mu$m. Adding to this effect the Λ^4 dependence of the relative compensating-to-fixed grating efficiency, we find that the compensating grating increases like $\Lambda^{5.3\sim 6.1}$, and this is more or less verified experimentally.[56] The revealed grating itself has maximum diffraction efficiency at $\Lambda \approx 11.2\ \mu$m.

These observations depend critically on the dynamics of the formation of the ferroelectric domain grating. Recently, it was shown that the fixing/revealing fields also have a strong effect on the fixing efficiency.[57] Applying a small amplitude negative pulse in general has the effect of creating a weak domain grating. In that case, the fixed hologram has appreciable efficiency, but the revealed hologram is weak. On the contrary, a high-amplitude negative pulse creates a strong domain grating, which results in a very weakly diffracting fixed grating (since all or most of the space-charges are screened through compensating the domain grating), and a very strong revealed grating (up to 50% of the diffraction efficiency of the grating before fixing). These results are summarized in the data of Table 1.

	Relative diffraction efficiency	
	fixed hologram	revealed hologram
$\mathbf{E}_a = -2.2\hat{\mathbf{c}}$ kV/cm	$0.74\eta_0$	$0.06\eta_0$
$\mathbf{E}_a = -3.0\hat{\mathbf{c}}$ kV/cm	$< 1 \times 10^{-3}\eta_0$	$0.40\eta_0$

Table 1: Fixing efficiency versus fixing pulse amplitude. η_0 denotes the diffraction efficiency of the hologram before fixing ($\sim$ 30% in most experiments) for grating spacing $\Lambda = 10$ μm. The data show that there is a sharp threshold at approximately 2.6kV/cm; below, fixing is very inefficient, as shown by the low-efficiency revealed hologram. Above, the opposite happens, since a single positive pulse is sufficient to reveal significant portion of the original hologram. After 12 pulses, as much as $0.5\eta_0$ was revealed in the same experiment.

The understanding of the physics of electrical fixing made it possible to fix angularly multiplexed holograms in SBN:75.[57] The multiplexing setup is very similar to that of Fig. 1. The angle θ between the reference and signal beams has to be in the order of a few degrees for efficient fixing to occur (the optimal $\Lambda = 7$ μm corresponds to $\theta = 1.6°$). This requirement leads to the following trade-offs: *(1)* the number of holograms increases as θ is made larger, because the angle selectivity $\Delta\theta$ goes approximately like $\Delta\theta \approx \lambda/n\theta$; *(2)* the signal-to-noise ratio improves as θ increases, because the reconstruction is less affected by scatter noise coming from the reference and crosstalk noise coming from partial reconstruction of overlapping holograms; *(3)* the individual capacity of the holograms increases with θ because the available angular bandwidth becomes larger. Very recently, up to 1,000 holograms were angularly multiplexed and electrically fixed in an SBN specimen, using a similar experimental arrangement.[58]

We described in detail one method of domain fixing in ferroelectric materials. Variants of this idea also exist. For instance, applying a negative field *during* recording has been reported in both SBN[54] and barium titanate.[59] In the former case, the diffraction efficiency of the fixed grating was found to be higher than if the fixing pulse were applied after recording. It is even possible to obtain fixed holograms in SBN:75 by applying the negative pulse *before* recording, i.e. by preliminary partial de-poling of the crystal.[58]

One potential limitation in the long term stability of fixed domain gratings is the electrostatic energy stored in the domain walls, which occurs because of the non-zero polarization gradient $\nabla \cdot \mathbf{P}$ (the spontaneous polarization is along the direction of the grating vector). By recording gratings perpendicular to the $\hat{\mathbf{c}}$-axis while applying an electric field anti-parallel to the $\hat{\mathbf{c}}$-axis, Horowitz et. al. demonstrated a new domain fixing technique based on the screening field.[60] The screening field cooperates with the external field in regions of high light intensity to cause domain reversal and works even though the domains are perpendicular to the grating vector. This method can be applied to fix patterns without the presence of a reference beam, and also to weakly photorefractive crystals. According to the authors, features as small as 1 μm can be fixed. To our knowledge, no multiplexing results using this method have been reported to-date.

3 Two-lambda readout

The motivation for using different wavelengths of light for recording and read-out is simple: If a crystal has an absorption spectrum with a substantial variation as a function of wavelength, then we record at a wavelength λ_1 at which the crystal is highly sensitive and read out at a second wavelength λ_2 at which the crystal is relatively insensitive. In this way we reduce the decay of the gratings caused by the read-out illumination. The main problem with using a different wavelength for read-out than was used for recording is the difficulty of achieving Bragg-match over the entire image. While we can easily Bragg-match a single grating, when a hologram of an image consisting of many plane wave components is recorded, it is generally impossible to match the entire spectrum simultaneously using a single plane wave read-out reference. Recording an image consisting of many plane-wave components can be represented in k-space by a cone of signal vectors that interferes with the reference beam to record a cone of grating vectors, as shown in Figure 7. When we attempt to reconstruct the signal with a reference at λ_2, only the gratings that lie on the circle of intersection between the two k-spheres is exactly Bragg-matched. Hence, only an arc of the signal cone will be reconstructed.

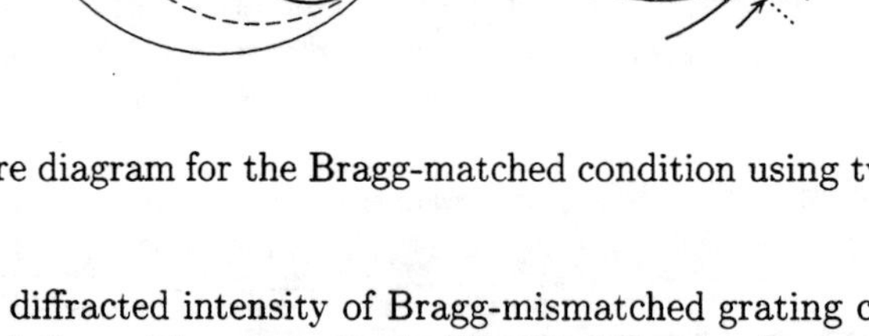

Figure 7: k-sphere diagram for the Bragg-matched condition using two wavelengths

However, the diffracted intensity of Bragg-mismatched grating components follows a *sinc* modulation with respect to the degree of Bragg-mismatch, so there will still be substantial diffraction from slightly mismatched grating components, and we can consider the reconstructed arc to be a strip of finite thickness. The effect on the reconstructed image of limiting the bandwidth of the signal cone depends on whether we record in the Fourier plane or image plane. When we record in the

Fourier plane (see Figure 8a, each plane wave component of the signal beam entering the crystal corresponds to a spatial location (pixel) on the input image. Hence, if we reconstruct only a limited angular bandwidth of the signal cone, we expect to reconstruct a strip of the image, limited in the x-dimension. An example of such a reconstruction is shown in Figure 8c.

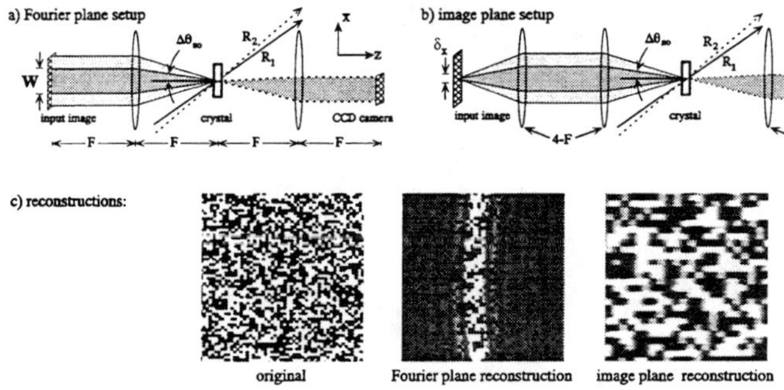

Figure 8: Basic setups for (a) Fourier plane and (b) image plane recording, with sample reconstructions (c) using the two-lambda method.

Recording in the image plane (Figure 8b) is analagous to recording in the Fourier plane, except that in place of the input image we would have its Fourier transform. Therefore, instead of reconstructing a strip of the image, we reconstruct a "strip" or band of the frequency spectrum of the image. If we position the read-out reference to Bragg-match the DC component of the image, the resulting reconstruction will be a low-pass filtered version of the original in the x-dimension, also shown in Figure 8c.

A number of solutions have been proposed for the Bragg-mismatch problem of the two-wavelength scheme. One option is to use Fourier-plane recording and spherical read-out beams[39] in order to Bragg-match a larger range of the signal cone. Another approach is to interleave strips from adjacent holograms,[40] also with Fourier-plane recording. However, we can recover all of the necessary information by recording in the image plane, without the added complexity of the above methods, if we simply adjust the system parameters according to the resolution of the images that we wish to store.

The width of the reconstructed strip of the signal cone can be maximized by reducing the crystal thickness, L, using wavelengths λ_1 and λ_2 that are close together, or reducing the angle between the signal and reference beams. The second and third options may not always be viable solutions because the wavelengths are partially determined by the crystal absorption spectrum, and angular multiplexing needs can limit the range of usable beam angles. However, we can change the crystal thickness. By using a sufficiently thin crystal and the peristrophic multiplexing[18] technique, a large number of holograms can be stored and recalled with a single plane wave reference.

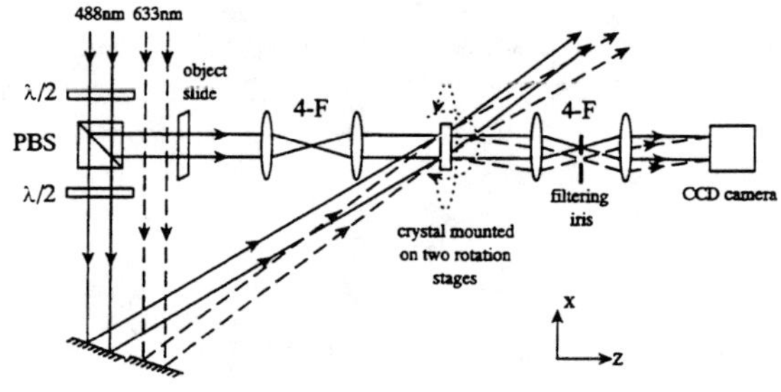

Figure 9: Experimental setup for 1000-hologram experiment using the two-lambda architecture.

3.1 Experiment

Figure 9 shows the experimental setup used for the two-lambda image-plane architecture. It consists of two 4-F systems to image the input object through the crystal and onto the CCD detector, with two offset beam paths for the recording and read-out reference arms. We used λ_1'=488nm polarized out-of-plane for recording and λ_2'=633nm polarized in-plane for read-out, provided by an argon ion and He-Ne laser, respectively. The photorefractive crystal was a $LiNbO_3$:Fe (0.015%) crystal, 4.6mm thick, cut from a boule from CrystalTech. The crystal was mounted on two rotation stages – one to provide angular tilt and the other for peristrophic tilt. We combined peristrophic multiplexing with angle multiplexing to compensate for the loss in angular selectivity from using thinner crystals.

Using 50 angular locations and 20 peristrophic locations, we recorded 1000 holograms using a 150μm-pixel random bit pattern mask as input. Figure 10 shows the resulting comb function recorded using an exposure schedule. The diffraction efficiencies are fairly uniform at about 6×10^{-6}. Of the 1000 holograms, we visually inspected about 100 for uniformity and randomly chose five for SNR and probability-of-error analysis. Figure 11 shows the original input image as seen through the crystal, a sample reconstruction at 488nm, and two reconstructions at 633nm. For hologram #263, we also show histograms for the 488nm and 633nm reconstructions below their respective images. For the reconstructions that were read out with the original reference at 488nm, the SNR ranged from approximately 3.0 to 4.0, with corresponding probabilities of error from 10^{-3} to 10^{-4}. The results for the reconstructions at 633nm were better than those at 488nm, despite the low-pass filtering effect of the two-lambda image-plane readout. For the 633nm reconstructions, the SNR varied from 3.5 to 5.5, with probabilities of error from 10^{-5} to 10^{-7}.

There are a few reasons for the apparent improvement in reconstruction quality using 633nm as opposed to 488nm. One reason is the method used for averaging pixel values in the reconstructions; the program that we used to calculate SNR averages CCD pixels only within a margin of each image pixel – edge values are discarded. Hence, any blurring effect at the edges of pixels in the x-dimension

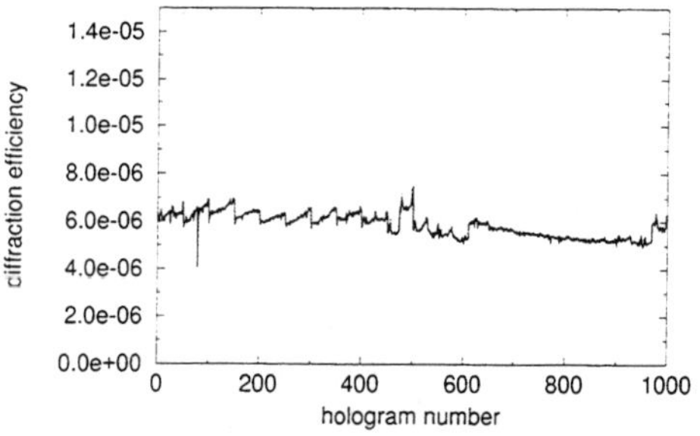

Figure 10: Experimental comb function for 1000 holograms.

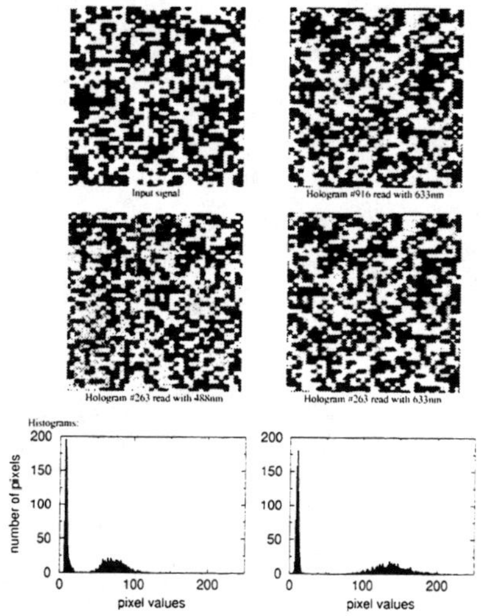

Figure 11: Sample reconstructions and histograms from the 1000-hologram experiments.

becomes less of a factor for the SNR and error calculations. Another reason is that the 488nm reference was out-of-plane (ordinary) polarized, whereas the 633nm reference was in-plane (extraordinary) polarized, resulting in a higher diffraction efficiency for the 633nm readout. Yet another factor is that the crystal has a larger effective scattering cross-section for blue light than for red.

To examine how well the two-lambda method reduced the read-out decay rate, we recorded two holograms with the same exposure at 488nm, and erased one with a non-Bragg-matched beam at 488nm and the other with an equal intensity (21.4mW/cm^2) beam at 633nm, periodically monitoring the grating strength by scanning for the reconstruction peak with a 633nm read-out beam. The decay rate was also measured with no erasure beam in order to determine the decay contribution from dark conductivity as well as from the monitoring beam. The decay characteristics are plotted in Figure 12b. The measured erasure time constant (which includes the effects of dark erasure) for the 488nm erasure was $\tau'_{e,488} =$ 3.21hrs, while that for the 633nm erasure was $\tau'_{e,633} = 35.6$hrs, giving a reduction in the read-out decay rate by a factor of 11.1. However, after factoring out the dark decay ($\tau_{e,dark} = 194$hrs), we find the actual contribution caused by the illumination to be $\tau_{e,488} = 3.27$hrs and $\tau_{e,633} = 43.6$hrs, corresponding to a ratio of 13.3.

The absorption spectrum for this crystal is shown in Figure 12a. For the two wavelengths used in our experiment, the absorption coefficients were $\alpha_{488} =$ 0.55cm^{-1} and $\alpha_{633} = 0.26$cm^{-1}. The ratio of these coefficients is 2.1, so the ratio of 13.3 in erasure time constants was larger than expected. At this point we haven't developed a theoretical model to predict relative erasure times from the absorption spectrum. However, it should not be entirely surprising that the ratio of absorption coefficients is different from that of the erasure time constants, since not every photon that is absorbed will contribute toward the erasure of the hologram. For example, some energy will be absorbed by the lattice, or an electron may be excited

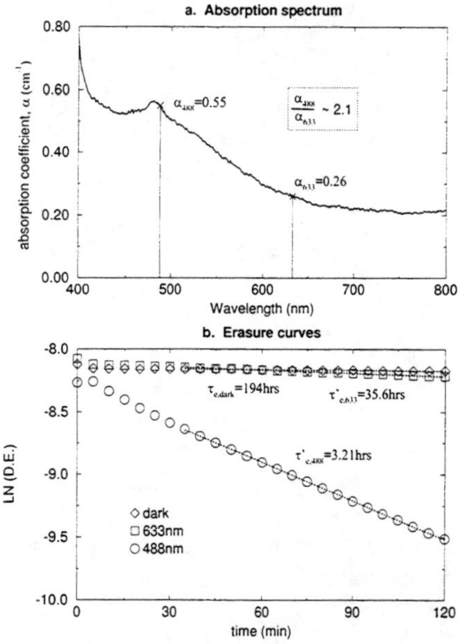

Figure 12: a) Absorption spectrum and b) erasure curves for $LiNbO_3$:Fe sample.

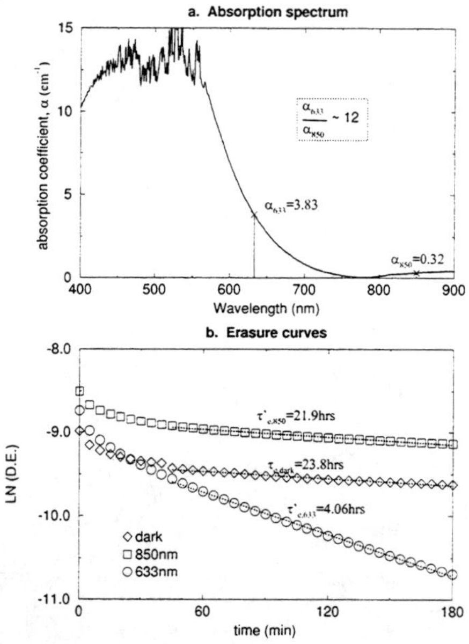

Figure 13: a) Absorption spectrum and b) erasure curves for $LiNbO_3$:Fe:Ce sample.

into the conduction band but immediately be trapped again. Still, we expect that larger ratios in absorption coefficients will be reflected in larger ratios in erasure time constants.

One important note is that the choice of crystal and wavelengths used in this experiment was not optimized for the two-lambda architecture, but was based simply on what was available in the lab. The crystal was one that we have had good results with in the past, and 488nm and 633nm were laser wavelengths that were readily at hand with good power output. However, to maximize the benefits of the two-lambda scheme, we can certainly be more selective in our choice of crystal and system parameters.

For instance, Figure 13a shows the absorption spectrum for a doubly-doped $LiNbO_3$:Fe:Ce (0.05% Fe, 0.03% Ce) crystal from Deltronic that we have begun to use in experiments. This crystal exhibits a much wider range of absorption coefficients over the same range of wavelengths than the crystal that we used for the 1000 hologram experiment. We tested this crystal using $\lambda_1' = 633$nm for recording and $\lambda_2' = 850$nm for read-out, corresponding to absorption coefficients of $\alpha_{633} = 3.83\text{cm}^{-1}$ and $\alpha_{850} = 0.32\text{cm}^{-1}$, respectively. This gives a ratio of 12.0 of absorption coefficients, nearly six times that for the previous crystal and wavelengths.

Figure 13b shows the erasure characteristics that we measured for the two wavelengths, using equal intensities of 15.7mW/cm^2, as well as for the dark erasure. From the data we obtain the erasure time constants (including dark effects), $\tau'_{e,633} = 4.06$hrs and $\tau'_{e,850} = 21.9$hrs. Again, factoring out the dark erasure ($\tau_{e,dark} = 23.8$hrs) to get the true decay contributions due to the illumination, we get $\tau_{e,633} = 4.79$hrs and $\tau_{e,850} = 273.4$hrs, giving a reduction in the erasure rate by a factor of 57.1.

These results are encouraging because they show that we can indeed greatly reduce the read-out decay by using a second wavelength for read-out at which the crystal is much less absorptive. However, it also illustrates a fundamental limitation of the two-lambda scheme, and that is the problem of dark erasure. While it may be true that the read-out wavelength slows the erasure *due to the read-out illumination* by a factor of nearly 60, unless something can be done to reduce the dark decay we will not observe this improvement in practice. Instead, we will remain limited by the time constants with dark erasure, $\tau'_{e,633}$ and $\tau'_{e,850}$, which only offer a reduction by a factor of 5.5 in erasure time. In this case, the first crystal would be a better option due to its lower dark conductivity.

4 Two-photon recording

In 1974, von der Linde et al. reported the recording of holograms with high sensitivities that can be read without erasure.[42] They recorded holograms with considerable diffraction efficiencies in $LiNbO_3$:Cu, $LiNbO_3$:Fe, and nominally pure $LiNbO_3$. Their experimental results were compatible with the assumption of two-photon absorption. Later on they reported three orders of magnitude improvement in recording sensitivity by using potassium tantalate niobate (KTN) instead of $LiNbO_3$,[61] but the storage time was only about 10 hours due to the dark conductivity of KTN. The main problem in these experiments was the high intensity required for the two-photon processes[61] ($I \sim 200MW/cm^2$ for $LiNbO_3$ and $I \sim 2MW/cm^2$ for KTN). One year later, they reported a drastic decrease in the required recording intensity by properly doping the crystal.[62] They used $LiNbO_3$:Cr^{3+} as the recording medium. The main process, called the two-step process, was then the sensitization of the crystal by promoting the electrons from the valance band to a real state within the band gap by one photon and recording the hologram by the second photon interacting with the electrons in that state or vice versa (i. e. recording holograms in a real state within the band gap using the first photon and then, using the second photon to send the electrons from these intermediate states to the conduction band to record the final hologram). The major problem in this case was the short dark storage time which was only about 20 hours.[62]

More recent experiments suggest that the holograms can be recorded efficiently with light intensities smaller than those needed for two-photon processes using the so-called secondary photorefractive centers. Due to the importance of this concept, we will consider it here in more detail.

4.1 Secondary photorefractive centers

Several groups had reported intensity-enhanced photorefractive effects with the enhancement factor of typically[63–65] 1.2 before Chen et al. demonstrated, for the first time, that this nonlinear effect can constitute the dominant mechanism for efficient hologram recording.[66] They recorded holograms in $LiNbO_3$:Fe using nanosecond optical pulses ($I \sim 1MW/cm^2$) with 10 times higher photorefractive sensitivity compared to cw (continuous wave) recording ($I < 1W/cm^2$) with the same exposure. The intensity in their experiment was far less than that required for multi-photon absorption processes to become important ($I \sim 1GW/cm^2$).[42] Therefore, they explained their experimental results by assuming that the nonlinear intensity-dependence was due to the presence of secondary photorefractive centers, first proposed by Kim et al.[65] in 1978. These centers are shallow traps within the band gap that are highly ionized in the dark, and become important in charge transport only at high light intensities. The key assumption, proposed by Chen et al.[66] for the first time, was that the transition of electrons between the secondary photorefractive centers and the deep traps ($Fe^{2+/3+}$ in $LiNbO_3$:Fe) is allowed.

The presence of secondary photorefractive centers is the only plausible argument that can explain the observed nonlinear intensity-dependent phenomena. Different aspects of these phenomena including light-induced absorption,[67–70] photoconductivity,[68,69] light-induced refractive index changes,[71] and holographic recording[66,69,72–74] in different photorefractive materials with different types and amounts of dopants have been used to support this theory. The high intensity needed for these experiments could be obtained by laser pulses,[66,73–75] by focused cw laser beams,[71,76] or in optical waveguides.[77–79] The results could not be explained by simple linear models (for example, see[80–82]) which are based on rather well understood charge transport properties in photorefractive crystals at usual cw laser intensities.[83] However, the two-center model proposed by Jermann and Otten,[70] which is based on the existence of the secondary photorefractive centers[65] (shallow traps) and direct charge transfer between these centers and deep traps[66] (for example, $Fe^{2+/3+}$ in $LiNbO_3$:Fe), can explain most of the results successfully. In the two-center model, a system of five equations (two rate equations for the concentrations of the ionized deep and shallow traps, current density equation, continuity equation, and Poisson's equation) in five unknowns (concentrations of the ionized deep and shallow traps, current density, density of the electrons in the conduction band, and the space charge field) is solved. The resulting space charge field is then used to calculate the change in the index of refraction through the electro-optic effect.

Although the origin of the secondary photorefractive centers in $LiNbO_3$ (like other photorefractive crystals) is not very well understood, the experimental data suggest that these centers are Nb_{Li} polarons[84] or at least are correlated with them. There are several reasons to accept this suggestion.[70] The concentration of Nb_{Li} is assumed to be[85,86] in the order of 10^{26}, similar to that of secondary centers in the two-center model necessary for explaining the experimental results.[70] Furthermore, it has been shown that the above-mentioned polaron states can be involved in light-induced charge transport[84] and bulk photovoltaic effect[87] in a way compatible with that expected from the secondary centers in the two-center model.[70] Also, doping $LiNbO_3$ with magnesium (Mg) or zinc (Zn) reduces the optical damage sensitivity[88] and diminishes the intensity-dependent parts of holographic quantities, which are the results of the secondary centers, in holographic experiments.[89] This can be contributed to the effect of Nb_{Li} polarons as secondary photorefractive centers if we assume that Mg and Zn occupy lithium sites,[88-90] and therefore decrease the concentration of Nb_{Li} polarons (which are caused by niobium ions occupying lithium sites). Similar results can be obtained by increasing the lithium content of the $LiNbO_3$ samples instead of doping them with Mg or Zn.[91,92]

Using the two center model, we consider the holographic recording experiments at high light intensities. Although this kind of recording is usually called two-photon recording, the reader must remember that the main recording mechanism is not a quantum mechanical two-photon process via a virtual state.

4.2 Two-photon holographic recording using high intensity pulses

Single holograms have been recorded efficiently using high intensity laser pulses. Since most of the experiments have been done using $LiNbO_3$ crystals,[66,69,73,74] we concentrate here on this holographic medium. The holograms were recorded using high intensity output pulses of Q-switched Nd :YAG lasers. Recording in doped $LiNbO_3$ crystals using green pulses only ($\lambda = 532nm$)[66,69] and both green and infrared pulses($\lambda = 1064nm$)[73,74] have been reported. In all experiments the results could not be explained by simple linear models assuming only deep traps (for example Fe^{2+}/Fe^{3+} model in $LiNbO_3$:Fe), but they are compatible with the two center model.

Figure 14 shows the basic processes involved in two-photon infrared holographic recording in $LiNbO_3$:Fe. The green pulses (in general, sensitizing pulses) sensitize the crystal by sending the electrons from the Fe^{2+} levels (deep traps) to the X centers (shallow traps). The lifetime of these shallow traps (or secondary photorefractive centers) is in the range of milliseconds.[73,74] Then, two interfering coherent infrared beams (in general, recording beams) record the hologram by sending the electrons from the shallow traps to the conduction band where they move mainly under the influence of drift and photovoltaic[64] forces. The electrons finally recombine with the ionized deep traps (Fe^{3+} levels) resulting in the build up of a space charge field which causes a change in the index of refraction through the electro-optic effect.

The infrared pulses must illuminate the crystal simultaneously or (with a short delay) after the green pulses for efficient hologram recording. The intensities used by Buse et al.[74] are $I_{532} = 110GW/m^2$ for the sensitizing (green) pulses and $I_{1064} = 125GW/m^2$ for each one of the recording (infrared) pulses. The interfering coherent pulses and the spatially homogeneous pulses illuminate the crystal

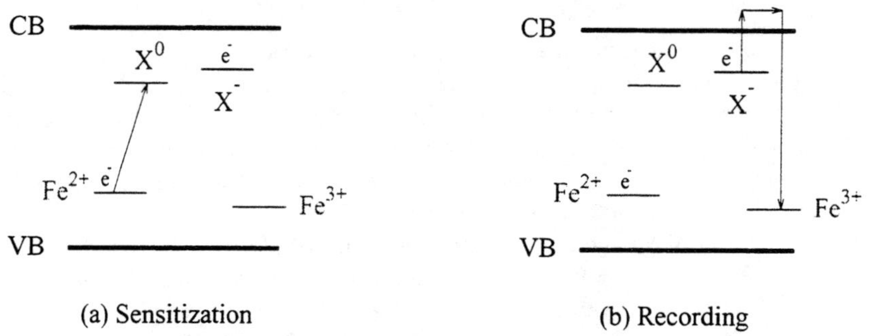

Figure 14: Sensitization and recording processes in two-photon infrared recording in $LiNbO_3$:Fe : (a) Homogeneous green pulses sensitize the crystal by sending the electrons from the deep traps (Fe^{2+} centers) to the shallow traps (X centers), (b) Infrared pulses record the hologram by sending the electrons from the shallow traps to the conduction band.

simultaneously.[74] The crystal will be very insensitive for holographic recording if the sensitizing (green) pulses are absent or they illuminate the crystal after the recording (infrared) pulses. Since the intensities involved are much smaller than those required for two-photon processes[42] and efficient hologram recording is possible when the recording pulses illuminate the crystal after the sensitizing pulses, the explanation suggested by Fig. 14 seems plausible.

After recording the hologram, it can be read by Bragg-matched infrared light without appreciable erasure. Nevertheless, the hologram can be erased by illuminating the sample with spatially homogeneous green pulses.[73,74] The recording and erasure curves (diffraction efficiency vs. exposure time) can by described by[73,74]

$$\begin{aligned} &\text{Recording:} \qquad \Delta n = \Delta n_s[1 - exp(-t/\tau_w)] \\ &\text{Erasure:} \qquad \Delta n = \Delta n_w exp(-t/\tau_e) \end{aligned} \tag{3}$$

where Δn_w and Δn_s denote the values of the refractive index change at the beginning of the erasure and after saturation, respectively, while τ_w and τ_e represent the recording and erasure time constants, respectively. The values of these parameters are usually found by fitting Equations 3 to experimental data. Note also that t in Equations 3 represents the exposure time (the time when the recording pulse is on), not the real continuous time.

Buse et al.[73,74] have investigated the variations of Δn_s, τ_w, and τ_e (among other holographic quantities) with the intensities of the sensitizing and recording pulses (I_{532} and I_{1064}, respectively) and the concentrations of the ionized ($C_{\mathrm{Cu}^{2+}}$ or $C_{\mathrm{Fe}^{3+}}$) and unionized ($C_{\mathrm{Cu}^{+}}$ or $C_{\mathrm{Fe}^{2+}}$) impurities in doped $LiNbO_3$ crystals. Since we need these variations in the discussion of multiplexing holograms, we summarize their results for the case of $LiNbO_3$:Fe here. Similar results could be obtained for other impurities if we replace $C_{\mathrm{Fe}^{3+}}$ and $C_{\mathrm{Fe}^{2+}}$ by the concentrations of ionized and unionized impurities, respectively.

$$\Delta n_s \sim C_{\mathrm{Fe}^{3+}} I_{1064}$$

$$\tau_w{}^{-1} \sim (C_{\mathrm{Fe}^{2+}}/C_{\mathrm{Fe}^{3+}}) I_{532} \quad (4)$$

$$\tau_e{}^{-1} \sim F(C_{\mathrm{Fe}^{2+}}, C_{\mathrm{Fe}^{3+}})(I_{532} + A I_{532}{}^2)$$

where A is some constant. Note that the variation of $\tau_e{}^{-1}$ with $C_{\mathrm{Fe}^{3+}}$ and $C_{\mathrm{Fe}^{2+}}$ has not been mentioned in the literature. Therefore, we included a general function (F) here. We will consider the possible form of this function later.

Finally, it should be mentioned that holograms have been recorded in doped $LiNbO_3$ crystals using green pulses for both sensitizing and recording.[66,69,70] The basic processes are similar to those depicted in Fig. 14.

4.3 Multiplexing holograms using two-photon recording

In this part, we will give a general discussion about the potential of two-photon recording for multiplexing many holograms. For clarity, we assume that the medium is a $LiNbO_3$:Fe crystal, and the sensitization and recording is done by green and infrared pulses, respectively.

For every multiplexing method, we need a recording schedule which, when used with correct erasure time, results in holograms of equal diffraction efficiencies.[93] The diffraction efficiency of each hologram varies inversely with the square of the number of holograms (M) as

$$\eta = \left(\frac{A_0 \tau_e / \tau_w}{M}\right)^2 = \left(\frac{M/\#}{M}\right)^2 \quad (5)$$

where A_0 is the square root of the saturation value of the diffraction efficiency which is approximately linearly related to the saturation value of the refractive index change (Δn_s) by Kogelnik's formula,[94] and τ_e and τ_w are as defined previously. We also used the definition of $M/\#$ (M-number), first given by Burr et al.[95] as

$$M/\# = \frac{A_0 \tau_e}{\tau_w}. \quad (6)$$

It can be used as a system metric[36] for holographic memory systems and can be measured from the recording and erasure of a single hologram.[96] The main problem in applying the above results to two-photon recording is that these results were obtained assuming constant τ_e and τ_w, while in two-photon recording these time constants are intensity-dependent. However, if we assume that the crystal is so thin that we can neglect the absorption of pulses inside it and the intensities of the sensitizing and recording pulses are constant throughout the recording, τ_e and τ_w

will be constant during the recording and Equation 5 also holds for the diffraction efficiency of the holograms in two-photon recording. One should be careful in designing a recording schedule for the two-photon recording, since the recording time of each hologram is discrete (integer multiples of the recording pulse length).

The main contribution to erasure in two-photon hologram multiplexing is the presence of the sensitizing beam. It is needed for recording each hologram, but at the same time causes erasure of the previous holograms through three different mechanisms depicted in Fig. 15(a). The erasure is caused by the excitation of the electrons from the deep traps to the conduction band by one sensitizing photon directly, by two sensitizing photons through the shallow traps, and by one sensitizing photon and one recording photon through the shallow traps (the third mechanism for erasure of the previous holograms is the same as that responsible for recording the present one). Therefore, we can expect that the total rate of erasure ($1/\tau_e$) would be the sum of three rates representing the above-mentioned mechanisms, i. e.

$$\frac{1}{\tau_e} = \frac{1}{\tau_{e(g)}} + \frac{1}{\tau_{e(g,g)}} + \frac{1}{\tau_{e(g,ir)}} \tag{7}$$

where g and ir in the subscripts refer to green and infrared photons, respectively.

A simple comparison of two-photon recording and conventional (single-photon) recording is depicted in Fig. 15. A single mechanism is responsible for both recording and erasure during multiplexing in the conventional recording, while the recording mechanism plus two other ones are responsible for the erasure in the two-photon recording. Therefore, one could expect the contribution of $\frac{\tau_e}{\tau_w}$ to the $M/\#$ (Equation 6) to be smaller for two-photon recording. Fortunately, the experimental results[73,74] suggest that $\frac{1}{\tau_{e(g,ir)}}$ in Equation 7 is larger than the other two terms by at least a factor of two. This might make it possible to compensate for the smaller contribution of $\frac{\tau_e}{\tau_w}$ to the $M/\#$ of two-photon recording by increasing A_0 using efficient mechanisms.

We also did an approximate calculation using the data in the literature[73,74] to get $M/\# \sim 0.2$ for two-photon recording in a $LiNbO_3$:Cu crystal. Also, our experimental measurements approximately resulted in $M/\# \sim .25$ for two-photon recording in a thin $LiNbO_3$:Fe crystal without any optimization, while the best $M/\#$ we could get using the same crystal in conventional cw recording was less than 0.4.

To get an idea about the dependence of the $M/\#$ on the doping density of the crystal and sensitizing and recording intensities, we can use Equations 4 to get

$$M/\# \sim \frac{C_{\mathrm{Fe}^{3+}} I_{1064}}{1 + AI_{532}} \tag{8}$$

where we simply assumed that the dependence of τ_e on $C_{\mathrm{Fe}^{3+}}$ and $C_{\mathrm{Fe}^{2+}}$ is the same as that of τ_w. This makes sense because the two major erasure mechanisms are those involving two photons (two sensitizing or one sensitizing and one recording). These mechanisms are exactly the same as that responsible for recording. It is also interesting to note that using experimental photoconductivity data for $LiNbO_3$:Fe,[82,97] we

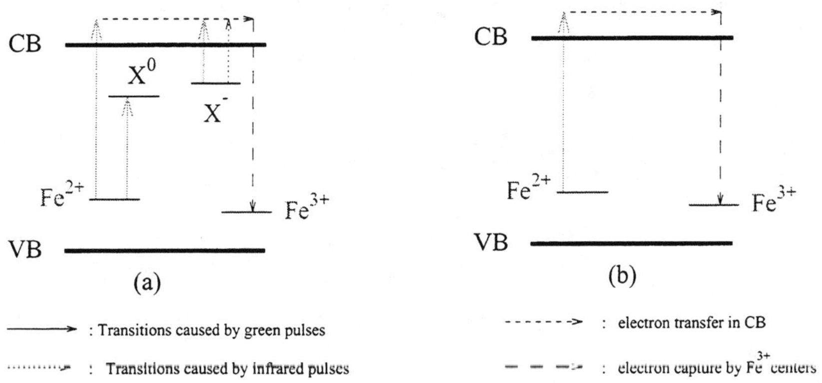

Figure 15: Erasure mechanisms in holgram multiplexing using two-photon and conventional (single-photon) recording : (a) Three mechanisms are involved in two-photon recording, one of which is the same mechanism responsible for recording, (b) Only one mechanism is responsible for both erasure and recording in the conventional case.

can get the same $(C_{Fe^{3+}}/C_{Fe^{2+}})$ dependence for the time constants using conventional, lower (cw) intensities. This also suggests that our assumption is acceptable.

So, using crude experimental data (Equation 8) one expects that to improve $M/\#$, the highest possible recording intensity (I_{1064}), the lowest possible sensitizing intensity (I_{532}), and the highest possible Fe^{3+} concentration $(C_{Fe^{3+}})$ should be used (note that $C_{Fe^{3+}}$ and $C_{Fe^{2+}}$ in a $LiNbO_3$:Fe crystal can be varied using oxidation treatments[98]). One should take note of the limitations of this analysis. The experimental data we used to get Equation 8 represent only a limited range of the values for I_{532}, I_{1064}, $C_{Fe^{3+}}$, and $C_{Fe^{2+}}$. Values out of those ranges will not necessarily give the same results. For example, using very high I_{1064} may result in saturation, or reducing I_{532} increases the time constants and therefore the recording time and, if reduced beyond certain limits, will diminish the strength of two-photon recording due to the lack of enough sensitization. Finally, one would expect that there is some optimum value for $C_{Fe^{3+}}$ (and $C_{Fe^{2+}}$) for a given total concentration of iron impurities, since with all iron in either Fe^{2+} or Fe^{3+} state, the recording sensitivity of the crystal will be highly reduced. The optimization of the $M/\#$ using the correct ratio of $C_{Fe^{3+}}/C_{Fe^{2+}}$ for conventional cw recording has been addressed by Burr et al.[99] Therefore, more sophisticated optimization studies, including the consideration of other impurities or even the use of undoped crystals, are necessary to obtain the best possible $M/\#$ in two-photon recording.

To summarize, two-photon recording seems to be a promising way for recording many holograms that can be read without appreciable erasure. Further studies are necessary for better understanding of the physical mechanisms and modeling them to obtain the optimum conditions for recording in order to improve the $M/\#$.

5 Long-term readout with a positive feedback loop

Copying is another method for achieving the nonvolatile storage of holograms in photorefractive crystals while preserving the system as a read/write memory. The basic idea is to periodically refresh the stored holograms in order to counteract the gradual decay in grating strength due to subsequent illumination for readout or recording. Unlike other methods for non-volatile storage which simply attempt to prevent or minimize the decay of the stored information, the copying technique is capable of actually strengthening the holograms beyond their initial diffraction efficiencies.

A number of copying techniques have been demonstrated experimentally,[43,100–102] such as using a thermoplastic media for temporary storage,[43] copying between two photorefractive crystals,[100] and copying in a single crystal with an optoelectronic feedback loop.[101] A system utilizing an optical feedback architecture was demonstrated by Qiao and Psaltis.[102] This system uses a single photorefractive crystal to store Fourier-transform holograms. The holograms are initially recorded in the crystal with a spatial light modulator (SLM) as input and a liquid-crystal light valve for thresholding. When a hologram becomes weak as a result of repeated illumination, that hologram can be read out, amplified and binarized through the LCLV, and then reintroduced into the crystal via the feedback loop to refresh the hologram while it is being read.

A basic problem that must be addressed in any system with such a feedback loop is that of instability; otherwise, any slight misalignment or perturbation of the setup will be amplified with each successive pass through the loop, resulting in a gradual smearing of the hologram with each copying operation. In this system, the feedback loop is stabilized by using a 128 $\times$ 128 microlenslet array to sample the image on the reading side of the LCLV. The reconstructed image that is fed back to the writing side of the LCLV is defocused by placing the LCLV slightly before the image plane in order to magnify each pixel to a larger spot. As a result of the sampling array, any misalignment between the input and output of the LCLV will be corrected as long as the readout sampling spot lies within the area of the defocused pixel on the writing side of the LCLV. Multiple holograms are angle-multiplexed by changing the angle of the reference beam with a mirror mounted on a rotation stage. These holograms can then be dynamically enhanced by sequentially reading out and refreshing each hologram, and repeating this cycle as necessary.

To demonstrate the system, a 5mm $\times$ 5mm $\times$ 5mm $BaTiO_3$ crystal with its c-axis 45° from the cut face was used. The writing beams of wavelength λ=514.4nm were extraordinary polarized and provided by an argon ion laser. The stabilization effect of the sampling array is demonstrated by recording and then repeatedly copying the letter "A". When sampling was used, the reconstructed image in the steady-state hologram maintained its integrity and remained virtually unchanged throughout the copying process. However, when the sampling was removed, the hologram gradually smeared during copying, eventually becoming unrecognizable.

The capability of the system to dynamically enhance/equalize holograms as well as selectively erase/rewrite them was tested by angularly multiplexing ten holograms and performing a series of experiments that we will briefly review here. The ten holograms were initially recorded with equal exposure times and their diffracted signal powers measured with an oscilloscope. As expected, the diffraction efficiencies of the earlier holograms were lower due to illumination and erasure by the recording beams for later holograms. However, after performing dynamic copying for 27 cycles, the holograms became enhanced and equalized. The fifth hologram was then erased by introducing a 180° phase shift in its recording reference beam with an electrically controlled liquid-crystal phase retarder and writing a hologram that was out of phase with the original. The fifth hologram was almost completely erased, while the other nine were also partially erased from the illumination. By repeating the dynamic copying cycles, the holograms were refreshed to their previous levels except for the fifth hologram, since its diffraction was below the threshold of the LCLV. Finally, a new hologram was rerecorded at the fifth position, and all the holograms were again equalized. The final diffraction efficiency of each hologram was approximately 0.12%.

6 Discussion and conclusions

Fixing techniques for photorefractive holograms are relatively well studied and robust. For instance, 1000 holograms have been fixed successfully using thermal fixing (section 2.1) and electrical fixing (section 2.2). In the case of thermal fixing, the drop in SNR was not significant, despite some loss (in the order of 70%) in diffraction efficiency. Thermal fixing therefore seems promising as one solution for high capacity HRAM applications. Electrical fixing is also promising, but it suffers from the low recording angle requirement, which imposes limitations in both the capacity and quality of the holograms. In addition, both fixing techniques require a relatively cumbersome experimental setup, either a furnace of a high-voltage power supply.

The two-lambda architecture can significantly reduce the decay of holograms due to readout, and does not require any special equipment except for a second laser. However, such an architecture does introduce new complexities and problems to the system, some of which we addressed in section 3 while others still require further investigation. On the system side, we have shown that the Bragg-mismatch problem of the two-lambda scheme can be minimized by properly adjusting system parameters, primarily the thickness of the crystal. We also combine peristrophic with angle multiplexing to counter the poorer angular selectivity of thin crystals. Also, by recording holograms in the image plane, we can retrieve entire data pages at a time with a simple plane-wave readout beam, without the added complexity of using spherical beams or interleaving holograms.

Two main problems remain, however. The first is the problem of dark conductivity. As was shown earlier, dark erasure can severely limit the effectiveness of the second wavelength at reducing the decay rate. Furthermore, the dark erasure prevents a two-lambda architecture from truly maintaining a constant grating strength, since at best it can only eliminate the erasure from the readout process. In contrast, periodic copying (section 5) for instance can restore the strength of holograms regardless of the cause of the holographic decay – whether it is due to readout or dark conductivity. Hence, a two-lambda scheme will probably be most useful when used in conjunction with some other process, such as copying, so as to expand the time frame over which we can refresh the holograms. It may be possible to affect the dark conductivity by changing the crystal temperature, impurity doping level, or oxidation/reduction state.

The second problem is the density limitation. Because of the $\delta_{x,min}$ constraint, the two-lambda system will restrict the storage density of the system except where very thin crystals are being used. Alternatively, if the crystal exhibits sharp changes in absorption behavior for small changes in wavelength, then the density can approach that of a one-wavelength system. Otherwise the two-lambda system will be most useful for storing large numbers of low-density holograms.

Two-photon recording (section 4) seems a promising alternative that sidesteps most of the problems associated with the two-lambda technique, however it does impose the complication of using a high-intensity pulsed laser for material sensitization. From a system viewpoint, usage of the two photon method results in a slight decrease of the $M/\#$, however, several system optimizations and improvements in the understanding of the physics are still needed.

This research was supported by the Air Force Office of Scientific Research and the Air Force Rome Laboratories. Ernest Chuang is funded by a National Science Foundation graduate fellowship. We are grateful to John Hong, Tallis Chang, Jian Ma, Chuan Xie, Geoffrey Burr, Allen Pu, and Jean-Jacques Drolet for their assistance, and to Yayun Liu for technical support.

7 References

[1] P.J. van Heerden, "Theory of optical information storage in solids", *Appl. Opt.*, 2(4):393–400, 1963.

[2] F. H. Mok, G. W. Burr, and D. Psaltis, "Angle and space multiplexed random access memory (HRAM)", *Optical Memory and Neural Networks*, 3(2):119–127, 1994.

[3] G. W. Burr, F. H. Mok, and D. Psaltis, "Angle and space multiplexed storage using the 90° geometry", *Opt. Commun.*, 117(1-2):49–55, 1995.

[4] D. Psaltis, "Parallel optical memories", *Byte*, 17(9):179, 1992.

[5] A. Pu and D. Psaltis, "High density recording in photopolymer–based holographic 3D disks", *Appl. Opt.*, 35(14):2389–2398, 1996.

[6] L. d' Auria, J. P. Huignard, C. Slezak, and E. Spitz, "Experimental holographic read-write memory using 3-D storage", *Appl. Opt.*, 13(4):808–818, 1974.

[7] J. F. Heanue, M. C. Bashaw, and L. Hesselink, "Volume holographic storage and retrieval of digital data", *Science*, 265(5173):749–752, 1994.

[8] D. Psaltis and F. Mok, "Holographic memories", *Sci. Am.*, 273(5):70–76, 1995.

[9] E. N. Leith, A. Kozma, J. Upatnieks, J. Marks, and N. Massey, "Holographic data storage in three-dimensional media", *Appl. Opt.*, 5(8):1303–1311, 1966.

[10] D. L. Staebler, J. J. Amodei, and W. Philips, "Multiple storage of thick holograms in $LiNbO_3$", In *VII International Quantum Electronics Conference*, Montreal, 1972.

[11] F. H. Mok, "Angle-multiplexed storage of 5000 holograms in lithium niobate", *Opt. Lett.*, 18(11):915–917, 1993.

[12] G. W. Burr, X. An, F. H. Mok, and D. Psaltis, "Large–scale rapid–access holographic memory (Paper F2.3)", In *SPIE Optical Data Storage Topical Meeting*, San Diego, 1995.

[13] I. McMichael, W. Christian, D. Pletcher, T. Y. Chang, and J. Hong, "Compact holographic storage demonstrator with rapid access", *Appl. Opt.*, 35(14):2375–2379, 1996.

[14] S. Yin, H. Zhou, F. Zhao, M. Wen, Y. Zang, J. Zhang, and F. T. S. Yu, "Wavelength–multiplexed holographic storage in a sensitive photorefractive crystal using a visible–light tunable diode–laser", *Opt. Commun.*, 101(5-6):317–321, 1993.

[15] G. A. Rakuljic, V. Levya, and A. Yariv, "Optical data storage by using orthogonal wavelength–multiplexed volume holograms", *Opt. Lett.*, 17(20):1471–1473, 1992.

[16] C. Denz, G. Pauliat, and G. Roosen, "Volume hologram multiplexing using a deterministic phase encoding method", *Opt. Commun.*, 85:171–176, 1991.

[17] H. Lee, X.-G. Gu, and D. Psaltis, "Volume holographic interconnections with maximal capacity and minimal cross talk", *J. Appl. Phys.*, 65(6):2191–2194, 1989.

[18] K. Curtis, A. Pu, and D. Psaltis, "Method for holographic storage using peristrophic multiplexing", *Opt. Lett.*, 19(13):993–994, 1994.

[19] D. Psaltis, M. Levene, A. Pu, G. Barbastathis, and K. Curtis, "Holographic storage using shift multiplexing", *Opt. Lett.*, 20(7):782–784, 1995.

[20] G. Barbastathis, M. Levene, and D. Psaltis, "Shift multiplexing with spherical reference waves", *Appl. Opt.*, 35:2403–2417, 1996.

[21] K. Curtis and D. Psaltis, "Characterization of the Du-Pont photopolymer for 3-dimensional holographic storage", *Appl. Opt.*, 33(23):5396–5399, 1994.

[22] D. von der Linde and A. M. Glass, "Photorefractive effects for reversible holographic storage of information", *Appl. Phys.*, 8:85–100, 1975.

[23] F. S. Chen, J. T. LaMacchia, and D. B. Fraser, "Holographic storage in lithium niobate", *Appl. Phys. Lett.*, 15(7):223–225, 1968.

[24] J. B. Thaxter and M. Kestigian, "Unique properties of SBN and their use in a layered optical memory", *Appl. Opt.*, 13(4):913–924, 1974.

[25] K. Bløtekjaer, "Limitations on holographic storage capacity of photochromic and photorefractive media", *Appl. Opt.*, 18(1):57–67, 1979.

[26] G. C. Valley and M. B. Klein, "Optimal properties of photorefractive materials for optical data processing", *Opt. Eng.*, 22(6):704–711, 1983.

[27] S. M. Silence, R. J. Twieg, G. C. Bjorklund, and W. E. Moerner, "Quasinondestructive readout in a photorefractive polymer", *Phys. Rev. Let.*, 73(15):2047–2050, 1994.

[28] K. Meerholz, B. L. Volodin, B. Sandalphon Kippelen, and N. Peyghambarian, "A photorefractive polymer with high optical gain and diffraction efficiency near 100 percent", *Nature*, 371(6497):497–500, 1994.

[29] G. G. Malliaras, V. V. Krasnikov, H. J. Bolink, and G. Hadjiioannou, "Photorefractive polymer composite with net gain and subsecond response at 633 nm", *Appl. Phys. Lett.*, 65(3):262–264, 1994.

[30] N. V. Kukhtarev, V. B. Markov, S. G. Odulov, M. S. Soskin, and V. L. Vinetskii, "Holographic storage in electrooptic crystals, I. Steady state", *Ferroelectrics*, 22:949–960, 1979.

[31] T. J. Hall, R. Jaura, L. M. Connors, and P. D. Foote, "The photorefractive effect – a review", *Progress in Quantum Electronics*, 10(2):77–145, 1985.

[32] N. V. Kukhtarev, V. B. Markov, S. G. Odulov, M. S. Soskin, and V. L. Vinetskii, "Holographic storage in electrooptic crystals, II. Beam coupling - light amplification", *Ferroelectrics*, 22:961–964, 1979.

[33] Pochi Yeh, *Introduction to photorefractive nonlinear optics*, Wiley & Sons, 1993.

[34] D. Psaltis, D. Brady, and K. Wagner, "Adaptive optical networks using photorefractive crystals", *Appl. Opt.*, 27(9):1752–1759, 1988.

[35] A. C. Strasser, E. S. Maniloff, K. M. Johnson, and S. D. D. Goggin, "Procedure for recording multiple-exposure holograms with equal diffraction efficiency in photorefractive media", *Opt. Lett.*, 14(1):6–8, 1989.

[36] F. Mok, G. W. Burr, and D. Psaltis, "A system metric for holographic memory systems", *Opt. Lett.*, 21(12):896–898, 1996.

[37] J. J. Amodei and D. L. Staebler, "Holographic pattern fixing in electro-optic crystals", *Appl. Phys. Lett.*, 18(12):540–542, 1971.

[38] F. Micheron and G. Bismuth, "Electrical control of fixation and erasure of holographic patterns in ferroelectric materials", *Appl. Phys. Lett.*, 20(2):79–81, 1972.

[39] H. C. Külich, "A new approach to read volume holograms at different wavelengths", *Opt. Commun.*, 64(5):407–411, 1987.

[40] D. Psaltis, F. Mok, and H.Y.-S. Li, "Nonvolatile storage in photorefractive crystals", *Opt. Lett.*, 19(3):210–212, 1994.

[41] G. Barbastathis and D. Psaltis, "Shift-multiplexed holographic memory using the two-lambda method", *Opt. Lett.*, 21(6):429–431, 1996.

[42] D. von der Linde, A. M. Glass, and K. F. Rodgers, "Multiphoton photorefractive processes for optical storage in $LiNbO_3$", *Appl. Phys. Lett.*, 25(3):155–157, 1974.

[43] D. Brady, K. Hsu, and D. Psaltis, "Periodically refreshed multiply exposed photorefractive holograms", *Opt. Let.*, 15(14):817–819, 1990.

[44] D. L. Staebler, W. J. Burke, W. Phillips, and J. J. Amodei, "Multiple storage and erasure of fixed holograms in Fe-doped $LiNbO_3$", *Appl. Phys. Lett.*, 26(4):182–184, 1975.

[45] B. F. Williams, W. J. Burke, and D. L. Staebler, *Appl. Phys. Lett.*, 28:224, 1976.

[46] W. Bollmann and H. J. Stohr, "Incorporation and mobility of oh^- ions in $linbo_3$ crystals", *Phys. Sta. Sol. (A)*, 39:477–484, 1977.

[47] H. Vormann, G. Weber, S. Kapphan, and E. Kratzig, "Hydrogen as origin of thermal fixing in $LiNbO_3$:Fe", *Solid State Communications*, 40(5):543–545, 1981.

[48] W. Meyer, P. Wurfel, R. Munser, and G. Muller-Vogt, "Kinetics of fixation of phase holograms in $LiNbO_3$", *Phys. Stat. Sol. (A)*, 53(1):171–180, 1979.

[49] W. Bollmann, "Diffusion of hydrogen (oh^- ions) in $linbo_3$ crystals", *Phys. Sta. Sol. (A)*, 104:855, 1987.

[50] M. Carrascosa and F. Agullo-Lopez, "Theoretical modeling of the fixing and developing of holographic gratings in $LiNbO_3$", *J. Opt. Soc. Am. B*, 7(12):2317–2322, 1990.

[51] A. Yariv, S. Orlov, G. Rakuljic, and V. Leyva, "Holographic fixing, readout, and storage dynamics in photorefractive materials", *Opt. Lett.*, 20(1):1334–1336, 1995.

[52] M. A. Krainak and A. W. Wu, "Electron compensation in a lithium niobate:fe holographic optical filter", In *OSA spring topical meeting, Boston*, 1996.

[53] F. Micheron and G. Bismuth, "Field and time thresholds for the electrical fixation of holograms recorded in $(Sr_{0.75}Ba_{0.25})Nb_2O_6$ crystals", *Appl. Phys. Lett.*, 23(2):71–72, 1973.

[54] Y. Qiao, S. Orlov, D. Psaltis, and R. R. Neurgaonkar, "Electrical fixing of photorefractive holograms in $(Sr_{0.75}Ba_{0.25})Nb_2O_6$", *Opt. Lett.*, 18(12):1004–1006, 1993.

[55] S. Orlov, D. Psaltis, and R. R. Neurgaonkar, "Dynamic electronic compensation of fixed gratings in photorefractive media", *Appl. Phys. Lett.*, 63(18):2466–2468, 1993.

[56] S. Orlov, D. Psaltis, and R. R. Neurgaonkar, "Spatial and temporal characteristics of electrically fixed holograms in photorefractive strontium-barium niobate", *Appl. Phys. Lett.*, 64(7):824–826, 1994.

[57] G. Barbastathis, D. Psaltis, T. Chang, J. Hong, and R. R. Neurgaonkar, "Electrical fixing of angularly multiplexed holograms in SBN:75", In *OSA Annual Meeting '95*, Portland, Oregon, 1996.

[58] J. Ma, T. Chang, J. Hong, R. R. Neurgaonkar, G. Barbastathis, and D. Psaltis, "Electrical fixing of 1,000 holograms in SBN:75", In *Optical Data Storage '96*, Maui, Hawaii, 1996.

[59] R. S. Cudney, J. Fousek, M. Zgonik, P. Günter, M. H. Garrett, and D. Rytz, "Photorefractive and domain gratings in barium titanate", *Appl. Phys. Lett.*, 63(25):3399–3401, 1993.

[60] M. Horowitz, A. Bekker, and B. Fischer, "Image and hologram fixing method with $(Sr_xBa_{1-x})Nb_2O_6$ crystals", *Opt. Lett.*, 18(22):1964–1966, 1993.

[61] D. von der Linde, A. M. Glass, and K. F. Rodgers, "High sensitivity optical recording in KTN by two-photon absorption", *Appl. Phys. Lett.*, 26(1):22–24, 1975.

[62] D. von der Linde, A. M. Glass, and K. F. Rodgers, "Optical storage using refractive index change induced by two-step excitation", *J. Appl. Phys.*, 47(1):217–220, 1976.

[63] F. S. Chen, "Optically induced change of refractive indecies in $LiNbO_3$ and $LiTaO_3$", *J. Appl. Phys.*, 40(8):3389–3396, 1969.

[64] A. M. Glass, D. von der Linde, and T. J. Negran, "High voltage bulk photovoltaic effect and the photorefractive process in $LiNbO_3$", *Appl. Phys. Lett.*, 25(4):233–235, 1974.

[65] D. M. Kim, Jr. J. G. Gallagher, T. A. Rabson, and E. K. Tittel, "Intensity enhanced bulk photovoltaic effects in $LiNbO_3$", *Appl. Phys.*, 17(4):413–416, 1978.

[66] C. T. Chen, D. M. Kim, and D. von der Linde, "Efficient hologram recording in $LiNbO_3$:Fe using optical pulses", *Appl. Phys. Lett.*, 34(5):321–324, 1979.

[67] G. A. Brost, R. A. Motes, and J. R. Rotge, "Intensity-dependent absorption and photorefractive effects in barium titanate", *J. Opt. Soc. Am.*, 5(9):1879–1885, 1978.

[68] L. Holtmann, K. Buse, G. Kuper, A. Groll, H. Hesse, and E. Krätzig, "Photoconductivity and light-induced absorption in $KNbO_3$:Fe", *Appl. Phys. A*, 53(1):81–86, 1991.

[69] F. Jermann and E. Krätzig, "Charge transport processes in $LiNbO_3$:Fe at high intensity laser pulses", *Appl. Phys. A*, 55(1):114–118, 1992.

[70] F. Jermann and J. Otten, "Light-induced charge transport in $LiNbO_3$:Fe at high light intensities", *J. Opt. Soc. Am. B*, 10(11):2085–2092, 1993.

[71] O. Althoff and E. Krätzig, "Strong light-induced refractive index changes in $LiNbO_3$", *Proc. SPIE: Nonlinear Optical Materials III*, 1273:12–19, 1990.

[72] K. Buse, L. Holtmann, and E. Krätzig, "Activation of $BaTiO_3$ for infrared holographic recording", *Opt. Commun.*, 85(2):183–186, 1991.

[73] K. Buse, F.Jermann, and E. Krätzig, "Infrared holographic recording in $LiNbO_3$:Cu", *Appl. Phys. A*, 58(3):191–195, 1994.

[74] K. Buse, F.Jermann, and E. Krätzig, "Infrared holographic recording in $LiNbO_3$:Fe and $LiNbO_3$:Cu", *Opt. Mat.*, 4(1):237–240, 1995.

[75] V. E. Wood, N. F. Hartman, and C. M. Verber, "Two-photon photorefractivity in pure and doped $LiNbO_3$", *Ferroelectrics*, 27(1–4):237–240, 1980.

[76] P. A. Augustof and K. K. Shvarts, "The temperature and light intensity dependence of photorefraction in $LiNbO_3$", *Appl. Phys.*, 21(2):191–194, 1980.

[77] T. Fujiwara, A. Terashima, and H. Mori, "Photorefractive effect in Ti-diffused channel waveguides using $LiNbO_3$ substrate with reduced optical absorption", *Appl. Phys. Lett.*, 25(26):2718–2720, 1989.

[78] R. Göring, A. Rasch, and W. Karthe, "Quantitative investigation of photorefractive effects in $LiNbO_3$ channel waveguides", *Proc. SPIE: Electro-Optic and Magneto-Optic Materials II*, 1274:18–21, 1990.

[79] R. Göring, Z. Yuan-Ling, and S. Steinberg, "Photoconductivity and photovoltaic behavior of $LiNbO_3$ and $LiNbO_3$ waveguides at high optical intensities", *Appl. Phys. A*, 55(1):97–100, 1992.

[80] A. Motes and J. J. Kim, "Beam coupling in $BaTiO_3$ photorefractive crystals", *Opt. Lett.*, 12(3):199–201, 1987.

[81] A. Motes and J. J. Kim, "Intensity-dependent absorption coefficient in photorefractive $BaTiO_3$ crystals", *J. Opt. Soc. Am. B*, 4(9):1379–1381, 1987.

[82] E. Krätzig and R. Sommerfeldt, "Influence of dopants on photorefractive properties of $LiNbO_3$ crystals", *Proc. SPIE: Nonlinear Optical Materials III*, 1273:2–11, 1990.

[83] P. Günter and J. P. Huignard, editors, *Photorefractive materials and their applications I and II, Topics Appl. Phys. 61–62*, Springer, Berlin, Heidelberg, 1988-1989.

[84] O. Schirmer and D. von der Linde, "Two-photon and X-ray induced Nb^{4+} and O^- small polarons in $LiNbO_3$", *Appl. Phys. Lett.*, 33(1):35–38, 1978.

[85] S. Abrahams and P. Marsh, "Defect structure dependence on composition in lithium niobate", *Acta Crystallogr Sect. B*, 42(1):61–68, 1986.

[86] O. Schirmer, O. Thiemann, and M. Wöhlecke, "Defects in $LiNbO_3$ -I. experimental aspects", *J. Phys. Chem. Solids*, 52(1):185–200, 1991.

[87] O. Schirmer, S. Juppe, and J. Koppitz, "ESR, optical, and photovoltaic studies of reduced undoped $LiNbO_3$", *Cryst. Lattice Defects Amorph. Mat.*, 16(1–4):353–357, 1987.

[88] T. Volk and N. Rubinina, "Nonphotorefractive impurities in lithium niobate: magnesium and zinc", *Sov. Phys. Solid state*, 33(4):674–680, 1991.

[89] M. Simon, F. Jermann, and E. Krätzig, "Intrinsic photorefractive centers in $LiNbO_3$:Fe", *Appl. Phys. B*, 61(1):89–93, 1995.

[90] B. C. Grabmaier and F. Otto, "Growth and investigation of MgO-doped $LiNbO_3$", *J. Cryst. Growth*, 79(1–3):682–688, 1986.

[91] G. Malovichko, V. Grachev, E. Kokonyan, O. Schirmer, K. Betzler, B. Gather, F. Jermann, S. Klauer, U. Schlarb, and M. Wöhleke, "Characterization of stoichiometric $LiNbO_3$ grown from melts containing K_2O", *Appl. Phys. A*, 56(2):103–108, 1992.

[92] F. Jermann, M. Simon, and E. Krätzig, "Photorefractive properties of congruent and stoichiometric lithium niobate at high light intensities", *J. Opt. Soc. Am. B*, 12(11):2066–2070, 1995.

[93] D. Psaltis, D. Brady, and K. Wagner, "Adaptive optical networks using photorefractive crystals", *Appl. Opt.*, 27(9):1752–1759, 1988.

[94] H. Kogelnik, "Coupled wave theory for thick hologram gratings", *Bell Syst. Tech. J.*, 48(9):2909–2947, 1969.

[95] G. W. Burr and D. Psaltis, "Effect of the oxidation state of $LiNbO_3$:Fe on diffraction efficiency of multiple holograms", *Opt. Lett.*, 21(12):893–895, 1996.

[96] A. C. Strasser, E. S. Maniloff, K. M. Johnson, and S. D. D. Goggin, "Procedure for recording multiple-exposure holograms with equal diffraction efficiency in photorefractive media", *Opt. Lett.*, 14(1):6–8, 1989.

[97] E. Krätzig and R. Orlowski, "Light induced charge transport in doped $LiNbO_3$ and $LiTaO_3$", *Ferroelectrics*, 27(1–4):241–244, 1990.

[98] W. Phillips and D. L. Staebler, "Control of the Fe^{2+} concentration in iron-doped lithium niobate", *J. Electron. Mat.*, 3(2):601–617, 1974.

[99] G. W. Burr and D. Psaltis, "Optimization of the oxidation state of $LiNbO_3$ for large scale holographic storage", to appear in *Opt. Lett.*.

[100] H. Sasaki, Y. Fainman, J. E. Ford, and S. H. Lee, "Dynamic photorefractive optical memory", *Opt. Let.*, 16(23):1874–1876, 1991.

[101] S. Boj, G. Pauliat, and G. Roosen, "Dynamic holographic memory showing readout, refreshing, and updating capabilities", *Opt. Let.*, 17(6):438–440, 1992.

[102] Y. Qiao and D. Psaltis, "Sampled dynamic holographic memory", *Optics Letters*, 17(19):1376–1378, 1992.

Optical neural networks using electron trapping materials

Alastair D. McAulay and Junqing Wang

Lehigh University, Bethlehem, Pa 18015.

ABSTRACT

The characteristics of electron trapping optical material are reviewed and equations for modeling the behavior developed. Classical supervised and unsupervised learning algorithms, suitable for optical implementation, are discussed. The principles, optical set up, experimental laboratory results and limitations are described for three optical learning demonstration systems: a supervised Hebbian optical learning associative memory, a supervised Perceptron classifier, and an unsupervised Hebbian optical learning novelty detector. Results show the potential and limitations of electron trapping optical material for optical neural network systems that use optical learning.

Keywords: Optical neural networks, rare earth phosphors, electron trapping material, thin film memory, optical learning, Hebbian learning, Perceptron learning.

1. INTRODUCTION

Neural networks provide a learning methodology for moving computer technology to a new powerful dimension in which machines or networked computer systems adapt according to circumstances and usage. The uncharged nature of photons, as compared with electrons, allow high speed, high density and remote addressing with minimal interference. Electron trapping optical material, ETOM, enables high speed, high density, optical addressability, and adjustable storage of a large number of weights and input/outputs for a neural network. In this paper we review and critique several demonstration neural network demonstration systems [12], [14], [17], that use optical learning with electron trapping material.

Several researchers including ourselves have demonstrated optical neural networks using optical supervised learning with ETOM, [9], [14], [17], [18], [28], [30], [31]. Unsupervised optical learning was demonstrated with ETOM in reference [12]. It is possible to read and write an array on a 2-D piece of ETOM simultaneously. Therefore, neural network weights stored in the material can be continuously raised or lowered in size by controlling the relative spatial intensities of the read and write beams [27].

Section 2 discusses the characteristics of electron trapping material and provides equations for modeling their operation. Hebbian and Perceptron learning approaches, suitable for optical learning, are discussed in section 3. Section 4, 5 and 6 describe the principles, optical set up, experimental results, and limitations for optical learning demonstration systems that use ETOM for learning. Optical supervised Hebbian learning is used for a heteroassociative memory in section 4.

Optical supervised Perceptron learning is used for an optical classifier system in section 5. Optical unsupervised Hebbian learning is used in section 6 for a novelty detector.

2. ELECTRON TRAPPING OPTICAL MATERIAL

Electron optical trapping material (ETOM) has proven to be a viable material for thin film optical memory [25] and is suitable for optical neural networks. We describe the operation, characteristics and modeling equations.

2.1 Material description

A thin film of Calcium or Strontium Sulfide is doped with rare earth elements Europium and Samarium [3], [15] and deposited on a glass or saphire substrate. The number of stored electrons at a point in the 2-D ETOM material can represent a weight in a neural network. The weights are selectively increased by illuminating with a blue image for which the intensity at each point represents the amount by which the weight at that point is to be increased. An IR image can be used to selectively reduce the spatially arranged weights. The output luminescence represents element by element multiplication between two matrices, one stored as trapped electrons and the other as the IR image.

Blue light excites electron transitions from the ground state of the Eu^{2+} to an excited state. If there are Sm^{3+} atoms sufficiently close to the Eu^{2+} atoms, a tunneling transition occurs which eventually results in increasing the electrons residing in samarium ground state energy levels. This means that some Eu^{2+} has converted to Eu^{3+} and Sm^{3+} to Sm^{2+} as a consequence of the charge exchange. The transferred electrons add to those already there and can remain at the samarium for very long times because the energy level is deep and transitions to the valence band are apparently forbidden. The number of additional electrons trapped at each point in the 2-D material depends on the intensity of the blue light at that point up to a saturation level.

Reducing the spatially arranged weights is accomplished with a near infrared image for which the intensity at each point represents the amount of weight reduction desired at that point. Near infrared light exposure is capable of exciting the electrons trapped at the samarium to an excited state of the samarium from whence the electrons transfer back to the excited states of the europium. In a very short time, the electrons fall to the lowest energy level of the Eu^{2+} excited states and then make a transition back to the ground state of the Eu^{2+} accompanied by the emission of orange light. With this near IR light exposure, the material returns to the state existing prior to blue light exposure, $Eu^{3+} \rightarrow Eu^{2+}$ and $Sm^{2+} \rightarrow Sm^{3+}$. Given the nature of the processes involved, complete reversibility exists, making it possible to cycle the material indefinitely. In fact, we have been reading and writing from the same material since 1988. Transition times will be fast since either tunneling processes or highly allowed electric dipole transitions are involved. We have demonstrated nanosecond response times. We have also shown that trapping occurs in better than 300 fs and that the energy required does not depend on pulse length [13]. There is a 20 ns recovery time after increasing or decreasing weights before the same spot can be addressed again.

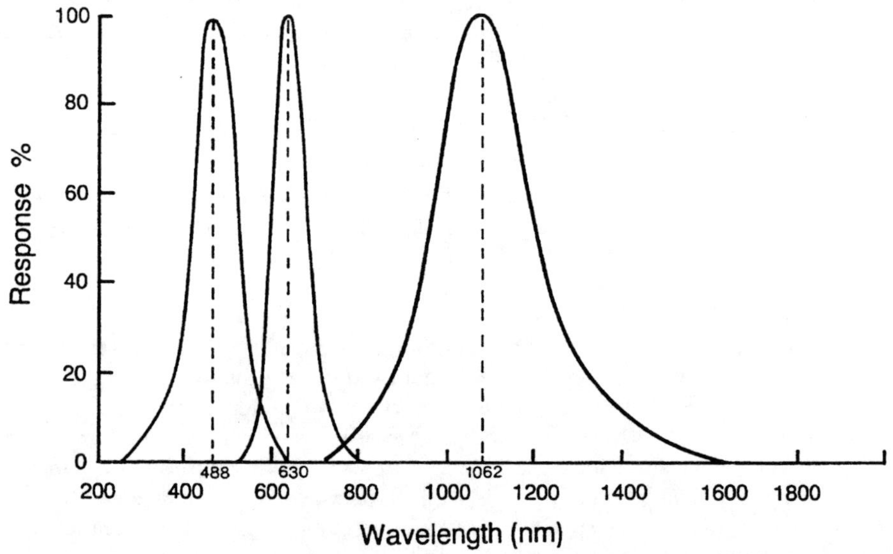

Figure 1: Spectral dependence of the emission, visible, and IR sensitivity of SrS:Eu,Sm

The spectral sensitivities and emission of SrS:Eu,Sm are seen in figure 1 [15]. For some of our samples, the most efficient wavelength for generating the trapped electrons at samarium sites is 488nm and the most efficient wavelength for producing the orange emission is 1062nm. Approximately 4 mJ/cm^2 of 488nm light is required to saturate the electron traps in the electron trapping 8μm thin film. Approximately 10 times this energy at near IR of 1.0μm is needed to deplete the electron traps to one percent of the saturated value. A method of fabrication is described in reference [12].

The resolution achievable depends on the thickness of the film. 1μm resolution has been achieved with 8μm thick film. In a previous experiment we stored 100 by 100 matrices in 1 cm square 12lp/mm material and used it in analog matrix-vector multiplication [17], [19]. The dynamic range can be several order of magnitude, even for the thin film and our earlier results verify the linearity over a significant range [17].

2.2 ETOM equation modeling

When we write a blue image on a 2-D piece of material, the number of electrons trapped at element (x,y) at time t grows exponentially up to a saturation level n_s:

$$n_{x,y}(t) = n_s \left(1 - exp\left[-\frac{\alpha}{n_s} B_{x,y}(t)t\right]\right) \tag{1}$$

where $B_{x,y}(t)$ is the intensity of the blue light image and α is a write sensitivity factor. If we write with constant intensity $B_{x,y}(T_W)$ for time T_W such that the material is not close to saturation, we can use a binomial approximation for the exponent to obtain:

$$n_{x,y}(T_W) = \alpha B_{x,y}(T_W)T_W \tag{2}$$

This shows that below saturation the number of trapped electrons is proportional to the write energy $B_{x,y}(T_W)T_W$.

Information is read off the material in a 2-D array by illuminating with infrared light at 1.06 μm. Orange luminescence is produced as electrons return to the Europium ground state. The intensity of orange luminescence at time t is written:

$$G_{x,y}(t) = \beta n_{x,y}(t)R_{x,y}(t) \tag{3}$$

where $R_{x,y}(t)$ is the infrared image at time t and β is a read sensitivity factor. As we read information out, the number of electrons falls according to the initial number of electrons $n_{x,y}(0)$ and the intensity of the IR read light $R_{x,y}(t)$:

$$n_{x,y}(t) = n_{x,y}(0)exp[-\gamma R_{x,y}(t)t] \tag{4}$$

where γ is the material decay sensitivity factor. An equation for the decaying intensity of orange luminescence is obtained by substituting equation (4) into equation (3):

$$G_{x,y}(t) = \beta n_{x,y}(0)R_{x,y}(t)exp[-\gamma R_{x,y}(t)t] \tag{5}$$

If we use a constant intensity read pulse of length T_R, significantly shorter than that required to read all the energy out, integration of equation (5) provides an orange luminescence energy of:

$$E_{x,y}(T_R) = \beta n_{x,y}(0)R_{x,y}(T_R)T_R \tag{6}$$

This shows that the output energy as a function of x and y is proportional to the product of elements in the stored trapped image $n_{x,y}(0)$ and the energy associated with the IR pulsed image $R_{x,y}(T_R)T_R$.

In the demonstrated optical learning system in section 4, 5 and 6, the trapped electron image $n_{x,y}(0)$ represents the weights $W_{x,y}$ and the IR read matrix $R_{x,y}(T_R)$ represents an input image $A_{x,y}$. Normally we sum all the elements in the orange luminescence to form an output y (not to be confused with the y in the x,y transverse direction):

$$y = \sum_x \sum_y A_{x,y}W_{x,y} \tag{7}$$

Figure 2 shows an optical implementation in which the summation is performed with a convex lens. The resulting operation of multiplication and summation is that of correlation.

2.3 Advantages and disadvantages of ETOM

The material advantages are fast write/read response, high resolution, endless read/write cycles, fast recovery, long term storage, room temperature operation, and optical write and read addressing. Optical access allows light to be focussed from a distance. Uniform material deposition allows the system to determine the

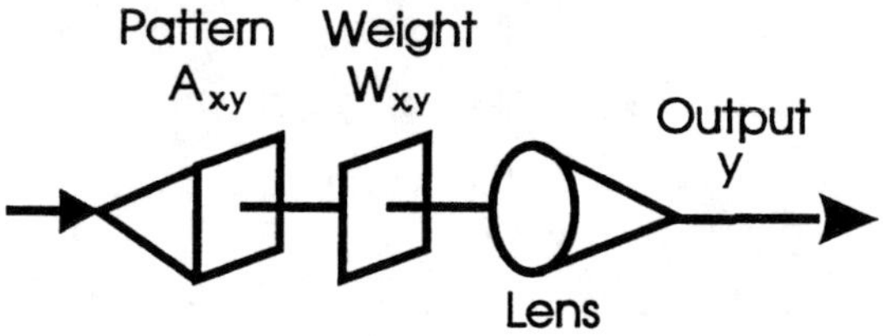

Figure 2: Optical correlator

pixel size at the time of writing. ETOM is cost effective because it requires only thin film deposition.

Disadvantages of ETOM are that the output is incoherent, the efficiency is low, and the material is not readily usable in 3-D [28]. The incoherent output makes it hard to collect or amplify the output light. The low efficiency makes it difficult to achieve high resolution, high speed, high dynamic range and low cost simultaneously. Thin film, needed for high resolution, can trap fewer electrons than thick film resulting in decreased output luminescence intensity and dynamic range. The number of photons available for detection at high speed is further reduced because of the shorter integration time for shorter pulses.

A further problem results from using light intensity. Correlation between two images for matching purposes requires each image have a zero mean. Therefore, positive and negative numbers are needed. Unfortunately, intensity cannot be negative. This may be overcome by correlating between both the images and the complements of the images. Equivalently, a dual rail coding can be used in which $-a \rightarrow (a, 0)$ and $a \rightarrow (0, a)$. The systems in this paper overcome this problem in different ways. For the Perceptron classifier in section 5 we showed earlier [14] that the Perceptron algorithm still converges for positive weights and that results improve if both positive and complementary images are correlated. For the unsupervised learning system in section 6 we remove a small percentage of the trapped electrons at each iteration to avoid the need to perform computations with the complementary images.

3. LEARNING

Learning enables computer systems and networked computer systems to adapt according to circumstances and environmental usage. Most of the demonstrations of optical learning have used photorefractive crystals and supervised learning [11], [21], [22], [24], and [26]. At present these approaches are slow because of the slow speed of photorefractive materials at reasonable power levels. Learning algorithms used in electronic computers are often too complex to implement optically, though researchers have experimented with optical backpropagation [15], [22]. The simplest and earliest learning algorithms for supervised learning are the Perceptron and Hebbian [5], [15]. They are used in this paper because their simplicity makes them good candidates for optical learning systems.

3.1 Supervised learning

Supervised learning [6], [7] is more common in artificial neural networks than unsupervised learning. In supervised learning a supervisor instructs the neural system during training by providing the correct output corresponding to a given input. The error between the neural system output and the correct output provided by the supervisor is used to adapt the system to improve future performance. The resulting adaption of weights during training is referred to as supervised learning. After training the system can be switched to a second mode for application in which opportunities to learn are normally removed.

In sections 4 and 5 we demonstrate two systems that use two different classical learning algorithms that are thought to be similar to those used in biological intelligence. It is well know that biological intelligence relies heavily on associative memory. In other words we remember things by associating them with other things. Our teaching makes use of this when we show a child a picture of a chair and the word chair for his or her association. Hebbian learning is a simple model of biological learning and in section 4 we use Hebbian learning for associative memory. In section 5 we use the Perceptron algorithm for classification of images into groups.

Supervised systems have limitations. They cannot cope with changes after training is switched off. Further, a persistent observer can learn to manipulate a trained system. Finally, the system cannot gain more knowledge than contained in its supervisor who in turn is limited by that of his supervisor and so on. This leads us to consider unsupervised learning.

3.2 Unsupervised learning

We normally associate learning with a supervising teacher. However, biological intelligent beings tend to learn even when not instructed. For example we learn as much or more from trying ourselves and the observation of others, movies, TV, or reading than we do from direct supervised instruction. Unsupervised systems like this are provided with an input but there is no supervisor to provide a corresponding output for adaption or to indicate whether what is learned is good or bad. Unfortunately it is not unusual to learn undesirable characteristics and behaviors this way. However, this does not mean that the learning is not goal driven. Typical examples of unsupervised learning in computer applications involve clustering of data into groups according to some specified goals. For example the clustering of features from previous mechanical designs in such a manner as to provide machine setups for new related designs [2]. Other types of unsupervised systems perform competitive learning and topological preservation [10].

Unsupervised learning is less demanding in terms of being able to implement a complex learning algorithm, giving it an advantage over supervised learning when considering optical implementations. However, the field is not well understood and it is more difficult to configure goals so as to accomplish defined tasks than to provide direct supervision. In section 6 we demonstrate a simple unsupervised optical Hebbian learning system for image novelty detection.

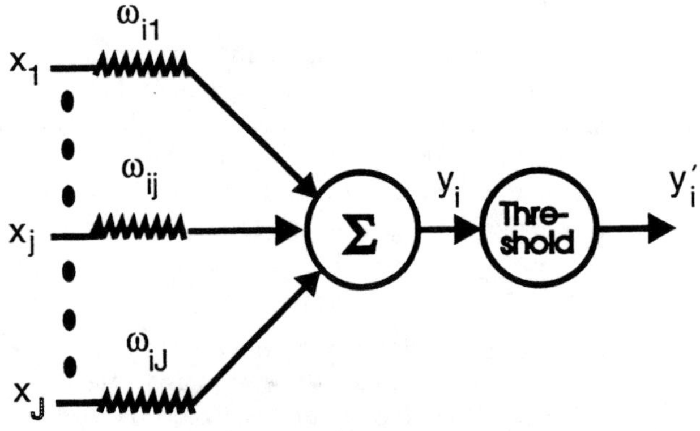

Figure 3: Neuronal model

4. OPTICAL HETEROASSOCIATIVE MEMORY

Our first neural system illustration involves a linear network that stores a set of associations. For example, peoples names may be associated with their phone numbers. Learning is accomplished in one step for each set of associations. For illustration we assume that the input (names) are orthogonal functions, meaning there is no overlap between them. Reference [15] and [17], which provide more details, use the name spatial light rebroadcaster (SLR) for a 2-D piece of ETOM. A spatial light rebroadcaster stores patterns for later rebroadcasting as incoherent information.

4.1 Hebbian learning for orthogonal heteroassociative memory

Hebb [5] introduced a biologically motivated concept of learning in neurons. His idea was that when an axonal input to a neuron (arriving via a synapse) helps cause that neuron to immediately emit a pulse, then the efficacy of that axonal input is increased. Efficacy refers to the ability to help the neuron produce pulses in the future. The model is supported by every day observation that after struggling to retrieve a name it is generally much easier to recall that name in the near future. This shows that exercising neurons for an associative memory causes these neurons to activate more easily in the future.

A McCulloch and Pitts type simple neuronal model [20] is shown in figure 3. For the ith neuron, the efficacy of the jth input is represented by the weight $w_{i,j}$. Inputs x_j are weighted by multiplying with weights $w_{i,j}$, summed, and thresholded. Neglecting the threshold, the operation performed can be written:

$$y_i = \sum_j \omega_{ij} x_j \tag{8}$$

Hebbian learning can be implemented by increasing a weight by an amount $a_{i,j}$ in proportion to the product of the input value x_j and the output value y_i:

$$a_{i,j} = \eta y_i x_j \tag{9}$$

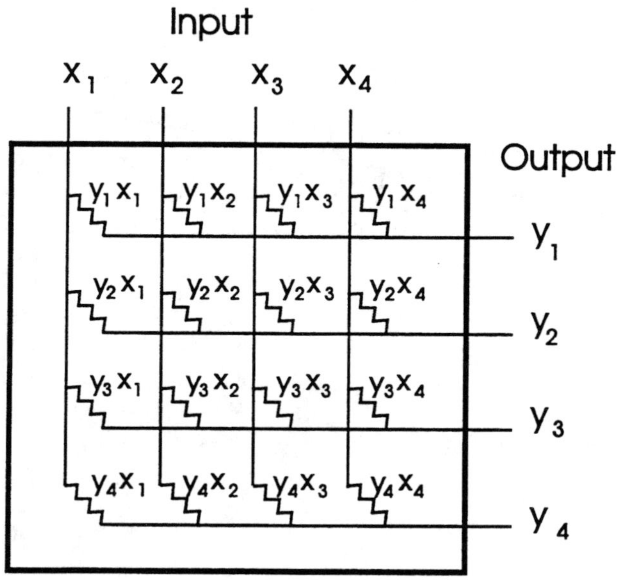

Figure 4: Outer product associative memory

where η is a proportionality factor. A larger weight at the neuronal input makes the input more effective in influencing the output of the neuron.

It turns out that if we consider two vectors $\mathbf{x}$ and $\mathbf{y}$ the outer product for these vectors $\mathbf{xy^T}$ or equivalently $\mathbf{yx^T}$ is equal to the matrix $\mathbf{A}$ with elements $a_{i,j}$ for $\eta = 1$. The matrix $\mathbf{A}$ in some sense accumulates the learning for the neural system. Therefore storing associations between vectors $\mathbf{x}$ and $\mathbf{y}$ in an outer product form is representative of Hebb's model. Information may be recovered by a single matrix-vector computation $\mathbf{y} = \mathbf{Ax}$ if the input vectors are orthogonal.

For the orthogonal case we can construct a heteroassociative memory that associates the vector pairs $\mathbf{x}_i$ and $\mathbf{y}_i$, i = 1 to J. We can envisage $\mathbf{x}_i$ to be the name of the ith person and $\mathbf{y}_i$ to be their phone number. When queried with the vector $\mathbf{x}_i$ (the name) the memory responds with the vector $\mathbf{y}_i$ (the phone number). The autoassociative memory, when $\mathbf{x}_i = \mathbf{y}_i$, is a special case. Figure 4 shows the process diagrammatically for a single associative pair. A vector $\mathbf{x}$, representing a person's name is entered at the top. The corresponding $\mathbf{y}$ for their phone number is entered from the right to form the outer product weight matrix, the elements of which are labeled in the box. A second vector $\mathbf{x}$ for another person and the corresponding $\mathbf{y}$, phone number, is used to compute another outer product matrix to be added to the first. This is repeated for all the names. The capacity for such networks, that is the maximum number of associations that may be stored, has been computed [7]. If one of the vectors $\mathbf{x}_i$, a name, is entered at the top, the matrix-vector multiplication by the weight matrix $\mathbf{W}$ will generate the corresponding phone number $\mathbf{y}_i$.

Mathematically, the vector associations are stored by taking the outer product

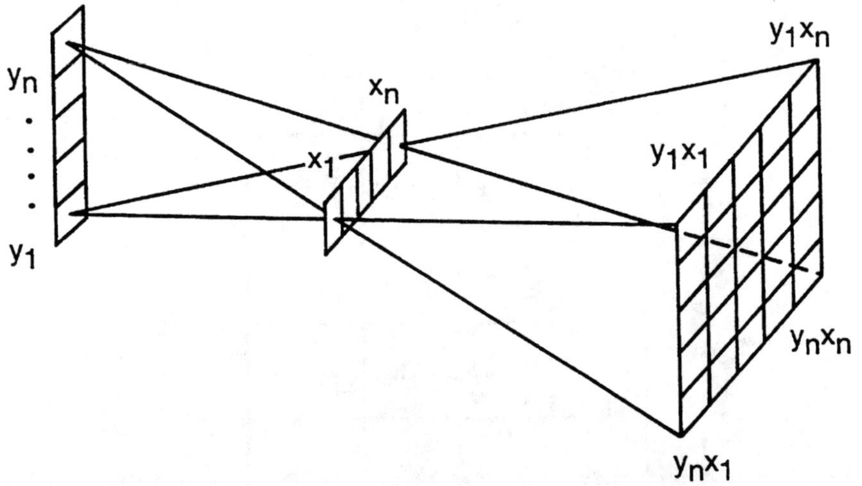

Figure 5: Illustrates optical computation of outer product matrix

for each pair and summing the resulting matrices. The dimension of the matrix is m by n where m is the dimension of the first vector and n that of the second vector. If there are J pairs of vectors to be stored:

$$\mathbf{A} = \mathbf{y}_1\mathbf{x}_1^T + \mathbf{y}_2\mathbf{x}_2^T + ... + \mathbf{y}_J\mathbf{x}_J^T. \tag{10}$$

If the input vectors $(\mathbf{x}_1, \mathbf{x}_2, ...\mathbf{x}_J)$ are orthogonal then $\mathbf{x}_i^T\mathbf{x}_j$ will be equal to 1 for $i = j$, otherwise $\mathbf{x}_i^T\mathbf{x}_j$ will be equal to 0. Given an input vector $\mathbf{x}_i$ it can be multiplied by the matrix $\mathbf{A}$ to recall the corresponding associated output vector $\mathbf{y}_i$ from the set $(\mathbf{y}_1, \mathbf{y}_2, ...\mathbf{y}_j)$. This is illustrated using equation (10) and separating the ith term.

$$\mathbf{A}\mathbf{x}_i = \mathbf{y}_i(\mathbf{x}_i^T\mathbf{x}_i) + \sum_{j \neq i} \mathbf{y}_j(\mathbf{x}_j^T\mathbf{x}_i) \tag{11}$$

The orthogonality condition

$$\begin{cases} (\mathbf{x}_j^T\mathbf{x}_i) = 0 & j \neq i \\ (\mathbf{x}_j^T\mathbf{x}_i) = 1 & j = i \end{cases} \tag{12}$$

leads to

$$\mathbf{A}\mathbf{x}_i = \mathbf{y}_i \tag{13}$$

Thus an output vector $\mathbf{y}_i$ (phone number) is obtained by multiplying the weight matrix $\mathbf{A}$ by the input vector $\mathbf{x}_i$ (the name).

4.2 Optical set up for orthogonal heteroassociative memory

Hebbian learning is associated here with the formation of an outer product matrix. Figure 5 shows how this is performed optically. Light shining through y_1 picks up the intensity associated with the transmittance. The light is spread

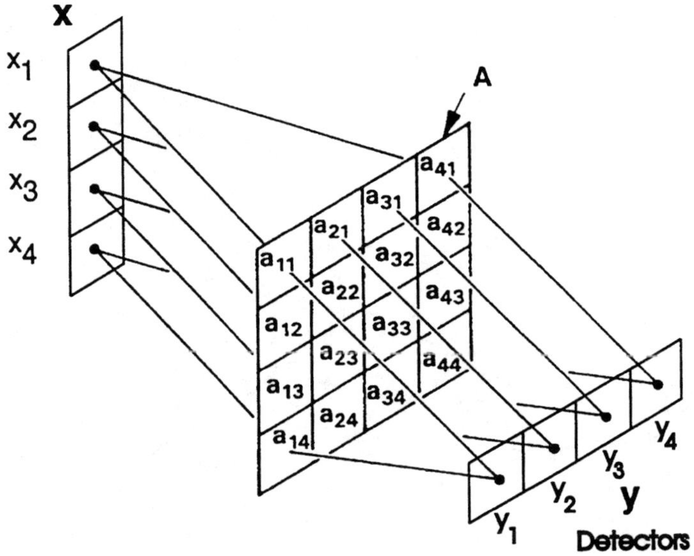

Figure 6: Illustrates optical matrix-vector multiplication

so as to pass through all elements of the vector **x** to generate the top row of the outer product matrix. The other rows of the outer product matrix are generated similarly.

The recall operation involves matrix-vector multiplication. Figure 6 shows how this may be accomplished optically. Light passing through an array at **x** is imaged onto the plane of the matrix in the vertical direction. In the horizontal direction the light is expanded to cover rows. The light is collected by columns on the receiving side. The computation performed is:

$$\begin{bmatrix} y_1 \\ y_2 \\ y_3 \\ y_4 \end{bmatrix} = \begin{bmatrix} a_{11} & a_{12} & a_{13} & a_{14} \\ a_{21} & a_{22} & a_{23} & a_{24} \\ a_{31} & a_{32} & a_{33} & a_{34} \\ a_{41} & a_{42} & a_{43} & a_{44} \end{bmatrix} \begin{bmatrix} x_1 \\ x_2 \\ x_3 \\ x_4 \end{bmatrix} \tag{14}$$

or:

$$\mathbf{y} = \mathbf{A}\mathbf{x} \tag{15}$$

Figure 7 shows the optical set up for the generation of a 32-bit heteroassociative memory using optical outer product computation for learning and optical matrix-vector product computation for recall. Vectors **x** and **y** form an outer product at the SLR composed of ETOM. An Argon laser beam is collimated and focused by means of cylindrical lens S1 onto one column of the film located in the plane labeled vector **y**. The column represents one of the output vectors, **y**, involved in the association. After passing through the vector **y** plane, the light spreads horizontally. There is a horizontal line of light corresponding to every clear spot in the column on the film representing vector **y**. Clear spots are considered logical

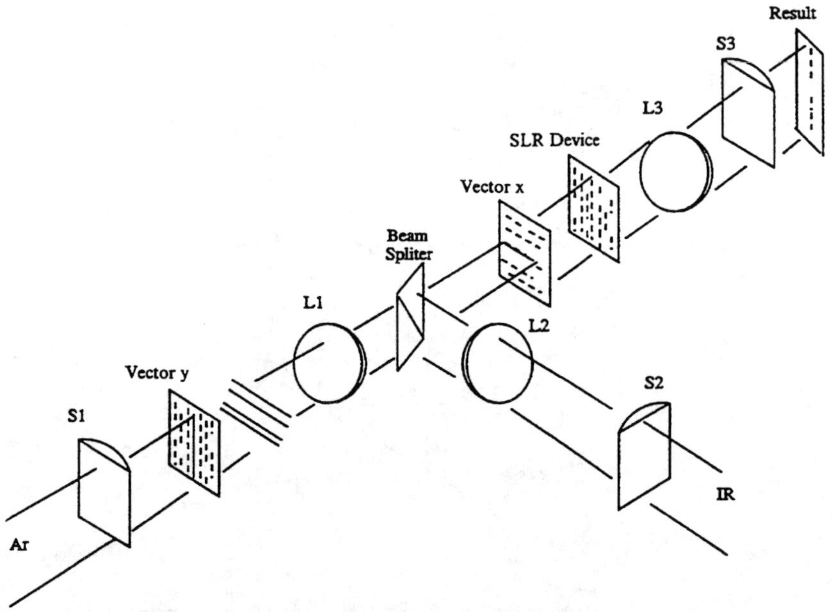

Figure 7: Optical set up for neural network associative memory with SLR device

ones and opaque spots are considered logical zeros. A spherical lens L1 focusses these lines on to one horizontal line in the plane labeled vector **x**. Film with the vector **x** to be associated with **y** is placed at this position. The light passing through the vector **x** expands vertically so as to produce the outer product of **x** and **y** at the spatial light rebroadcaster (SLR). Note that vertical columns are passed only where the vector **x** has ones. The SLR stores this outer product.

The film at the position labelled vector **y** is moved horizontally in order to perform an outer product for another associated pair. The film at the position labelled vector **x** is moved vertically to provide the **x** that corresponds with the **y** selected. The resulting outer product for the new associated pair is stored on top of that previously stored on the SLR. This implements equation (10). For the purpose of illustration, vector pairs were selected so that the resulting matrix has only ones or zeroes.

If the input vectors are orthogonal then a simple matrix-vector multiplication [4] can recall an output vector **y** from an input vector, **x**. The optical set up is arranged in such a way that the Argon laser and IR laser come from different angles, which reduces the complexity of the experimental procedure (Figure 7).

Cylindrical lens S2 and spherical lens L2 act in the same manner as S1 and L1, except that the light is not modulated. They cause a single horizontal line to be illuminated in the plane labelled vector **x**. Film with the input vector **x** whose association is sought is placed at this row. The illumination is such that the light is spreading vertically to illuminate those columns of the SLR corresponding to

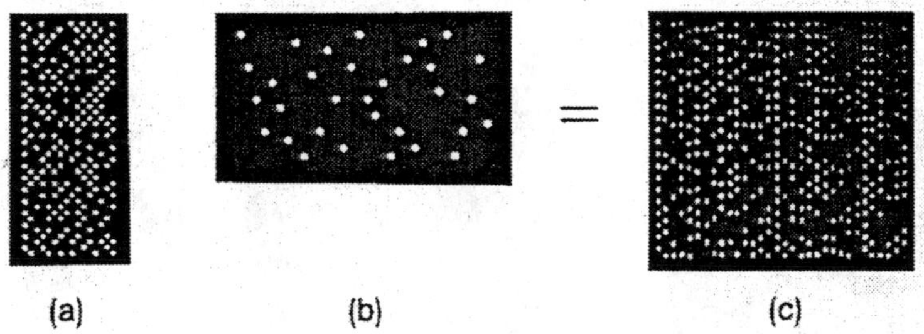

Figure 8: Experimental result of orthogonal neural network associative memory: (a) vector **y**, (b) vector $\mathbf{x}^T$, (c) associative memory matrix

positions in vector **x** that have ones. The resulting luminescence is imaged onto the results plane with leans L3. A cylindrical lens S3 smears the incoherent light in the horizontal plane. A single vertical column is detected on film or a detector array. The vector resulting from the matrix-vector multiplication is the output **y** associated with the input **x**.

4.3 Experimental results for orthogonal heteroassociative memory

Figure 8 shows the experimental results of an orthogonal neural network heteroassociative memory. Figure 8(a) shows a matrix **y** whose 16 columns each represent one of the output vectors. Each vector has 40 bits. The input vectors corresponding to these outputs are the 16 rows in figure 8(b). Each row has 32 bits. The first column in figure 8(a) corresponds to the first row of figure 8(b) and so on. The two matrices were placed on film and the two figures were produced by shining collimated light through each film separately.

The **y** and **x** matrices were placed in the positions labelled vector **y** plane and vector **x** plane respectively in the optical set up shown in figure 7. The outer product was generated and stored as described previously by shifting the **y** horizontally and **x** vertically until the outer products for all 16 pairs were generated and stored on the SLR. The SLR was then illuminated to produce luminescence corresponding to the sum of the outer products and this was photographed to produce figure 8(c). The intensity of the IR controls the initial intensity of luminescence. Maximum output energy for the photograph is achieved if sufficient IR energy is used so that all the energy stored in the SLR is read out. However, in this case the matrix must be reconstructed for further use.

Heteroassociative memory recall is illustrated in figure 9. The sum of outer product matrix stored in the SLR is shown in figure 9(a). Figure 9(b) shows the film used with 16 input vectors arranged as columns. This film is placed at position labelled vector **x** in figure 7 such that the vectors are horizontal. One of the columns labelled p, q, or r, in figure 9(b) is placed at the active row position in the optical set up. The output obtained at the result position in figure 7 was

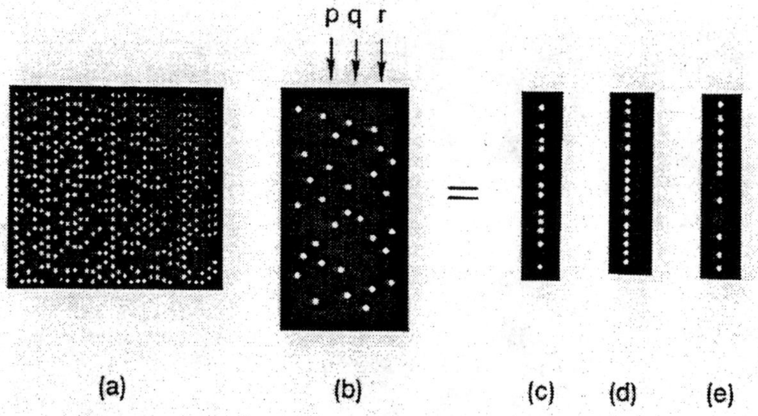

Figure 9: Recall from associative memory for three different input vectors: (a) outer product matrix storing associative memory, (b) array containing input vectors **p**, **q**, **r**, (c) output vector for matrix times vector **p** (6 th column of matrix, $\mathbf{y}_6$), (b) output vector for matrix times vector **q** (10th column of matrix, $\mathbf{y}_{10}$), (d) output vector for matrix times vector **r** (14th column of matrix, $\mathbf{y}_{14}$)

photographed for each case, p, q, or r.

The cases were taken so that the reader can verify the correction operation by careful examination of the result. The sixth column in figure 9(b) is labelled p and was used first. It has a one in the tenth entry from the top and no other ones. Hence, on premultiplying by the matrix, this should result in the selection of the tenth column of the matrix in figure 9(a). The actual result photographed, shown in figure 9(c), is seen to be a replica of the tenth column of the matrix as required. Placing the column q at the active row position in figure 7 produced the result in figure 9(d). This column has a one at position six so that the result is seen to match the sixth column of the matrix. In the same way the third input vector r, which has a one in the seventh position, is shown to replicate the seventh column of the matrix at the output. Other input vectors, having several ones also work satisfactorily but were not included because it is much harder for an observer to verify correct operation.

4.4 Extensions and limitations for orthogonal heteroassociative memory

Orthogonality, implying a cross product of zero between inputs, is a severe constraint. Moreover, noisy inputs cannot provide a correct association. For this reason, a nonlinear feedback may be used to cut out weak associations and allow strong ones to feed back. Such networks have been studied extensively and are often referred to as Hopfield networks [8]. These networks are related to those for noisy input autoassociative memory [15] which involve a sequence: correlation, selection of the peak correlation, and reestablishment of the selected vector.

We also comment that learning for this network is a matter of computing an outer product and accumulating these products. Some researchers feel that the word learning should be reserved for cases where repetitions of input/output associations are needed in order for the system to iteratively improve weights. In section 6 we use an iterative Hebbian learning algorithm in this manner.

5. PERCEPTRON OPTICAL CLASSIFIER

One of the earliest iterative learning algorithms is the Perceptron which lends itself to classification of input data sets into groups. For example, with a video camera we could distinguish between aircraft friend and foe. Prior size, shift, and orientation transformations are desirable [15]. More details can be found in reference [14].

5.1 Perceptron Algorithm

Neural networks tend to use recursive algorithms for learning [6] [7] [15]. These may be viewed as gradient methods that seek the minimum of a criterion function J. The weights at the $(k+1)$th iteration are obtained from those at the (k)th iteration from:

$$\mathbf{w}(k+1) = \mathbf{w}(k) - \mu \nabla_{\mathbf{w}} J|_k \tag{16}$$

where

$$\nabla_{\mathbf{w}}^T = \left[\frac{\partial}{\partial w_1}, \frac{\partial}{\partial w_2} \cdots \right] \tag{17}$$

Equation (16) searches for the minimum of the criterion function J by adjusting weights in proportion to the negative slope. When the slope is uphill (positive) the weights are adjusted to move back downhill. When the slope is downhill (negative) the weights are adjusted to move further downhill. As the minimum is approached the slope gets smaller causing smaller adjustments to the weights. At the minimum the slope is zero and no more adjustment is need in equation (16).

For methods that rely on linear discrimination between classes, the minimum of J must satisfy the equation:

$$\mathbf{w}\tilde{\mathbf{y}} > 0 \tag{18}$$

where for a two class system:

$$\tilde{\mathbf{y}} = \begin{cases} \tilde{\mathbf{x}} & \text{if } \tilde{\mathbf{x}} \text{ in class } C_1 \\ -\tilde{\mathbf{x}} & \text{if } \tilde{\mathbf{x}} \text{ in class } C_2. \end{cases} \tag{19}$$

and $\tilde{\mathbf{x}}$ is an input vector augmented by 1, e.g. $\tilde{\mathbf{x}} = \{x_1, x_2, x_3, \cdots, 1\}$. The use of $\tilde{\mathbf{y}}$ simplifies the equations by allowing a single decision, equation (18), to separate correct and incorrect classification. The "one" augmentation from $\mathbf{x}$ to $\tilde{\mathbf{x}}$ allows the discriminant function plane to be shifted away from the origin. For a 3-D example, $D(x) = w_1 x_1 + w_2 x_2 + w_3 = w^T \tilde{x}$ for $w = (w_1, w_2, w_3)$ and $\tilde{x} = (x_1, x_2, 1)$

The well known Perceptron algorithm [23],[29] is obtained by selecting:

$$J_1 = \frac{1}{2} \left[|\mathbf{w}^T \tilde{\mathbf{y}}| - \mathbf{w}^T \tilde{\mathbf{y}} \right] \tag{20}$$

In this case, $\nabla_W J_1 = 0$ if $\mathbf{w}^T \tilde{\mathbf{y}} \geq 0$ and $\nabla_W J_1 = \tilde{\mathbf{y}}$ if $\mathbf{w}^T \tilde{\mathbf{y}} \leq 0$.

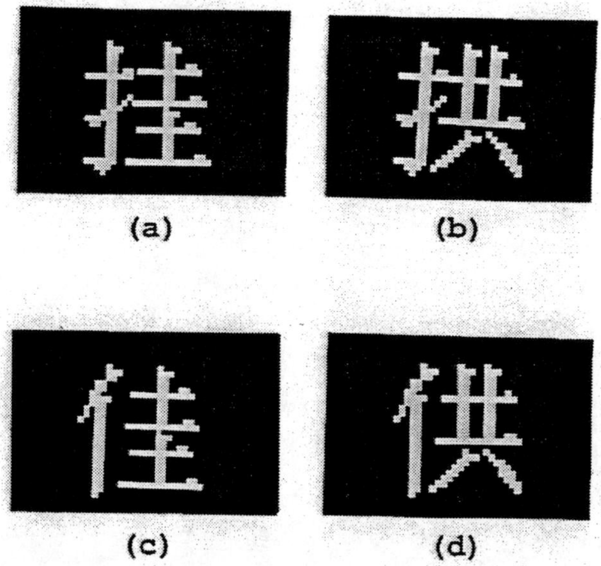

Figure 10: Patterns used for optical learning demonstration: (a) Hang, (b) Arch, (c) Good, (d) Offer

In this case, the learning rule may be written:

$$\mathbf{w}(k+1) = \mathbf{w}(k) + \mu \begin{cases} 0 & \text{if } \mathbf{w}^T\tilde{\mathbf{y}} > 0 \\ \tilde{\mathbf{y}} & \text{if } \mathbf{w}^T\tilde{\mathbf{y}} \leq 0 \end{cases} \tag{21}$$

where the top alternative condition corresponds to correct classification and the bottom alternative condition to incorrect classification. Also, the overrelaxation constant $\mu > 0$.

Note that if a new value is correctly classified during training, the weight is left unchanged. If the new value is incorrectly classified during training, an update to the weights is made. From equation (19), the update involves adding the input vector for C_1 and subtracting the input vector for C_2. The Perceptron has been shown to converge in a finite number of steps [7] [6] [23] [29]. In reference [14], we show that the Perceptron algorithm converges if only positive weights are permitted, a restriction due to light intensity being only positive (see section 2.3).

5.2 Optical setup for Perceptron classifier

Four binary chinese characters were used as shown in figure 10. Much simpler patterns were used earlier for optical demonstrations of Perceptron learning using photorefractive crystals [22] and SLRs [16]. In the first experiment, we wish to classify patterns (a) and (b) in figure 10 as class C_1 by providing a high output at a detector, and patterns (c) and (d) as class C_2 by providing a low output at the detector. In order to show the versatility of the system, we then perform a second experiment in which the patterns (a) and (c) are classified as class C_1 by providing

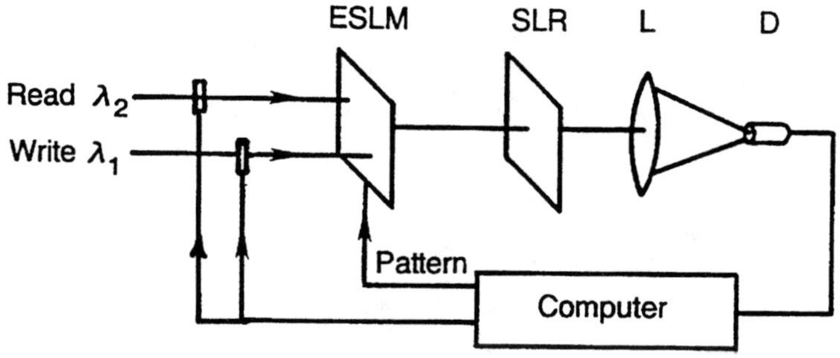

Figure 11: Optical set up for Perceptron binary classification experiment with an SLR

a high output at a detector and patterns (b) and (d) as class C_2 by providing a low output at the detector.

Figure 11 shows an optical set up. The computer controls shutters for timing writing and reading onto the *SLR*. We start by writing uniformly with low intensity on the SLR. Then we set pattern (a) on ESLM. This pattern is multiplied pixel by pixel by the weights on *SLR* and the resulting array is spatially integrated to a detector by lens L in the same manner as shown earlier, figure 2, and described by equation (7). The computer reads the detector. If the output is correct during training, the computer controls the write shutter to open only long enough to compensate for energy lost from the SLR during read. This corresponds to no change in the algorithm, equation (21). If the detector output is opposite to that desired during training for class one, then the computer opens the write shutter for a longer time in order to add information selectively on the SLR. If the detector output is opposite to that desired during training for class two, then the computer opens the read shutter to remove some of the information selectively on the SLR.

The electronic threshold is not a bottleneck in the system because the optics performs a 2-D inner product while only a single number needs to be thresholded electronically. Classification occurs fast because light travels a foot in a nanosecond and the response time of the SLR is in the order of nanoseconds.

5.3 Experimental results for Perceptron classifier

We start by writing a small intensity on the SLR to represent small weights in normalized terms $\mathbf{w} = 0.1$. The computer rotates through all the patterns using the Perceptron algorithm until there are no more changes in the weights stored on the SLR. The intensities used are 0.32 mW/cm^2 for the Argon write and 0.72 mW/cm^2 for the IR read. The exposure times are approximately 0.5 seconds for write and 1 second for read. Long times were used in order to operate with low power levels. The learning curve is shown in figure 12. After 10 cycles through all the patterns, the patterns (a) and (b) provide the desired high output and patterns (c) and (d) the desired low output.

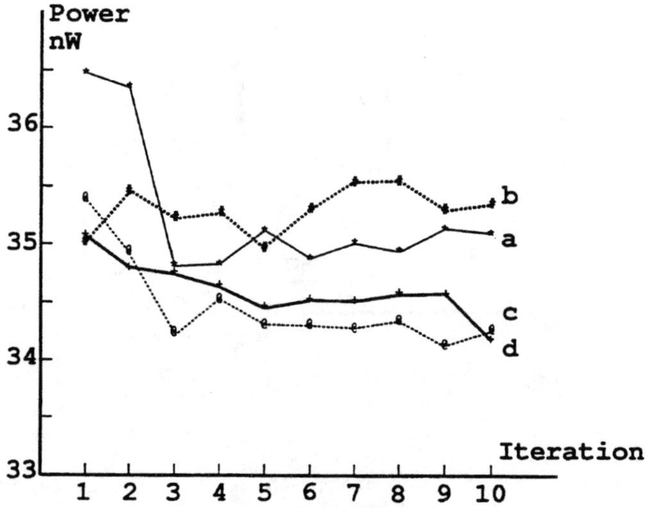

Figure 12: Learning curve for optical learning for (a) and (b) in class one and (c) and (d) in class two

As a result of this training, if one of the input patterns is set on ESLM, the correct output will be seen at the detector. For example, (a) and (b) give a high output, indicating class one and (c) and (d) give a low output, indicating class two. The final weight pattern on the SLR for the thick film and the thin film are shown in figure 13(a) and (b) respectively. Note that it has weights equal to '1' in regions at the right side of the vertical bar at the left in the figure, where patterns (a) and (b) have '1' but patterns (c) and (d) have '0'. Three identical cross bars on the left vertical part are also present on both patterns (a) and (b) and absent on patterns (c) and (d). Therefore, the cross bars also show up on the final weight pattern. When an inner product is taken between pattern (a) or (b) with the weight pattern, a high results. When an inner product is computed between the

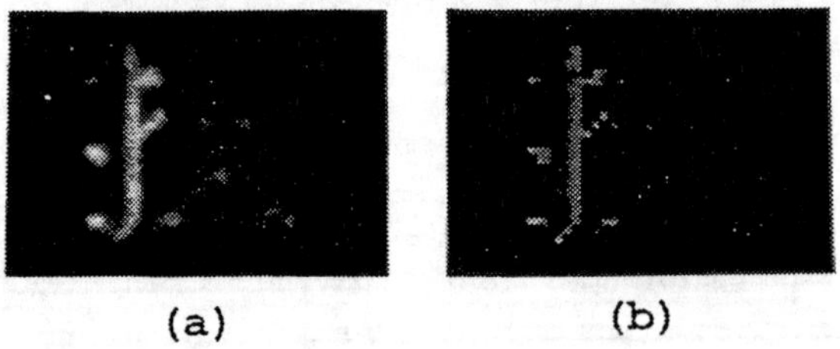

Figure 13: Weight patterns for optical learning with (a) and (b) in class one and (c) and (d) in class two: (a) final pattern for thick film, (b) final pattern for thin film

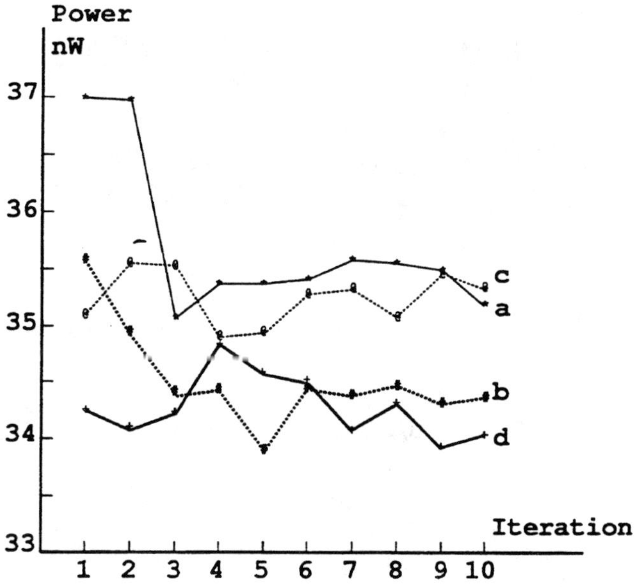

Figure 14: Learning curve for optical learning for (a) and (c) in class one and (b) and (d) in class two

weight matrix and pattern (c) or (d), a zero results.

In a second experiment, we rearranged the classes with the same patterns. This time we set patterns (a) and (c) in figure 10 to be in class one and patterns (b) and (d) to be in class two. After 10 iterations, the classes are learned as shown in figure 14. The final weight pattern on the SLR for the thick film and the thin film for this class selection are shown in figure 15(a) and (b) respectively. Note that it has weights equal to '1' in the tree like pattern at the right side where both patterns (a) and (c) have '1's. The thin film show a higher resolution as expected. Therefore, when an inner product is taken between pattern (a) or (c) with the weight pattern, a high output results. When an inner product is computed between the weight matrix and pattern (b) or (d), a low output results.

5.4 Extensions and limitations for Perceptron classifier

As mentioned earlier (section 2.3, 5.1), the use of positive weights allows convergence but reduces performance in the correlation operation. In reference [14] we show that results are improved by simultaneously performing the Perceptron algorithm on the complementary images which is equivalent to negative weights. The results for the positive and negative weights are merged to provide improved results.

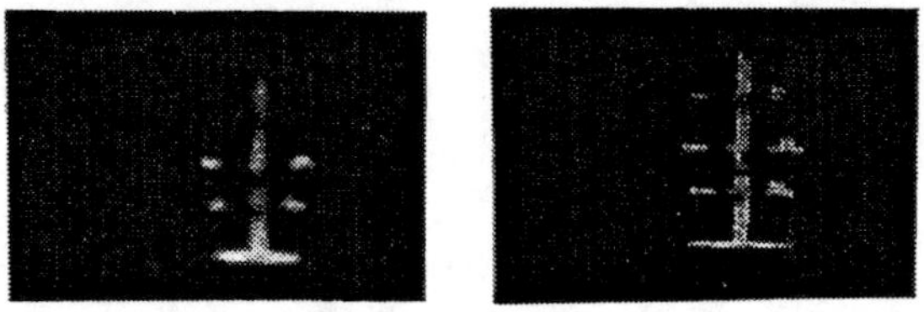

Figure 15: Weight patterns for optical learning with (a) and (c) in class one and (b) and (d) in class two: (a) final pattern for thick film, (b) final pattern for thin film

6 UNSUPERVISED HEBBIAN OPTICAL NOVELTY DETECTOR

A novelty detector for images can be used as an alarm system because it detects changes in an input image. If the input settles to a new constant image, the detector learns that changes from this new images must now be detected. The detector uses some concepts from the systems described in section 4 and 5. It uses element by element matrix multiplication summing as in the Perceptron classifier in section 5. It uses Hebbian learning as in the heteroassociative memory in section 4. However, in this case the Hebbian algorithm is iterative. Further it uses an all-optical loop which is a potential way to increase speed over what could be accomplished electronically. All-optical refers here to all devices having optical inputs and outputs. More details can be found in reference [12].

6.1 Equations for unsupervised Hebbian novelty detector

The classical McCulloch and Pitts neuronal model [20] was shown in figure 3 and equation (8) (see section 4.1). We can take an image $A_{x,y}$ as an input by recognizing that the input vector x_j can be written as a matrix by folding it into rows. The weight vector for a single neuron can also be written as a weight matrix $W_{x,y}$ by folding it into rows. Therefore, the vector $\mathbf{y} = \mathbf{x}^T\mathbf{w}$ can be written as a matrix equation (7). Previously, figure 2, we showed how the sum of the inner product elements between the input and the weights represented by equation (7) can be performed optically in 2-D.

We use interative Hebbian learning for the novelty detector in contrast to the non iterative Hebbian learning used in section 4.1. Hebbian learning, section 4.1, equation (9), requires that weights be updated by an amount proportional to the product of the output $\mathbf{y}$ and the input matrix $\mathbf{A}_{x,y}$. Therefore, we modify figure 2 to feed back the output $\mathbf{y}$ to multiply by the input $\mathbf{A}_{x,y}$ as shown in figure 16. The feedback loop effectively adjusts the weights by using the Hebbian learning rule of equation (9) recursively [7]

$$\omega_{ij}(t) = \omega_{ij}(t - \Delta T) + \eta\ y(t - \Delta T)\ A_{x,y}(t - \Delta T) \tag{22}$$

The new weight is set based on values of the output at a time ΔT earlier.

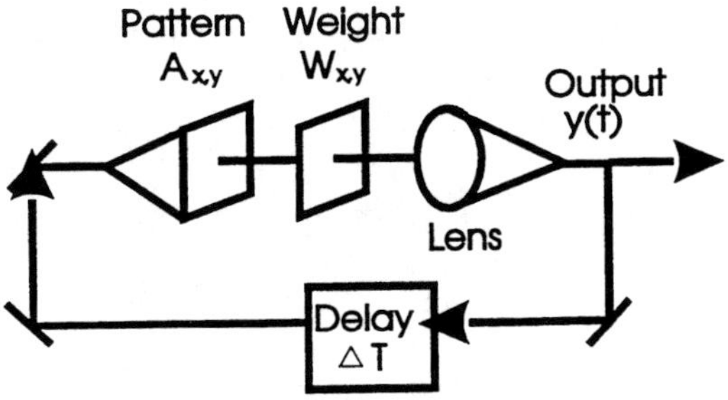

Figure 16: Hebbian learning with a correlator

The correlator model, equation (7) and the Hebbian learning model, equation (22) are used to derive our optical novelty detector.

The output $y(t)$ of equation (7), figure 16, is high when the input image correlates with the weight matrix and lower if it is less well correlated. Hence in a feedback recursive system any change in input image $A_{x,y}(t)$ will result in a drop of output. Here we use the analog output and omit the threshold operation of the neural network model, figure 3.

6.2 Optical setup for unsupervised Hebbian learning novelty detector

The system uses unsupervised learning as $A_{x,y}(t)$ is the only input and no output is provided for training. Figure 17 shows the optical set up for the unsupervised Hebbian learning novelty detector. In order to provide the delay of equation (22), we use two SLRs made of ETOM: SLR_1 stores the weights and SLR_2 the scalar output y. Because the luminescence is too weak to provide adequate feedback for the next stage, we use an image intensifier and an optically addressable liquid crystal light valve to boost the intensity and change the wavelength from orange to blue. We refer to the combination of image intensifier and liquid crystal light valve as OLCLV. In order to avoid the cost of two OLCLVs, we fold the system and alternate between two cycles [12]. The two SLRs share the OLCLV.

SLR_2 is first cleared. 7.5 μW of Argon light at wavelength 488 nm passes through SH_1 to write pattern $A_{x,y}$ at P onto SLR_1. Film is used for pattern $A_{x,y}$ at P for simplicity. Then shutters SH_1, SH_2, SH_5 and SH_7 are opened for the first read cycle to allow 0.233 mW of IR light from the right side to pass through shutter SH_1 and the pattern $A_{x,y}$ at P to read the weights $W_{x,y}$ at SLR_1 for one second. The benefit of a synchronized system, made possible with read/write material, is that we can run with low safe intensities and low speed for visual verification of the system. The potential speed of the system described is limited by the LCLVs and input LCDs. Lens L spatially integrates the output passing through shutter SH_5 to form the scalar output at the OLCLV according to the correlator neuronal model, equations (7). The output scalar is also measured by a photomultiplier tube and oscilloscope at D to provide the system output. Argon light at the left passes through shutter SH_2 to read the value at OLCLV by reflecting off the left side of the OLCLV and passing back through shutter SH_7 and BS_4 to write the

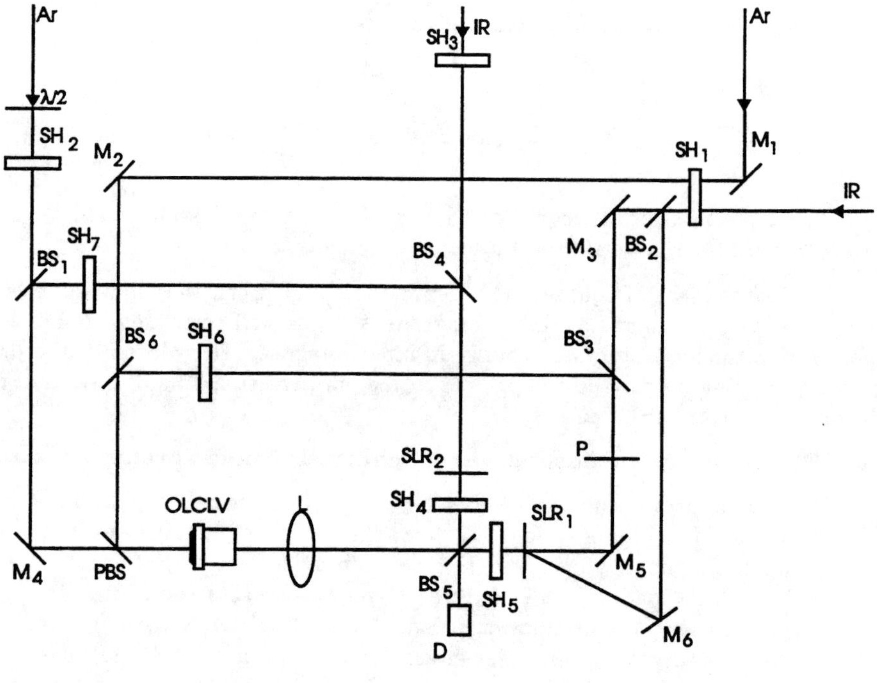

Figure 17: Optical set up used for unsupervised learning with two SLRs

Figure 18: Input patterns used for optical experiments: (a) X pattern, (b) O pattern

scalar y onto SLR_2. This completes the first cycle. The first set of shutters are closed and shutters SH_2, SH_3, SH_4 and SH_6 are opened after a delayed time ΔT. Opening shutter SH_3 allows IR in the center to read the scalar y at SLR_2. The output passes through shutter SH_4 and is spatially integrated by lens L on its way to OLCLV. The output is fed back using Argon light via SH_2, BS_6, SH_6 and BS_3 to pass through P for writing onto SLR_1. This completes the second cycle of updating the weights at SLR_1 according to the recursive Hebbian learning rule of equation (22). The process is repeated continuously to perform the alarm system function. Note that mirror M_6 provides the forgetting for SLR_1. It is also necessary to clear SLR_2 before writing on it in each cycle.

6.3 Experimental results for unsupervised Hebbian learning novelty detector

Figure 18(a) shows an 'X' that is placed at P. As long as it remains at P the output is high as shown at iteration '0' in the correlation output y in figure 19. The pattern stored as weights on SLR_1 is shown in figure 20(a). A second pattern 'O', shown in figure 18(b), is then placed at P. The weights at SLR_1 are shown in figure 20(b) where 'O' is effectively used to read a stored pattern of 'X'. Orange luminescence occurs at the overlap of 'O' and 'X'. As a result, the power at the output decreases as shown by the drop in figure 19 at iteration '1'. If the 'O' remains, the weights will gradually shift to that shown in figure 20(c) and the output y will increase back to a high level as shown in figure 19 after 11 iterations. This shows that a change of input will reflect in a drop of output y. Continuation of the new pattern will allow it to become the status quo from which change is detected.

6.4 Limitations of unsupervised Hebbian learning novelty detector

Thin film electron trapping material has a low efficiency of generating luminescence. Further, the luminescence is incoherent and the wavelength must be changed for a feed back loop [1]. Therefore we found it necessary to use an image intensifier and an optically addressable liquid crystal light valve. These devices reduce the resolution and speed. More efficient materials and an efficient gain mechanism would allow an all-optical system like this to iterate at high speeds.

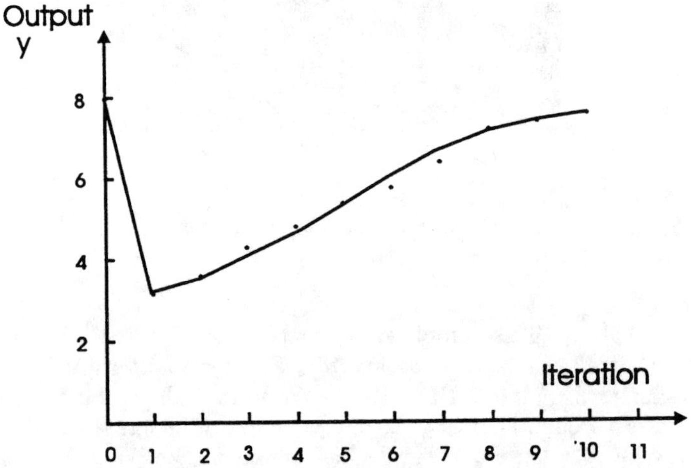

Figure 19: Output of unsupervised learning alarm system during experiments in switching from X to O

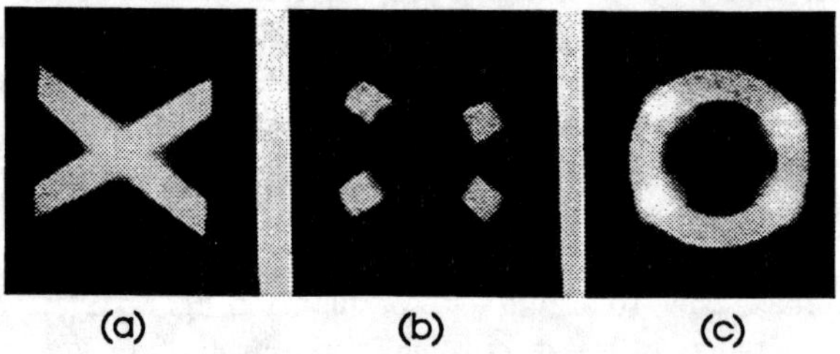

Figure 20: Weights on SLR during optical experiments: (a) after X present for some time, (b) immediately after X replaced by O, (c) after O now present for some time

7. CONCLUSION

Electron trapping optical material (ETOM) has many desirable characteristics, including: fast write/read response, high resolution, endless read/write cycles, long term storage, room temperature operation, and optical write and read addressing.

However, ETOM also has major limitations. The efficiency of luminescence is much less than 1% for thin film, which is very low. Further, luminescence is incoherent. The incoherent nature of the output and low signal levels make it difficult to directly use the optical output for further optical processing. An optical loop for optical learning cannot therefore be achieved without the use of other devices which in turn limit capability. An optical loop is desirable because of its potential to operate at speeds that would be difficult for electronics. The weak incoherent light output impacts the three demonstration systems differently.

In the supervised Hebbian learning system for orthogonal inputs, a photodetector array reads the output and electronic gain may be used. A useful extension is to a Hopfield network in which a noisy input is classified amongst those stored. This requires a nonlinear feedback loop and gain which are not readily accomplished optically for weak incoherent signals. One solution is to return to electronics for the feedback. In this event the whole system could have been electronic and the optics provides minimum advantage.

In the Perceptron learning system, the weights stored on ETOM are adjusted during training. A simple algorithm provides feedback based on whether the output is correct or incorrect during training. Without gain in the loop this is difficult to accomplish optically. While the training samples are input one at a time by the instructor it is easy for the same instructor to determine whether the system is providing the correct output. However, we would expect a fast source of inputs for training in a high speed system for which optics would be worthwhile.

In the unsupervised Hebbian learning system, we used an image intensifier and liquid crystal light valve for gain and wavelength changing in the optical loop. The overall performance is now decided by the resolution and speed of these devices. A remaining advantage of the ETOM is its storage capability. This enables us to construct a synchronous system, which is easier to operate than an asynchronous one. For example, the system can be run at arbitrarily slow speeds and low power.

In summary, three systems were successfully demonstrated using ETOM. At this time, the performance of these systems is limited by the low efficiency of the thin film material and the resulting difficulty associated with handling weak incoherent light signals.

6. REFERENCES

1. D. T. Brower and R. E. Revay, "Tuning of read/write/erase processes in electron trapping optical memory media," in *Optical Data Storage*, D. B. Carlin, D. B. Kay, Eds., Proc. Soc. Photo-Opt. Instrum. Eng. **1663**, 86-91 (1992).

2. C. L. P Chen and S. R. LeClair, " Integration of design and manufacturing: solving setup generation and feature sequencing using an unsupervised-learning approach," Computer Aided Design, **26**(1), 59-75 (1994).

3. J. Gasiot and P. Braunlich and J. P. Fillard, " Nanosecond infrared laser stimulation of luminescence in rare-earth doped sulphides," Applied Phys. Lett. **40**(5), 376-378 (1982).

4. J. W. Goodman, P. Kellman, and E. W. Hansen, Appl. Opt. **16**, 733 (1977).

5. D. O. Hebb, *The organization of behavior*, (John Wiley, New York, 1949, 1988).

6. R. Hecht-Nielsen, *Neurocomputing* (Addison-Wesley, New York, 1990).

7. J. Hertz, A. Krogh, and R. G. Palmer, *Introduction to the theory of neural computation* (Addison-Wesley, New York, 1991).

8. J. J. Hopfield and D. W. Tank, "Computing with neural circuits: a model," Science, **233**, 625-633 (1986).

9. S. Jutamulia, G. M. Storti, J. Lindmayer, W. Seiderman, "Use of electron trapping materials in optical signal processing 2: two-dimensional associative memory," Appl. Opt. **30**(20), 2879-28841 (1991).

10. T. Kohonen, *Self-organization and associative memory* (Springer Verlag 1989).

11. H. S. Li, Y. Q, D. Psaltis, "Optical network for real-time face recognition," Appl. Opt. **32**(26), 5026-5035 (1993).

12. A. D. McAulay and J. Wang, "Optical 2-D alarm system using electron trapping optical material," Optical Engineering, **35** (8), 1996.

13. A. D. McAulay and Junqing Wang, "Methods for addressing electron trapping optical material", SPIE Advances in Optical Information Processing VII, **2754-23**, April 1996.

14. A. D. McAulay, J. Wang, and X. Xu, "Optical Perceptron Learning for Binary Classier with Spatial Light Rebroadcasters," Applied Optics, **32**(8), 1346-1353 (1993).

15. A. D. McAulay, *Optical Computer Architectures*, Wiley, New York, 1991.

16. A. D. McAulay, J. Wang, X. Xu, and M. Zeng, "Demonstration of optical learning with luminescent materials for image classification," in *Proceedings of IEEE International Conference on Systems, Man, and Cybernetics* **1**, 237-241, (Oct. 1991).

17. A. D. McAulay, J. Wang, and C. T. Ma, "Optical heteroassociative memory using spatial light rebroadcasters (SLRs)," Appl. Opt. **29**(14), 2067-2073 (1990).

18. A. D. McAulay and J. Wang, "Spatial light rebroadcaster bit-slice word-addressable holographic memory," Appl. Opt. **30**(20), 2885-2889 (1991).

19. A. D. McAulay, "Logic and Arithmetic with Luminescent Rebroadcasting Devices," in *Advances in Optical Information Processing III*, D. R. Pape, ed., Proc. Soc. Photo-Opt. Instrum. Eng. **936**, 321-326 (1988).

20. McCulloch and Pitts, "A logical calculus of ideas immanent in nervous activity," Bulletin of Mathematical Biophysics **5**, 115-133 (1943).

21. Y. Owechko, G. J. Dunning, E. Maron and B. H. Soffer, " Holographic associative memory with nonlinearities in the correlation domain," Applied Optics, **26**(10), 1900-1910 (1987).

22. D. Psaltis, and D. Brady, and K. Wagner, "Adaptive optical networks using photorefactive crystals," Appl. Opt. **27**(9), 1752-1759 (1988).

23. Rosenblatt F., *Principles of Neurodynamics: Perceptrons and the Theory of Brain Mechanism*, (Spartan, Washington, 1961).

24. B. H. Soffer, G. J. Dunning, Y. Owechko, and E. Maron, " Associative holographic memory with feedback using phase-conjugate mirrors," Optics Letters, **11**(2), 118-120 (1986).

25. P. Asthana and B. Finkelstein, "Superdense optical storage,", IEEE Spectrum, Aug, 1995.

26. K. Wagner, and D. Psaltis, "Optical neural networks: an introduction by the feature editors," Applied Optics, **32**(8), 1261-1263 (1993).

27. Z. Wen, N. H. Farhat and Z. J. Zhao, "Dynamics of electron trapping materials for use in optoelectronic neurocomputing," Applied Optics, **32** (35), 7251-7265 (1993).

28. X. Yang, C. Y. Wrigley and J. Lindmayer, "Three-dimensional optical memory based on transparent electron trapping thin films," in *Photonics for Computers, Neural Networks and Memories*, W. J. Miceli, J. A. Neff and S. T. Kowel, Eds., Proc. Soc. Photo-Opt. Instrum. Eng. **1773**, 413-422 (1992).

29. T. Y. Young and T. W. Calvert, *Classification, estimation and pattern recognition*, (Elsevier, New York, 1974).

30. F. T. S. Yu and S. Jutamulia, *Optical Signal Processing, Computing and Neural Networks* (John Wiley, New York, 1992).

31. F. T. S. Yu, S. Wu, X. Yang, and S. Jutamulia, and G. M. Storti, "Optical neural network using novel electron trapping materials," Optical Memory and Neural Networks, **1**(1) 49-54 (1992).

A Review of Correlation Filters and Their Application for Scene Matching

Abhijit Mahalanobis
Hughes Missile Systems Company
Bldg. 805, MS F3
P.O. Box 11337
Tucson, Arizona 85734

Abstract

Recent advances in both processor technology as well as algorithms have brought the application of correlation filters closer to reality. In addition to object recognition, correlation filters can be used for scene matching applications. In particular, a form of the filters known as *unconstrained correlation filters* is shown to be optimal from a detection standpoint. Typical applications may include robot vision, autonomous land vehicle, and registering terrain images.

1. Introduction

Correlation filters are useful for the distortion invariant recognition of targets in a scene. A thorough summary of the key elements developed in the first ten years has been presented by Kumar[1]. However, advances have continued both in processor design as well as filter synthesis techniques. Some of the recent algorithm improvements based on new performance criteria related to class separation, distortion tolerance as well as the quality of the correlation plane are reviewed in this paper.

Section 2 is a brief review of the earlier correlation filters. For completeness, we outline the development of the original synthetic discriminant function (SDF)[2], the minimum variance SDF (MVSDF)[3], and the minimum average correlation energy (MACE) filter[4]. Section 3 describes the recent advances. The importance of new performance criteria is discussed. Among the new filters reviewed are the Optimal Trade-off (OT) SDF[5,7,8] filter which optimally trades off among various performance criteria, the mean square error (MSE) SDF[6] and the maximum average correlation height (MACH) filter[10] which has proved useful for target detection in clutter. The newest of the OT filters[9] and the MACH filters are examples of unconstrained correlation filters (UCFs)[10] which are also discussed in this section. We review a new approach to pattern recognition using correlation filters referred to as the distance classifier correlation filter (DCCFs)[11] which measure "distances" between targets and ideal references (instead of raw correlations). One fundamental difference between DCCFs and the other filters is that the resulting decision boundaries can be quadratic. In Section 4, we discuss an application of UCFs for scene matching. In particular, the form used is a variant of the MACH filter which has demonstrated significant promise for recognizing distorted views of an object in clutter. Our conclusions and directions for future work are given in Section 4.

2. Review of Earlier Correlation Filters

In this section, we provide a brief review of a few well known correlation filter design techniques that provide the background for the topics to be discussed in Section 3. The interested reader is referred to the tutorial by Kumar[1] for a more historically and mathematically complete summary. The notation followed in this paper is as follows: images in the space domain are denoted in lower case italics while upper case italics are used to represent the same in the frequency domain. Thus, a two dimensional (2D) image $x(m,n)$ has Fourier transform $X(k,l)$. Vectors are represented by lower case bold characters while matrices are denoted by upper case bold characters. Either $x(m,n)$ or $X(k,l)$ can be expressed as a column vector $\mathbf{x}$ by lexicographical scanning. The superscript ^{T} denoted the transpose operation, and $^+$ denotes the complex conjugate transpose of vectors and matrices.

Consider a two-dimensional image $x(m,n)$. Its correlation with a filter $h(m,n)$ is given by

$$g(m,n) = x(m,n) \otimes h(m,n) = \sum_i \sum_j x(i,j) h(i+m, j+n)$$

where $\otimes$ denotes the two-dimensional correlation operation. Typically, this operation is computed more efficiently via fast Fourier transform (FFT) as follows

$$g(m,n) = IFFT\left\{X(k,l) \cdot H^*(k,l)\right\}$$

where $H(k,l)$ is the FFT of $h(m,n)$, and IFFT is the "inverse" FFT function. Correlation filter theory suggests that filters based on the spectral characteristics and statistical properties of the patterns may be useful for recognizing these patterns, and that the resulting output can be made to demonstrate invariance to arbitrary changes in the input image. Perhaps the greatest advantage of correlation filters is that shift-invariance and the ability to process multiple objects in the image are inherently given. Therefore prior knowledge of the objects or segmentation processes are not required, and the recognition algorithms also offer an automatic tracking capability.

The synthesis of such filters for automatic target recognition has been the topic of investigation for many years. Even though significant advances have been made, and many formulations have been offered, it is not clear that a "best" filter has been found or that an unique solution exists which is optimum for all cases. In one of the earliest works, Hester and Casasent[2] proposed the first SDF filters to recognize more than one image of an object using a single filter. The SDF was intended to overcome the limited ability of the matched spatial filter (MSF) to handle arbitrary types of distortions (including changes in the pattern due to variations in scale and orientation). The strategy was to design the filter using *training images* chosen to cover the range of possible variations in the targets' signature. Thus, in principle, scale and aspect independence could be built into the filter by using suitable training images. In this approach, the filter is designed to yield a specific value at the origin of the correlation plane (also referred to as the "correlation peak") in response to each training image. This was achieved by solving a set of linear equations describing the constraints on the correlation peaks of the training images which can be expressed as

$$\mathbf{X}^T\mathbf{h} = \mathbf{u} \quad (1)$$

where $\mathbf{X} = [\mathbf{x}_1 \quad \mathbf{x}_2 \quad \cdots \quad \mathbf{x}_N]$ is a $d \times N$ matrix with the N training image vectors (each with d pixels) as its columns, and $\mathbf{u} = [u_1 \quad u_2 \quad \cdots \quad u_N]^T$ is a $N \times 1$ vector containing the desired peak values for the training images. However, since the number of training images N is generally much fewer than the dimension of the filters, the system of equations is under-determined and many filters exist which satisfy the conditions in Eq. (1). To find an unique solution, it was assumed by Hester and Casasent that **h** is a linear combination of the training images. This lead to the projection SDF solution

$$\mathbf{h} = \mathbf{X}(\mathbf{X}^T\mathbf{X})^{-1}\mathbf{u} \qquad (2)$$

where **h** is the $d \times$ filter vector. Unfortunately, the projection SDF was unsuccessful because its correlation plane was usually cluttered with sidelobes that often exceeded the filter's response to the target. This made it difficult to find the output when the target shifted and caused excessive false alarms. To overcome this problem, and make the true correlation peak easily detectable, several ad-hoc variations of the basic SDF idea were proposed which are summarized in [1].

The next major innovation in SDF design methodology came about with the development of the MVSDF[3]. Here, the filter **h** was designed to minimize the effect of additive noise on the correlation output. Assuming a zero mean input noise process with correlation matrix **C**, the output variance can be shown to be $\sigma^2 = \mathbf{h}^T\mathbf{C}\mathbf{h}$. The MVSDF was minimizes σ^2 while satisfying the conditions in Eq. (1) and was shown to be

$$\mathbf{h} = \mathbf{C}^{-1}\mathbf{X}(\mathbf{X}^T\mathbf{C}^{-1}\mathbf{X})^{-1}\mathbf{u} \qquad (3)$$

The projection SDF in Eq. (2) is a special case which is obtained if the noise is white (i.e. **C** is the identity matrix).

One difficulty in using the original MVSDF formulation in Eq. (3) is that **C** (a matrix of size $d \times d$) may be difficult to estimate and computationally difficult to invert. However, if the noise process is stationary, the inversion becomes trivial in the frequency domain since the matrix is diagonal. Like its predecessor, the MVSDF also controls only a single point in the correlation plane. Therefore, its output may contain large sidelobes which can cause errors and false alarms.

One way to control the entire correlation plane and reduce sidelobe levels is to minimize the correlation energy. Using Parseval's theorem, the necessary analyses are readily carried out in the frequency domain. Here, $\mathbf{x}_i$ (the columns of the matrix $\mathbf{X}$) are vector representations of the Fourier transform of the training images. For the N training images, the average correlation energy is given by

$$E_{ave} = \mathbf{h}^{+}\mathbf{D}\mathbf{h} \tag{4}$$

where $\mathbf{D} = \frac{1}{N}\sum_{i=1}^{N}\mathbf{X}_i^{*}\mathbf{X}_i$ is a diagonal matrix, and $\mathbf{X}_i$ is a diagonal matrix with the elements of $\mathbf{x}_i$ along its main diagonal. The MACE filter[4] minimizes E_{ave} subject to the constraints in Eq. (1), and is given by

$$\mathbf{h} = \mathbf{D}^{-1}\mathbf{X}\left(\mathbf{X}^{+}\mathbf{D}^{-1}\mathbf{X}\right)^{-1}\mathbf{u} \tag{5}$$

The solution in Eq. (5) is in the frequency domain and $\mathbf{h}$ is therefore the vector representation of the Fourier transform of the filter.

MACE filters have been shown to be effective for finding training images in background and clutter, and produce impressive correlation peaks. They appear to be the first filters which attempt to control the entire correlation plane. However, there are two main draw-backs. First, there is no built-in immunity to noise. Secondly, the MACE filters are often excessively sensitive to intra-class variations. None-the-less, the MACE filters have paved the way for the frequency domain analysis and development of correlation filters, and set the stage for the subsequent developments to be discussed in this paper.

The filters outlined here are only a few of the critical developments that provide the background for the advances to be discussed in Section 3. The interested reader is referred to a tutorial paper[1] by Kumar for a summary of many other interesting and technically important developments.

3. Advanced Correlation Filters

The SDF variations (except for the MACE filter) described in Section 2 were designed to control only one point in the correlation plane. Unfortunately, this strategy seems inadequate because either large sidelobes make this controlled point difficult to find, or its value is significantly different from the desired value specified in the training process. It has been suggested that perhaps performance can be improved if all points in the correlation plane are somehow taken into account. Indeed, the MACE filter affects the entire correlation plane by suppressing sidelobes everywhere. However, a more systematic development is needed to translate correlation plane control into improved discrimination and distortion tolerance. The advances discussed in Sections 3.1 - 3.6 are recent steps taken in this direction.

3.1 Mean Square Error SDFs

One approach is to find the filter whose correlation output would best approximate a specified desired shape by minimizing a mean square error criterion. Hence, this filter is referred to as the mean square error SDF (MSE-SDF)[6]. The linear constraints are still imposed on the central correlation value. The advantage is that a sculpted shape may be easier to find in the correlation plane rather than a single point. If the known shape is found then the relative position of the correlation peak can be better predicted (or the search can be narrowed).

Let $d(m,n)$ be the image of the desired shape. In principle, the choice of this shape can be arbitrary. However, later we will address the question what is the best shape to use for optimizing distortion tolerance and discrimination. We now summarize the theoretical development of the MSE-SDF. All analysis is conducted in the frequency domain. The vector **d** is obtained by lexicographically rearranging the rows of the Fourier transform of $d(m,n)$. Using Parseval's theorem, it can be shown that the "error" in the shape of the correlation surface for the *i*-th image is essentially

$$\mathbf{e}_i = \mathbf{X}_i^* \mathbf{h} - \mathbf{d} \tag{6}$$

where $\mathbf{X}_i$ is a diagonal matrix containing the samples of the Fourier transform of the i-th training image along the main diagonal. The mean square error averaged over all N training images is

$$MSE = \frac{1}{N}\sum_{i=1}^{N}\mathbf{e}_i^+\mathbf{e}_i = \frac{1}{N}\sum_{i=1}^{N}(\mathbf{X}_i^*\mathbf{h}-\mathbf{d})^+(\mathbf{X}\mathbf{h}-\mathbf{d}) \qquad (7)$$

With simple algebraic manipulations, it can be shown that

$$MSE = \mathbf{h}^+\left[\frac{1}{N}\sum_{i=1}^{N}\mathbf{X}_i^*\mathbf{X}_i\right]\mathbf{h} + \mathbf{d}^+\mathbf{d} - \mathbf{h}^+\left(\frac{1}{N}\sum_{i=1}^{N}\mathbf{X}_i^*\right)\mathbf{d} - \mathbf{d}^+\left(\frac{1}{N}\sum_{i=1}^{N}\mathbf{X}_i\right)\mathbf{h} \qquad (8)$$

$$= \mathbf{h}^+\mathbf{D}\mathbf{h} + E_d - \mathbf{h}^+\mathbf{p} - \mathbf{p}^+\mathbf{h}$$

where $\mathbf{D} = \frac{1}{N}\sum_{i=1}^{N}\mathbf{X}_i^*\mathbf{X}_i$ is the same diagonal matrix used in the formulation of the MACE filter, and $\mathbf{p} = \frac{1}{N}\sum_{i=1}^{N}\mathbf{X}_i^*\mathbf{d}$. We design the MSE-SDF to minimize MSE subject to the constraints in Eq. (1). Using the method of Lagrange multipliers, the solution can be found to be

$$\mathbf{h} = \mathbf{D}^{-1}\mathbf{p} + \mathbf{D}^{-1}\mathbf{X}(\mathbf{X}^+\mathbf{D}^{-1}\mathbf{X})^{-1}[\mathbf{u} - \mathbf{X}^+\mathbf{D}^{-1}\mathbf{p}] \qquad (9)$$

Note that here $\mathbf{X}$ is a $d \times N$ matrix with the Fourier transform of the training image vectors as its columns. An interesting situation occurs when $\mathbf{d}$ is the zero vector (i.e., the desired shape is zero everywhere). In this case $\mathbf{p}$ also is zero and the MSE-SDF in Eq. (9) reduces to

$$\mathbf{h} = \mathbf{D}^{-1}\mathbf{X}(\mathbf{X}^+\mathbf{D}^{-1}\mathbf{X})^{-1}\mathbf{u} \qquad (10)$$

and is identical to the MACE filter in Eq. (5). Thus, the MACE filter is a special case of the MSE-SDF which is obtained if the ideal shape for the correlation plane is to be zero everywhere except at the peak (i.e. a delta function). Clearly then, setting the ideal shape to zero and minimizing the energy of the correlation plane are equivalent.

The filter in Eq. (9) is the best choice in a mean square error sense to sculpt the correlation planes of all N training images to look like $d(m,n)$ on the average. Simultaneously, the central points of the correlation planes satisfy the conditions in Eq. (1). Further details about the MSE-SDF can be found in reference [6].

3.2 The choice of the best desired shape

As seen in Eq. (9), the properties of the MSE-SDF will depend on the desired correlation shape **d**. The questions arises then what is the best desired shape to use? It may make intuitive sense to choose geometrical shapes such as squares, circles, or even bell-shaped (Gaussian) patterns for $d(m,n)$. However, such choices are ad-hoc and may not be optimum for pattern recognition. An alternative approach would be find **d** such that the MSE criterion in Eq. (8) itself is minimized. Towards this end, we take the gradient of *MSE* with respect to **d** and set it to zero, i.e,

$$\nabla_{\mathbf{d}}(MSE) = 2\mathbf{d} - \left[\frac{1}{N}\sum_{i=1}^{N}\mathbf{X}_i^*\right]\mathbf{h} = \mathbf{0} \tag{11}$$

If $\mathbf{M} = \frac{1}{N}\sum_{i=1}^{N}\mathbf{X}_i$ is the mean of the Fourier transforms of the training images, then the optimum choice for the desired shape is

$$\mathbf{d} = \mathbf{M}^*\mathbf{h} \tag{12}$$

Therefore, the shape which minimizes *MSE* is the Fourier transform of the correlation surface produced by **h** in response to the mean image of the training set. Substituting this expression for **d** into Eq. (7) yields

$$\begin{aligned} MSE &= \frac{1}{N}\sum_{i=1}^{N}\left(\mathbf{X}_i^*\mathbf{h} - \mathbf{M}^*\mathbf{h}\right)^+\left(\mathbf{X}_i^*\mathbf{h} - \mathbf{M}^*\mathbf{h}\right) \\ &= \mathbf{h}^+\left[\frac{1}{N}\sum_{i=1}^{N}(\mathbf{X}-\mathbf{M})^*(\mathbf{X}-\mathbf{M})\right]\mathbf{h} \\ &= \mathbf{h}^+\mathbf{S}\mathbf{h} \end{aligned} \tag{13}$$

where $\mathbf{S} = \frac{1}{N}\sum_{i=1}^{N}(\mathbf{X}-\mathbf{M})^{*}(\mathbf{X}-\mathbf{M})$ is a diagonal matrix. The corresponding MSE SDF is thus obtained by minimizing $\mathbf{h}^{+}\mathbf{S}\mathbf{h}$ subject to the linear constraints in Eq. (1), and is given by

$$\mathbf{h} = \mathbf{S}^{-1}\mathbf{X}\left(\mathbf{X}^{+}\mathbf{S}^{-1}\mathbf{X}\right)^{-1}\mathbf{u}. \tag{14}$$

Applying Parseval's theorem to Eq. (13), it can be easily shown that

$$\mathbf{h}^{+}\mathbf{S}\mathbf{h} = \sum_{i=1}^{N}\left|\mathbf{H}^{*}\mathbf{x}_{i} - \mathbf{H}^{*}\mathbf{m}\right|^{2} \cong \sum_{i=1}^{N}\left|\mathbf{g}_{i} - \overline{\mathbf{g}}\right|^{2} \tag{15}$$

where $\mathbf{H}$ is diagonal matrix with the elements of $\mathbf{h}$ along its main diagonal, $\mathbf{m} = \frac{1}{N}\sum_{i=1}^{N}\mathbf{x}_{i}$ is the "average" or "mean" training image vector, and $\mathbf{g}_{i}$ and $\overline{\mathbf{g}}$ are vector representations of the correlation surfaces of the i-th training image and the mean image respectively. The quantity $\mathbf{h}^{+}\mathbf{S}\mathbf{h}$ is referred to as the *Average Similarity Measure* (ASM)[10] because it is a metric for the average similarity in the shapes of the in-class correlation surfaces.

We now discuss how ASM ($\mathbf{h}^{+}\mathbf{S}\mathbf{h}$) may serve as a metric for distortion. The distortions of the object in the input plane are represented by the training images $\mathbf{x}_{i}$. These distortions are reflected in the output as variations in the structure and shape of the corresponding correlation planes, $\mathbf{g}_{i}$, and are quantified by ASM as shown in Eq. (15). Another interpretation can be made by treating $\mathbf{H}$ as a linear transformation. Then, ASM is a measure of the distances of the training images from the class center under this transform which is also a metric for class compactness. Therefore minimizing ASM should lead to a compact set of correlation planes that resemble each other, and exhibit the least possible variations. Conversely, if a filter successfully reduces distortions, then distorted input images should yield similar output planes, leading to a small value of ASM. Thus minimizing ASM should increase distortion tolerance of the filter.

An interesting point is noted by observing that

$$\begin{aligned} ASM &= \mathbf{h}^{+}\mathbf{S}\mathbf{h} \\ &= \mathbf{h}^{+}\mathbf{D}\mathbf{h} \ - \ \mathbf{h}^{+}\mathbf{M}\mathbf{M}^{*}\mathbf{h} \\ &= ACE \ - \ \mathbf{h}^{+}\mathbf{M}\mathbf{M}^{*}\mathbf{h} \end{aligned} \tag{16}$$

where $ACE = \mathbf{h}^{+}\mathbf{D}\mathbf{h}$ is the criterion minimized by the MACE filter. Thus ASM is the difference between the average correlation energy of the training images and the correlation energy of the average image. Based on this relation, the minimization of ASM can be viewed as a generalization of the MACE criterion.

3.3 Maximum Average Correlation Height (MACH) Filter

The primary purpose of correlation filters is distortion-invariant recognition of objects in clutter. Traditionally, SDF filters have been designed by imposing linear constraints on the training images to yield a known value at a specific location in the correlation plane. However, such conditions do not explicitly control the filters' ability to generalize over the entire domain of the training images. Various filters exhibit different levels of distortion tolerance even with the same training set and constraints. Another reason to question the method of using hard constraints on the training images is that these conditions do not hold for the test images, and the outputs obtained from the filter in practice are almost certain to differ from the values specified during training. Based on these observations, in a major departure from conventional SDF filter design philosophy, we have removed the hard constraints on the peak values and no longer use Eq. (1) for filter design.

Consider the correlation surface $\overline{\mathbf{g}}$ produced by the mean image **m.** If $\overline{\mathbf{g}}$ exhibits a sharp well-defined peak, then minimizing ASM should cause the rest of the class to also follow this behavior. In other words, if ASM is minimum and $\overline{\mathbf{g}}$ is well shaped, then all true-class correlation planes are expected to resemble $\overline{\mathbf{g}}$ and to exhibit well-shaped structures. We therefore require **h** to have a high correlation with the mean image while making ASM small. This is achieved by maximizing the criterion

$$J(\mathbf{h}) = \frac{\mathbf{h}^{+}\mathbf{m}\mathbf{m}^{+}\mathbf{h}}{\mathbf{h}^{+}\mathbf{S}\mathbf{h}} \tag{17}$$

The optimum solution is found by setting the derivative of $J(\mathbf{h})$ with respect of $\mathbf{h}$ to zero and is given by

$$\mathbf{h} = \mathbf{S}^{-1}\mathbf{m} \tag{18}$$

The filter in Eq. (18) is referred to as the maximum average correlation height (MACH) filter[10] because it maximizes the height of the mean correlation peak relative to the expected distortions. To date, the MACH filter has demonstrated better performance than previous SDF filters. We attribute this to the inclusion of the ASM criterion which reduces the filters' sensitivity to distortions, and to the removal of hard constraints on the peak. The latter fact permits the correlation planes to adjust to whatever value best permits the optimization of performance criterion.

3.4 Optimal Trade-off SDF Filter Design

Since several different performance criteria exist, the question arises as to which among them is the best for correlation filter design. The answer may be that there is no clear winner and compromises must be made based on the particular application. None-the-less, it is desirable to find a general filter expression and the methodology for optimizing it for a given set of conditions. Soon after the MVSDF and the MACE filter had been reported, it was suspected[13] that noise variance and correlation energy could be linearly combined into a single performance criterion to obtain a filter that yielded sharp peaks as well as low output noise variance. This concept was independently and fully developed by Refregier[5] who introduced a formal technique for trading off between competing performance criteria. These filters have been referred to as the optimal trade-off (OT) SDF filters[5].

We have seen that unconstrained filters (such as the MACH filter) perform better than conventional SDF filters with hard constraints. We now review a recently published[9] technique for synthesizing unconstrained OT correlation filters. Filters can be described in terms of four figures of merit, namely, output noise variance (ONV), average correlation energy (ACE), average similarity measure (ASM), and average correlation height (ACH). The ideal filter should minimize ONV, ACE, and ASM and maximize ACH. The best approach in such a situation is the optimal trade-off approach introduced by Refregier[7]. In this approach, n -1 of the n criteria on interest are held constant, and the filter is optimized with respect to the one remaining criterion. Thus, no other filter could do better than the OT filter for all n criteria. As shown by Refregier, to determine the OT filter **h**, we minimize the following energy function, which is a weighted sum of the four performance measures:

$$\begin{aligned} E(\mathbf{h}) &= \alpha(ONV) + \beta(ACE) + \gamma\,(ASM) - \delta(ACH) \\ &= \alpha\ \mathbf{h}^{+}\mathbf{Ch} + \beta\ \ \mathbf{h}^{+}\mathbf{Dh} + \gamma\ \ \mathbf{h}^{+}\mathbf{Sh} - \delta\left|\mathbf{h}^{+}\mathbf{m}\right| \end{aligned} \tag{19}$$

where α, β, γ, and δ are non-negative, and $\alpha^2 + \beta^2 + \gamma^2 + \delta^2 = 1$. Note that this weighted sum is the result of optimally trading off between the four performance measures, not just arbitrarily combining them. The ACH appears in E(**h**) with a negative sign because it is to be maximized, whereas the other three criteria should be minimized. We can rewrite E(**h**) in Eq. (19) as

$$E(\mathbf{h}) = \mathbf{h}^{+}\mathbf{Ph} - \delta\left|\mathbf{h}^{+}\mathbf{m}\right| \tag{20}$$

where $\mathbf{P} = \alpha\ \mathbf{C} + \beta\ \mathbf{D} + \gamma\ \mathbf{S}$ is a diagonal matrix with positive diagonal elements. Minimizing $E(\mathbf{h})$ without any constraints leads to

$$\mathbf{h} = \frac{\delta}{2}\mathbf{P}^{-1}\mathbf{m} \tag{21}$$

The constant $\delta/2$ is not critical since it scales ONV, ACE and ASM and ACH identically. The OT filter in Eq. (21) is the provides the best balance between all of the above performance criteria for any choice of α, β, γ, and δ. The actual values of these parameters depend on the data and the application, and must be determined by searching all possible combinations.

3.5 Distance Classifier Correlation Filters

In Section 3.1, we discussed the MSE-SDF which was designed to sculpt the correlation planes of training images into arbitrary shapes. It was shown in Section 3.2 what the best shape for this purpose should be. The MACH filter in Section 3.3 was the first unconstrained filter to utilize this shape as a criterion for improving distortion tolerance. However, the possibility of comparing the full shapes of the correlation planes for decision making is still not completely addressed by the MACH filter. The distance classifier correlation filter[11] (DCCF) was therefore developed to directly measure the similarity between the sculpted shapes of the correlation plane and the ideal shape for that class. As we will now discuss, this is substantially different from the conventional approach to correlation filtering, and leads to quadratic decision boundaries.

All analysis is carried out in the frequency domain. Recall from the discussion on ASM in Section 3.2 that the filter can be viewed as a linear transform. Therefore, for our purposes the terms "transformation" and "filtering" are interchangeable. The distance classifier transform **H** is designed to optimally separate the classes while making them as compact as possible (i.e. for each class, the correlation planes should be as *similar* as possible to the transformed ideal reference shape for that class). For shift-invariance, this transform must be diagonal in the frequency domain. Multiplication of **x** by a diagonal **H** is equivalent to multiplying $X(k,l)$ by $H(k,l)$ or a convolution/correlation in space domain which provides shift-invariance. We may thus refer to the correlation plane as a result of the transformation. Figure 1 depicts a 3-class example where $\mathbf{m}_1$, $\mathbf{m}_2$ and $\mathbf{m}_3$ represent the class centers, and **z** is the unknown input to be classified. The transformation by **H** is expected to make the classes distinct by moving their centers apart while shrinking the boundaries so that $\mathbf{H}^*\mathbf{z}$ can be easily identified with its correct class (class 3 in the figure since d_3 is the smallest distance). Thus, the distance comparisons are between the entire filtered image and filtered ideal reference, which is equivalent to using the information in the full correlation plane.

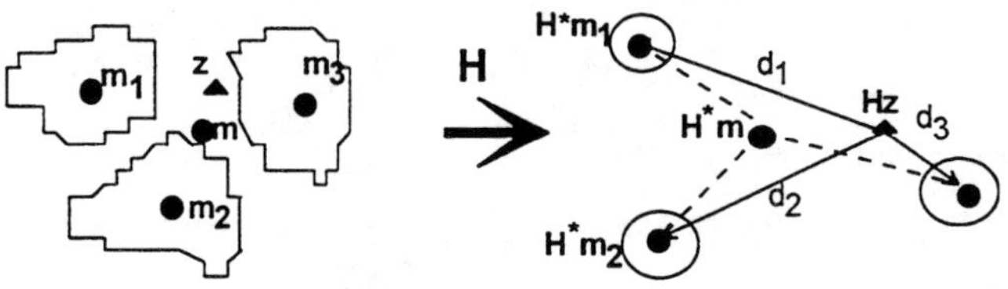

Figure 1: In the DCCF algorithm, transformation by **H** increase inter-class distances while making each class more compact to simultaneously improve distortion tolerance and discrimination

To make the "intra-class" separation between the correlation peaks very large, we make the mean peak values of the classes as different as possible. Towards this end, we formulate the criterion

$$A(\mathbf{h}) = \frac{1}{C}\sum_{k=1}^{C}\left|\mathbf{m}_k^{+}\mathbf{h} - \mathbf{m}^{+}\mathbf{h}\right|^2 = \frac{1}{C}\sum_{k=1}^{C}\mathbf{h}^{+}(\mathbf{m}-\mathbf{m}_k)(\mathbf{m}-\mathbf{m}_k)^{+}\mathbf{h} \tag{22}$$

$$= \mathbf{h}^{+}\mathbf{T}\mathbf{h}$$

where $\mathbf{T} = \frac{1}{C}\sum_{k=1}^{C}(\mathbf{m}-\mathbf{m}_k)(\mathbf{m}-\mathbf{m}_k)^{+}$ is a $d \times d$ full matrix of rank $\le C$ -1, and $\mathbf{m}_k$ and $\mathbf{m}$ are the mean image of the k-th class and the overall mean of the entire training set, respectively. If $A(\mathbf{h})$ in Eq. (22) is maximized, the class mean correlation peaks ($\mathbf{m}_k^{+}\mathbf{h}$) will differ greatly. The individual correlation peaks of the different classes will also become well separated provided ASM is simultaneously minimized. Our objectives are therefore met by maximizing the ratio of A(**h**) and ASM, i.e.

$$J(\mathbf{h}) = \frac{\mathbf{h}^{+}\mathbf{T}\mathbf{h}}{\mathbf{h}^{+}\mathbf{S}\mathbf{h}} \tag{23}$$

with respect to **h**. The optimum solution is the dominant eigenvector of $\mathbf{S}^{-1}\mathbf{T}$. The details for numerically obtaining this solution are in reference [12].

For testing purposes, the distance to be computed between the transformed input and the ideal shape for class k is

$$d_k = \left|\mathbf{H}^{*}\mathbf{z} - \mathbf{H}^{*}\mathbf{m}_k\right|^2 = p + b_k - 2\mathbf{z}^{+}\mathbf{h}_k \qquad 1 \le k \le C \tag{24}$$

where $\mathbf{z}$ is the input image, $p = \left|\mathbf{H}^{*}\mathbf{z}\right|^2$ is the transformed input image energy, $b_k = \left|\mathbf{H}^{*}\mathbf{m}_k\right|^2$ is the energy of the transformed k-th class mean, and

$$\mathbf{h}_k = \mathbf{H}\mathbf{H}^{*}\mathbf{m}_k \tag{25}$$

is viewed as the effective filter for class k. Since there are only C classes to which distances must be computed, we require only C such filters. The constants b_k do not depend on the input and can be pre-computed. Since p is independent of the class, it can be dropped if linear decision boundaries are sufficient. The third term in Eq. (24) is the only term that requires correlation. Essentially, we want the minimum value of d_k over all possible shifts between the input and the reference. Therefore, this term should be chosen as the maximum value in the cross-correlation of $\mathbf{z}$ and $\mathbf{h}_k$. The decision strategy is to label the input as belonging to the class to which the distance is minimum.

The DCCF approach has demonstrated promising performance on both infra-red (IR) [14] as well as synthetic aperture radar (SAR) [15,16] imagery. Test results showed that DCCFs outperform other correlation filters in recognizing targets while rejecting noise, clutter and other confusing objects.

4. Application of Correlation Filters for Scene Matching

Scene matching is useful for aligning an image with a data base. A wide variety of applications are possible including vision for robots and autonomous land vehicles, registration of satellite imagery, map reading, wide area screening and surveillance. The general case involves sensing an image in a window whose size is determined by the field of view (FOV) of the sensor. This sensed image must then be found and registered within a larger reference image of the same area. The traditional approach has been to construct a matched filter of the reference and correlate it with the sensed image. Limitations of earlier digital and optical processors have encouraged the development of binary correlation algorithms. These factors contribute to performance limitations which may be alleviated by the technique proposed here.

From a correlation filter standpoint, the scene matching problem differs from the traditional target recognition problem in one important way. Conventional filters are constructed from reference imagery of targets that are smaller than the sensed scene size. In the present case, the reference imagery is larger than the sensed image. Consequently the filter is larger than the sensed image. This situation is illustrated in Figure 2.

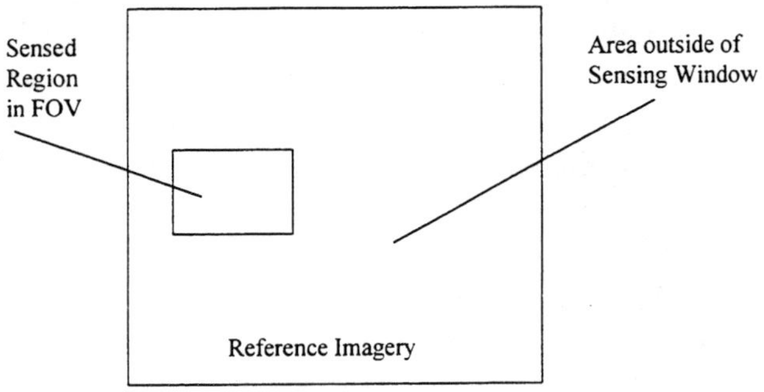

Figure 2: Sensed Image is smaller than Reference Imagery from which Filter must be made.

Let **w** be the vector representing the Fourier transform of the windowed image (i.e. values outside of the window are set to zero prior to applying the Fourier transform). In general, the sensor window may fall anywhere over the reference scene **x**. The objective is to determine its position (and hence the position of the sensor platform) from the location of the correlation peak.

We note that the expectation of **w** over all possible position of the window is $E\{\mathbf{w}\} = \mathbf{x}$. This is true because over a sufficient number of trials the window will have covered the entire scene. Therefore, if **h** is the filter vector, then the expectation of the correlation peak is

$$peak = E\{\mathbf{w}^{+}\mathbf{h}\} = E\{\mathbf{w}^{+}\}\mathbf{h} = \mathbf{x}^{+}\mathbf{h} \tag{26}$$

The detection strategy requires this peak to be as large as possible while suppressing the sidelobes caused by the regions outside the window with the filter as well as any effects of additive noise. As in the derivation of the MACE filter, using Parseval's theorem it can be shown that the expected value of the energy in the sidelobes is

$$Energy = E\left\{\left|\mathbf{X}^{*}\mathbf{h}\right|^{2}\right\} = \mathbf{h}^{+}\mathbf{D}\mathbf{h} \tag{27}$$

where **D** is a diagonal matrix with the power spectrum of reference image as its main diagonal. Similarly, the output variance caused by noise with covariance **C** is given by

$$variance = \mathbf{h}^{+}\mathbf{Ch} \tag{28}$$

Consequently, the optimum filter for enhancing the peak while suppressing sidelobes and the effects of additive noise is obtained by maximize the performance criterion

$$J(\mathbf{h}) = \frac{\left|\mathbf{x}^{+}\mathbf{h}\right|^{2}}{\mathbf{h}^{+}\mathbf{Dh} + \mathbf{h}^{+}\mathbf{Ch}} = \frac{\mathbf{h}^{+}\mathbf{xx}^{+}\mathbf{h}}{\mathbf{h}^{+}(\mathbf{D}+\mathbf{C})\mathbf{h}} \tag{29}$$

The solution to this is given by

$$\mathbf{h} = (\mathbf{D}+\mathbf{C})^{-1}\mathbf{x} \tag{30}$$

which is a variation of the MACH filter (a form of UCFs) described in Section 3. A simplification may be made by assuming the noise to be white with covariance $\sigma \cdot \mathbf{I}$ so that the filter becomes

$$\mathbf{h} = (\mathbf{D}+\sigma \cdot \mathbf{I})^{-1}\mathbf{x} \tag{31}$$

We now illustrate the application of the correlation filter in Eq. (31) for matching aerial imagery over an urban area. Consider the scene shown in Figure 3. This photograph represents the search area, and is the reference image from which the filter must be synthesized. An example of a sensed image over the same area is shown in Figure 4. As noted earlier, the sensed image is smaller than the reference image and represents an unknown windowed region. The objective of the filtering process is to identify the location where sensed scene matches the reference image. Typically, the two image are obtained under different conditions and can exhibit significant variations. For instance, a close examination will reveal that the shadows cast by the buildings suggest the images were collected at different times of day. Other variations are introduced by contrast and scale difference between the two images as well as distortions in the sensed image introduced by the optics of the camera.

The reference image was scanned and digitized to a resolution of 256 x 256 pixels and a filter of corresponding size was obtained per Eq. (31). The sensed scene was zero padded to the same size. The resulting correlation is show in Figure 5. The position of the peak reveals where the upper left corner of the sensed image occurs in the reference image. The peak is 11 standard deviations above the mean of the correlation plane. The sidelobes are well suppressed and do not contribute to any false alarms over the search area.

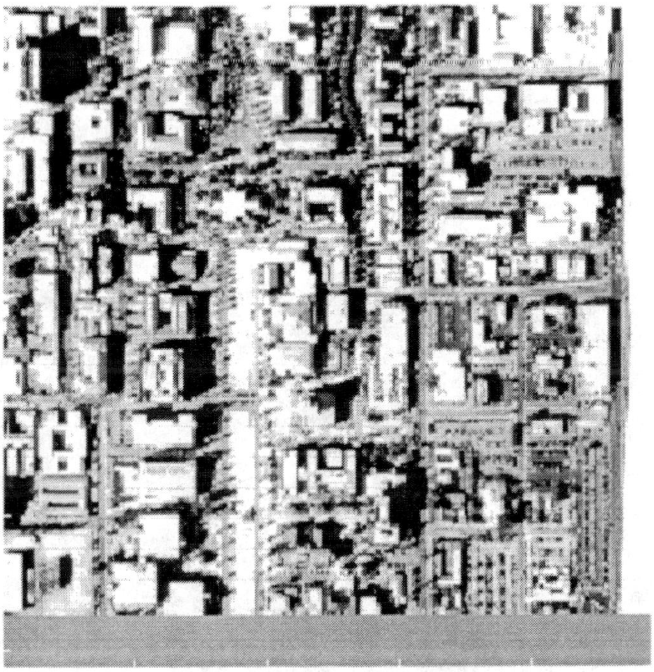

Figure 3: Aerial Reference Imagery of Urban Scene

Figure 4: Sensed Scene over same Urban Area shown in Figure 1.

5. Conclusion

The example presented here shows the potential of the filters to be use not only for recognizing vehicular targets, but also matching scenes. We are currently conducting extensive tests to ascertain where the limitations of correlation filters lie in terms of the signal to clutter ratio on the targets and the resolution requirements. The technique illustrated here provides a general means for finding a scene in a small window over a larger search area. The optimum correlation filter was shown to be the MACH filter, a variant of the UCF family. Work is currently underway to study the robustness of this approach, and to predict the performance under a variety of conditions. Future applications will include matching data from different sensors which operate in different regions of the electromagnetic spectrum.

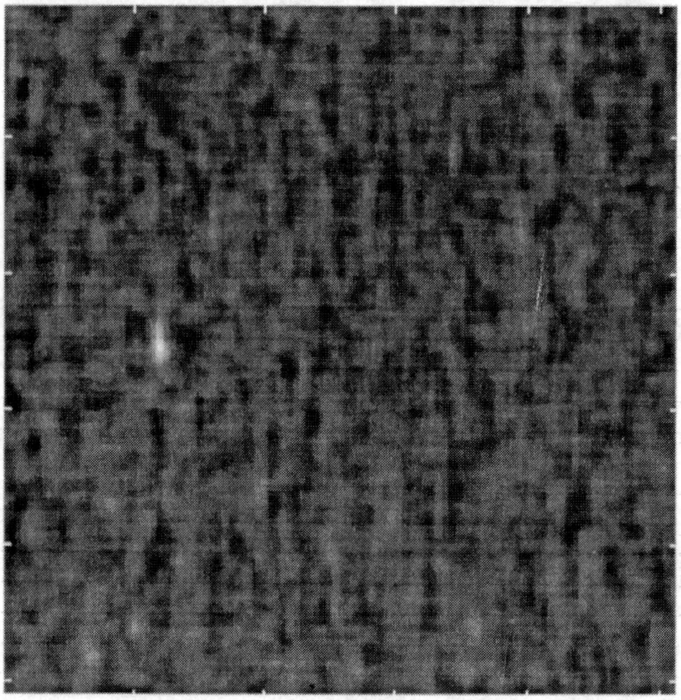

Figure 5: Correlation plane with strong peak showing the registration of the upper left corner of the sensed scene over the reference image.

References

1. B.V.K. Vijaya Kumar, "Tutorial survey of composite filter designs for optical correlators", *Applied Optics*, Vol. 31, pp. 4773-4801, (1992)
2. C.F. Hester and D. Casasent, "Multivariant techniques for multiclass pattern recognition", *Applied Optics*, Vol 19, pp. 1758-1761, (1980)
3. B.V.K. Vijaya Kumar, "Minimum variance synthetic discriminant functions", *JOSA*, Vol. A 3, pp. 1579-1584 (1986)
4. A. Mahalanobis, B.V.K. Vijaya Kumar, and D. Casasent, "Minimum average correlation energy filters", *Applied Optics*, Vol. 26, pp. 3633-3640 (1987)
5. Ph. Refregier, "Filter design for optical pattern recognition: multi-criteria optimization approach", *Optics Letters,* Vol. 15, pp. 854-856, (1990)
6. B.V.K. Vijaya Kumar, A. Mahalanobis, S. Song, S.R.F. Sims, and J. Epperson, "Minimum squared error synthetic discriminant functions", *Optical Engineering*, Vol. 31, pp. 915-922, (1992)
7. P. Refregier and J. Figue, "Optimal trade-off filter for pattern recognition and their comparison with Wiener approach", *Optical Computing Process.*, Vol. 1, pp. 3-10, (1991)
8. V. Laude and Ph. Refregier, "Multi-criteria characterization of coding domains with optimal Fourier spatial light modulator filters", *Applied Optics*, Vol. 33, pp. 4465-4471 (1994)
9. B.V.K. Vijaya Kumar, D.W. Carlson, and A. Mahalanobis, "Optimal trade-off synthetic discriminant functions filters for arbitrary devices", *Optics Letters*, Vol. 19, pp. 1556-1558, (1994)
10. A. Mahalanobis, B.V.K. Vijaya Kumar, S. Song, S.R.F. Sims, and J. F. Epperson, "Unconstrained correlation filters", *Applied Optics*, Vol. 33, pp. 3751-3759, (1994)
11. A. Mahalanobis, B.V.K. Vijaya Kumar, and S.R.F. Sims, "Distance classifier correlation filters for distortion tolerance, discrimination, and clutter rejection", *Proc. SPIE*, pp. 2026-74, San Diego ,(1993)
12. A. Mahalanobis, B.V.K. Vijaya Kumar, and S.R.F. Sims, "Generalized Distance Classifier Correlation Filters for Multi-Class Automatic Target Recognition", submitted to *Pattern Recognition* (1994)
13. A. Mahalanobis, "New correlation filters for symbolic rule-based pattern recognition", Doctoral dissertation, Carnegie Mellon University, Pittsburgh, pp. 205-215, (1987)
14. A. Mahalanobis, B.V.K. Vijaya Kumar, and S.R.F. Sims, "Performance evaluation of distance classifier correlation filters", *SPIE*, paper no. 2238-01, Hybrid Image and Signal Processing IV, Orlando, April 1994.
15. A. Mahalanobis, A. Forman, M. Bower, N. Day, R. Cherry, "A quadratic distance classifier for multi-class SAR ATR using correlation filters", *SPIE, OE/LASE '93*, Paper No. 1875-13, Los Angeles, January 1993

16. L. M. Novak, G.J. Owirka, and C. M. Netishen, "Radar target identification using spatial matched filters", Special issue of *Pattern Recognition* on correlation filters and neural networks, Vol. 27, pp. 607-617, 1994
17. R.D. Juday, "Optimal realizable filters and the minimum Euclidean distance principle", *Applied Optics*, Vol. 32, pp. 5100-5111 (1993)
18. Javidi, Ph. Refregier, and P. Willet, "Optimum receiver design for pattern recognition with spatially disjoint target and scene noise", *Proc. SPIE*, Vol. 2026, (1993)